Erosion of Geomaterials

Erosion of Geomaterials

Edited by
Stéphane Bonelli

iSTE

WILEY

First published 2012 in Great Britain and the United States by ISTE Ltd and John Wiley & Sons, Inc.

ISTE Ltd
27-37 St George's Road
London SW19 4EU
UK

www.iste.co.uk

John Wiley & Sons, Inc.
111 River Street
Hoboken, NJ 07030
USA

www.wiley.com

Library of Congress Cataloging-in-Publication Data

Erosion of Geomaterials / edited by Stéphane Bonelli.
p. cm.
Includes bibliographical references and index.
ISBN 978-1-84821-351-7
1. Sediment transport. 2. Soil erosion. 3. Levees--Protection. 4. Dam failures--Prevention. I. Bonelli, Stéphane.
TC175.2.E76 2012
627'.8--dc23

2012016138

British Library Cataloguing-in-Publication Data
A CIP record for this book is available from the British Library
ISBN: 978-1-84821-351-7

Printed and bound in Great Britain by CPI Group (UK) Ltd., Croydon, Surrey CR0 4YY

MIX
Paper from responsible sources
FSC® C013604

Table of Contents

Foreword . . . xiii
Paul ROYET

Introduction . . . xv
Stéphane BONELLI

Chapter 1. Introduction to the Process of Internal Erosion in Hydraulic Structures: Embankment Dams and Dikes . . . 1
Jacques FRY

1.1. Introduction . . . 1
1.2. The significance of internal erosion for hydraulic structures . . . 2
1.2.1. The set of hydraulic structures in France . . . 2
1.2.2. The vulnerability of hydraulic structures . . . 2
1.2.3. Erosion as a leading cause of failure . . . 3
1.2.4. Internal erosion: one failure per year in France . . . 4
1.3. The impact of incidents on embankment dams and dikes . . . 5
1.3.1. Terminology . . . 5
1.3.2. Initiation areas . . . 5
1.3.3. The importance of design . . . 8
1.3.4. Four mechanisms of erosion, classified according to their boundary conditions . . . 11
1.3.5. Triggering mechanisms . . . 13
1.4. Main results of erosion trials . . . 16
1.4.1. Which law of erosion? . . . 16
1.4.2. Concentrated leak erosion . . . 16
1.4.3. Backward erosion . . . 19
1.4.4. Contact erosion . . . 21
1.4.5. Suffusion . . . 22

1.5. Remarks on the applicability of erosion trials 24
1.5.1. Problem of passing on the results from the laboratory to the hydraulic structure 24
1.5.2. Scaling effect of outflows in the absence of similarity 25
1.5.3. Influence of the geostatic structure of the soil on the erosion threshold 26
1.5.4. Initiation of internal erosion in a cohesionless soil 31
1.5.5. Erodibility and researching erosion laws 32
1.6. Conclusion 35
1.7. Bibliography 36

Chapter 2. Suffusion, Transport and Filtration of Fine Particles in Granular Soil 39
Didier MAROT and Ahmed BENAMAR

2.1. Introduction 39
2.1.1. Chapter objectives 39
2.1.2. Terminology 40
2.2. Dominant parameters that influence suffusion 41
2.2.1. Parameters that modify the geometry of the porous medium 41
2.2.2. Parameters that modify the physicochemical characteristics of the medium 44
2.3. Main initiation criteria for suffusion 45
2.3.1. Grain-size distribution criteria 45
2.3.2. Confronting granular criteria 48
2.3.3. Hydraulic criteria 50
2.3.4. Summary and final remarks 52
2.4. An initiation criterion formulated using a geohydromechanical approach 53
2.4.1. Geometric criterion 53
2.4.2. The hydromechanical criterion 53
2.4.3. Summary and final remarks 54
2.5. The scaling effect and the energetic approach 55
2.5.1. Identifying the scaling effect 55
2.5.2. Energetic approach 55
2.5.3. Summary 58
2.6. Coupling the phenomena of suffusion and filtration-clogging 58
2.7. Processes causing filtration 59
2.7.1. Background knowledge 60
2.7.2. Theoretical background 61
2.8. Filtration modeling 64
2.8.1. Modeling in a continuous medium 64

2.8.2. Convection model – dispersion with deposition kinetics. 67
2.9. Confrontation between the laboratory filtration tests and the modeling 69
2.9.1. Validation of the model in the case of suspended particles 69
2.9.2. Results and preferential flow 70
2.10. Filtration and clogging 71
2.10.1. Phenomenological laws 71
2.10.2. Physicochemical aspects 72
2.10.3. Filtration and clogging 73
2.11. Conclusion 74
2.12. Bibliography 75

Chapter 3. The Process of Filtration in Granular Materials. 81
Eric VINCENS, Nadège REBOUL and Bernard CAMBOU

3.1. Introduction. 81
3.2. Fundamental characteristics of the filtering granular media 82
3.2.1. Pores 82
3.2.2. Constrictions. 83
3.3. The distribution of constriction size 84
3.3.1. The definition of pores and constrictions using a DEMDT approach 85
3.3.2. The influence that the density of the medium has on the size of the constrictions 89
3.4. A probabilistic approach of constriction sizes 92
3.4.1. Particle size distribution 93
3.4.2. Different geometrical configurations according to the density of the medium. 94
3.4.3. Distribution of constriction sizes: several models 96
3.5. Diameter of control constrictions 99
3.5.1. Probabilistic approach 100
3.5.2. Control diameter: several definitions 103
3.6. A continuous approach of the process of filtration. 105
3.7. Conclusion 110
3.8. Bibliography 111

Chapter 4. Contact Erosion between Two Soils. 115
Rémi BÉGUIN, Pierre PHILIPPE, Yves-Henri FAURE and Cyril GUIDOUX

4.1. Introduction. 115
4.2. Areas prone to CE in hydraulic structures 116
4.2.1. General characteristics of constitutive materials. 116
4.2.2. Heterogeneities within the concerned hydraulic structures 117

4.2.3. Possible configurations of CE within a hydraulic structure 118
4.3. Description of CE mechanisms on a local scale 120
4.3.1. CE in the general context of surface erosion 120
4.3.2. Local measures in the contact area between two soil layers 121
4.3.3. Flow regimes and bottom shear stress 122
4.3.4. Erosion threshold of the particles of fine soil 125
4.3.5. Kinetics of erosion and transport of the fine particles 128
4.4. CE of a fine soil under a coarse soil 131
4.4.1. Determination of the threshold of CE initiation 133
4.4.2. Influence of cohesion 139
4.4.3. CE kinetics 141
4.4.4. Modeling 143
4.5. CE of a fine soil on a coarse soil 146
4.5.1. Conditions for the occurrence of CE 146
4.5.2. Comparison with the configuration of a fine soil under a coarse soil 148
4.6. Possible scenarios that may lead to failure of a hydraulic structure 149
4.7. Conclusion and perspectives 151
4.8. Bibliography 152

Chapter 5. Concentrated Leak Erosion 155
Nadia BENAHMED, Christophe CHEVALIER and Stéphane BONELLI

5.1. Introduction 155
5.2. General points 156
5.2.1. Experimental reproduction of pipe flow with erosion 156
5.2.2. Erosion parameters: shear stress and erosion rate 157
5.2.3. Erosion law: shear stress threshold and erosion coefficient 158
5.3. The device and the protocol of the HET 160
5.3.1. Equipment 160
5.3.2. Preparation of the specimen 161
5.3.3. Test procedures 162
5.4. Methods of interpretation 164
5.4.1. Determining the shear stress 165
5.4.2. Interpretation by means of turbidity 165
5.4.3. Interpretation using the flow/pressure relation 169
5.5. Effect of different soil parameters on erosion 174
5.5.1. Trial campaign 174
5.5.2. The influence of clay content 176
5.5.3. The influence of the nature of the clay 178
5.5.4. The influence of the density and the moisture content 180
5.6. Importance of the erosion index for hydraulic structures 181

5.6.1. The case of earthfill dams 182
5.6.2. The case of dikes 183
5.7. Conclusion 183
5.8. Bibliography 184

Chapter 6. Modeling of Interfacial Erosion 187
Stéphane BONELLI, Frédéric GOLAY and Fabienne MERCIER

6.1. Introduction. 187
6.2. Modeling of a two-phase medium 189
6.2.1. Definition of volume quantities 189
6.2.2. Definition of interfacial quantities 192
6.2.3. Conservation of mass equations 193
6.2.4. Equations of motion 194
6.3. Modeling of the soil/fluid interface. 195
6.3.1. Simplifying assumptions 195
6.3.2. Rewriting of jump equations 196
6.3.3. The erosion law 196
6.4. Modeling of flow with erosion 198
6.4.1. Two-phase flow with erosion and transport 198
6.4.2. Analysis of the orders of magnitudes 199
6.4.3. Modeling of dilute flow with slow erosion 203
6.4.4. Turbulence modeling. 204
6.5. Numerical modeling. 205
6.5.1. Interface capturing: the fictitious domain method and the level set method 206
6.5.2. Interface capturing: Lagrangian approach and remeshing 208
6.6. Validation of numerical models 209
6.6.1. Simplified modeling and exact solutions in the case of pipe flows 209
6.6.2. Plane Poiseuille flow with erosion. 211
6.6.3. Axisymmetric Poiseuille flow with erosion 212
6.7. Illustrative examples 213
6.7.1. Scouring of a vertical cylinder 213
6.7.2. Modeling of Jet Erosion Test. 215
6.7.3. Modeling of Hole Erosion Test. 216
6.8. Conclusion 219
6.9. Bibliography 219

Chapter 7. Physics of Sediment and Aeolian Transport 223
Bruno ANDREOTTI and Philippe CLAUDIN

7.1. Introduction. 223
7.2. Static transport threshold 224

7.2.1. Shields number 224
7.2.2. The influence of longitudinal slope 226
7.2.3. Influence of soil cohesion 227
7.3. Aeolian transport 227
7.3.1. Qualitative description 227
7.3.2. Dynamic transport threshold 229
7.3.3. Focal point of the velocity profiles 230
7.4. Quantitative description of transport 232
7.4.1. Interface between the sediment bed and the fluid 232
7.4.2. Flux and conservation of matter 233
7.4.3. Saturated flux 235
7.4.4. Length of saturation 237
7.5. Linear stability analysis of a flat erodible bed 240
7.5.1. Qualitative description of instability 240
7.5.2. Linear stability analysis 242
7.5.3. A scaling law for elementary dunes 246
7.6. Conclusion 248
7.7. Acknowledgments 249
7.8. Bibliography 249

Chapter 8. Two-Phase Modeling of Bedload Transport 253
Pascale AUSSILLOUS, Elisabeth GUAZZELLI and Yannick PEYSSON

8.1. Introduction 253
8.2. Incipient motion 254
8.2.1. Shields number 256
8.2.2. Force balance on a grain 256
8.2.3. Force balance on a bed of particles 257
8.2.4. Shields curve 259
8.2.5. Laminar regime 261
8.2.6. Turbulent regime 263
8.3. Bedload transport 265
8.3.1. Particle flux 265
8.3.2. Two-phase modeling 267
8.3.3. Closure equations in laminar regime 269
8.3.4. Resolution for shear flows 271
8.3.5. Comparison with experiments in laminar regime 275
8.4. Conclusion and outlook 276
8.5. Acknowledgments 276
8.6. Bibliography 277

Chapter 9. Characterization of Natural Cohesive Sediments and Water Quality of Rivers 281
Fabien TERNAT, Patrick BOYER, Fabien ANSELMET and Muriel AMIELH

9.1. Introduction. 281
9.2. Behavior of pollutants in rivers 282
9.2.1. General balance equations. 287
9.2.2. Term of deposit flow and erosion 290
9.3. Erosion of fine cohesive sediments. 292
9.3.1. Erosion threshold of a non-consolidated layer. 292
9.3.2. Erosion threshold of a consolidated and saturated layer 293
9.4. Experimental characterization. 302
9.4.1. Characterization of sediments in the water column 302
9.4.2. Characterization of the sediment column 306
9.5. Example of operational application. 310
9.5.1. Simplified approach by a box model 310
9.5.2. Application to the radioactive contamination of the Techa River (Russia). 312
9.6. Conclusion 317
9.6.1. Review 317
9.6.2. Outlook. 318
9.7. Bibliography 319

Chapter 10. Sediment Transport and Morphodynamics in Nearshore Areas 325
Vincent REY and Damien SOUS

10.1. Introduction. 325
10.2. Marine sediments 326
10.2.1. Characteristics of sediments. 326
10.2.2. Settling velocity of sediments. 328
10.2.3. Rheology of suspensions and deposits. 330
10.3. Sediment transport. 331
10.3.1. Suspending the particles, the notion of threshold 331
10.3.2. Modes of transport 334
10.3.3. Equations and transport models. 335
10.4. Sediment structures and morphodynamic structures 336
10.4.1. Sediment structures outside the breaking area 336
10.4.2. Sediment structures in breaking zone 343
10.5. Local effects: scouring around the structures 351
10.5.1. Scouring around the pipelines. 352
10.5.2. Scouring around vertical cylinders 353
10.5.3. Scouring upstream from nearshore structures 353

10.6. Long-term morphodynamics of beaches 354
10.6.1. Evolution models of the profile of the beaches. 354
10.6.2. Shoreline evolution models 356
10.7. Conclusion 359
10.8. Bibliography 360

List of Authors 365

Index 369

Foreword

Soil erosion, be it internal erosion or surface erosion, is the main mechanism responsible for disorders or failures of civil engineering structures. These failures usually take place as soon as the soil, which constitutes either the entire engineering structure or only its foundation, comes into contact with water flows. The affected engineering structures are usually reservoir dams, canals, flood-protection dikes (structures that we will be referring to under the umbrella term "hydraulic structures"); as well as bridges, flood-protection dams, and engineering works carried out with the purpose of stabilizing coastlines.

This is the broad scope of this present compendium. Furthermore, this compendium is part of a twofold context: on the one hand, the need for improving the safety concerns regarding these engineering structures, and on the other hand, the significant research programs performed on erosion, whether these were recently carried out or are currently underway.

The regulations applied to hydraulic structures have been recently reinforced in France and they have been adapted according to the stakes involved in case of failure. These regulations insist on the necessity of assessing the safety level of these structures, particularly through risk analyses, as well as on strengthening their supervision. These regulations will apply in France from now on to 700 "large" dams, tens of thousands of small dams, approximately 8,000 km of canal dikes, and around 10,000 km of dikes and levees meant to protect against flooding or marine submergence.

The consequences that climate change is likely to have on the sea level as well as on continental hydrology will lead to increasing demands on coastline and fluvial dikes and levees. This, in turn, brings about the need for evaluating their safety and, should they fail that evaluation, the need to reinforce safety measures.

In recent years, these issues that are at stake have justified a strong mobilization on the part of the scientific community, which structured itself around several research projects, both on national and international levels. In France, we should mention the GdR MiDi and MeGe and the projects ERINOH, Dunes, and Carpeinter, to name only the most important projects.

This compendium is mainly concerned with the results of such research works; it aims to answer the queries of those responsible with the management of the civil engineering works that we have mentioned above.

Paul ROYET
June 2012

Introduction

The question of the natural risks involved in soil erosion remains a very active subject. The complexity of this issue is linked to multi-scale and multi-physics couplings, to the time durations involved, and to the highest quality required. This research is linked to the multidisciplinary character of many of the remaining questions. These subjects are dominated by safety and environmental protection concerns.

While an old issue, soil erosion remains so far an "open" problem. A great number of research works that deal with soil erosion are published regularly. This question is most often approached either on large spatial scales, from a hydrological and geomorphological viewpoint, or on a smaller scale, from a hydraulic perspective. On the contrary, the mechanical/physical approach of this subject, though it has been developed in scientific journals, has not yet motivated a reference work.

This compendium aims to deliver a significant part of the current French scientific progress on the problem of the erosion of geomaterials, with a focus on the mechanical/physical aspect, while targeting a twofold objective – coherence and pedagogy. This work does not pretend to be thorough on such an age-old, multi-disciplinary, and highly debated subject. In this respect, the contributions oscillate between a phenomenological outlook that is well grounded in experiments and a mechanical/physical approach that can offer a modeling framework.

The outline of this work is structured as follows: the first five chapters, are dedicated to the phenomenon of internal erosion. Once the internal erosion has been introduced within the framework of hydraulic structures (embankment dams and dikes), the basic mechanisms of internal erosion will be tackled one-by-one. The

---

Introduction written by Stéphane BONELLI.

last five chapters, deal with surface erosion. Interfacial surface erosion and sediment transport as well as bedload transport are examined from different perspectives and in different situations (aeolian, fluvial, coastline). Each chapter is concerned with two crucial aspects: the motion threshold, in connection with the critical stress force, and the flow of the eroded and transported matter, that sometimes includes a coefficient of erosion. This is the common thread that underpins this work.

The geomaterials we will be concerned with are mainly soils and sediments. The interactions are especially of mechanical origin, but several complex interactions, of physical-chemical origin are also considered. Throughout the chapters, the authors will highlight the different kinds of geomaterial erosion we may experience, at the same time trying to extract the practical consequences that are of interest to a practitioner.

Chapter 1, written by Jean-Jacques Fry, is an introduction to internal erosion as it presents itself in hydraulic structures: dams and dikes. The importance of this issue is proven by the feedback gathered from various incidents. The main results of erosion tests and their applicability are summarized and they set the tone for the next chapters.

In the second chapter, Didier Marot and Ahmed Benamar tackle a difficult problem that is likely to come up in hydraulic structures: suffusion, i.e. the transport and the filtration of fine particles in a granular soil. The significant parameters and the suffusion initiation criteria are analyzed and discussed. Subsequently, the migration of fine particles and their filtration is examined.

Eric Vincens, Nadège Reboul and Bernard Cambou are the authors of Chapter 3. They too deal with the process of filtration in granular materials, but this time from a micro-mechanical perspective, using a fine modeling scheme that is both theoretical and numerical, by means of the discrete element method. At the end of the chapter, by returning to the scale of a continuous medium, new perspectives are opened for the modeling of hydraulic structures.

In Chapter 4, Rémi Béguin, Pierre Philippe, Yves-Henri Faure and Cyril Guidoux devote their attention to a still under-researched mechanism of erosion, but which can be found in numerous dikes: contact erosion between two types of soil. The description of the mechanisms that take place at grain level allows for a better understanding of the conditions under which they occur as well as of the kinetics of erosion. Finally, several possible scenarios that might lead to a failure of hydraulic structures are suggested.

Chapter 5, written by Nadia Benahmed, Christophe Chevalier and Stéphane Bonelli, is concerned with flow erosion as it takes place in a pipe. In fact, this chapter thoroughly analyzes the "Hole Erosion Test", which was carried out for hydraulic structures in the United States and Australia, and recently in France. Several laboratory tests show the influence on erosion parameters of the clay nature, of the soil water content, and of its density. Finally, this chapter emphasizes the importance of the coefficient of erosion needed to evaluate the safety of hydraulic structures in relation to the risks of failure caused by pipe flow with erosion.

In Chapter 6, Stéphane Bonelli, Frédéric Golay and Fabienne Mercier present an original modeling of interfacial surface erosion. The modeling of a two-phase flow with erosion and transport is simplified, and two numerical models are proposed: one of them is based on the fictitious domain method and the level-set method, while the other is based on remeshing. Several examples of validation and illustration are presented, with wide applicability to laboratory tests.

In Chapter 7, we enter the field of sediment transport physics. Using a fundamental approach, Bruno Andreotti and Philippe Claudin open up new perspectives regarding the erosion law. The authors show us what are the scaling laws that govern transport thresholds, sand flux saturation length, and what is the wavelength at which an erodible bed is destabilized. It becomes apparent that, contrary to naturalistic classifications, we must take aquatic sand ripples, aeolian dunes and the giant Martian dunes to be of the same type.

In Chapter 8, Pascale Aussillous, Elisabeth Guazzelli and Yannick Peysson prefer a continuous two-phase medium approach for the modeling of bedload transport. Two essential aspects are examined: the setting into motion threshold and the transported sediment flow. The closure laws proposed in a laminary situation allow us to obtain a smooth agreement between modeling and experimentation. In particular, it is demonstrated that the motion threshold that corresponds to the critical number of Shields mainly depends on a friction coefficient of the granular medium and its compactness.

Chapter 9 is dedicated to natural cohesive sediments within the context of the quality of water streams. This chapter is written by Fabien Ternat, Patrick Boyer, Fabien Anselmet and Muriel Amielh. Having introduced the behavior of pollutants in water streams, the authors have carefully analyzed the erosion of fine cohesive sediments both theoretically and experimentally. Finally, for illustrative purposes, the authors provide us with the example of radioactive contamination of the river Techa in Russia.

The tenth and final chapter branches out to the crucial matter of sedimentary and morphodynamic transport in coastal areas and along the shoreline. In this chapter,

Vincent Rey and Damien Sous write successively about marine sediments, sedimentary transport, sedimentary and morphodynamic structures, as well as the scouring around these structures. This chapter ends with a presentation of the modeling of beach and coastline morphodynamics in the long-term range.

More than half of these chapters are a result of the research carried out over several years within the ERINOH Project – French Research Agency (ANR) Project and National Project – coordinated by the Institute for Applied Research and Experimentation in Civil Engineering (IREX). This compendium equally integrates the results from several other research projects: Dunes and Carpeinter, Zephyr (both financed by the ANR), PEA ECORS (SHOM) and MICROLIT (INSU-RELIEFS), Project Scale of GRR SER from the Haute Normandie Region of France. Finally, throughout the years, the two GdR CNRS MiDi and MeGe have brought together the majority of the present authors.

To conclude, I wish to express my gratitude to M. François Nicot, editor, for the confidence he has given me. I also wish to warmly thank the authors of these chapters, who have generously offered their time and shared their expertise in writing up this compendium.

Our aim in this book has been to provide a means for physicists and engineers to share with the reader their most recent findings in their field, while at the same time maintaining an accessible format. We hope that the reader finds in this compendium a well-documented information resource, and above all, a key for approaching the issue of erosion of geomaterials in an up-to-date manner.

Chapter 1

Introduction to the Process of Internal Erosion in Hydraulic Structures: Embankment Dams and Dikes

1.1. Introduction

The first part of this book presents the initial results of the National Project ERINOH (the acronym for *Erosion Interne des Ouvrages Hydrauliques*, i.e. Internal Erosion of Hydraulic Structures) and this chapter provides a general introduction to the problem. What is this lesser known pathology that can seriously damage the safety of such hydraulic structures as embankment dams, dikes, and canals? So far, there has been a strong imbalance in the way academics have presented the physical phenomena that govern the maintenance of hydraulic structures.

Relevant textbooks place the emphasis on the mechanical analysis of general stability, although stability is only marginally involved in the majority of incidents and failures. Consequently, this chapter aims to provide several basic elements that are necessary to understand and analyze the complex hydraulic phenomena that, under changeable circumstances, bring into play the interactions between water and porous media. The main objective of this chapter is to render familiar the knowledge we have obtained so far, thus offering a more global and easier reading of the chapters that follow.

Chapter written by Jacques FRY.

1.2. The significance of internal erosion for hydraulic structures

1.2.1. *The set of hydraulic structures in France*

France disposes of a significant stock of hydraulic structures. The linear lengths of dikes are roughly equivalent to 13 times the largest dimension of its territory, with more than 9,000 km of protection against flooding, 8,000 km of dikes for navigation canals, and 1,000 km of hydroelectric canals. The number of small embankment dams, whose height does not surpass 15 m, is around several tens of thousands, while the number of large dams approaches 600.

While the first characteristic of French hydraulic assets is their amplitude, the second characteristic is their age: most dikes are more than 100 years old and most dams are older than half a century. Finally, the predominance of natural materials used in the backfilling of these dams is the third characteristic.

The maintenance of such a wide patrimony, both old and built with local materials, requires a costly and decentralized upkeep, which is difficult to achieve in an economically restrictive context. This, in turn, poses a problem for exploitation safety, and hence the need for innovative and economical solutions.

Hydraulic structures are civil engineering structures whose function is to retain or transport water. Apart from being a natural resource, water is also a kind of "fluid" energy. Because it stands in the way of the water, a hydraulic structure must constantly fight against this energy, which can take advantage of the slightest fault in order to break loose. Consequently, following the example of the International Commission On Large Dams (ICOLD), we have regarded the loss of this main function of the hydraulic work as a failure.

1.2.2. *The vulnerability of hydraulic structures*

The dams are either built of natural materials without using any binder (backfill) or of materials that are reinforced using hydraulic binders (i.e. lime in the first gravity dams that were built using masonry, and then cement, fly ash, or slag cement in the case of concrete dams). The statistical data gathered by the International Commission on Large Dams (1995), presented in Table 1.1, emphasizes the fact that backfill dams make for more vulnerable structures than concrete or masonry dams. However, this table does not include the more fragile backfill dams that make up the flood-protection dikes whose failures hit the headlines during large floods.

Type of dam	Number of dams	Number of failures	Number of failures per 100,000 years*dams
Concrete or masonry	5,500	2	2
Loose material	16,500	26	9

Table 1.1. *Number of failures/year/dam in the world (excluding China) from 1970 to 1990 according to ICOLD (1995)*

1.2.3. *Erosion as a leading cause of failure*

Water is the main load as well as the main aging factor that the facility must hold against throughout its exploitation. Without maintenance, the weathering of its watertightness or a massive flood can easily trigger the failure of the entire hydraulic facility.

The forces that traverse the facility either go through the solid phase, in the form of intergranular stress forces (also called effective stress forces), or go through the liquid phase, i.e. water. Water, in turn, can either be found in a permanent and isotropic form, i.e. hydrostatic pressure, or in a transient and anisotropic form, i.e. hydrodynamic pressure. Consequently, there are two types of forces that are fundamental in every abnormal motion, which can trigger two types of completely different failures:

– a global mechanical shearing that separates the grains through tangential intergranular forces that are too elevated, throughout the entire sliding surface, called sliding or general instability;

– a local pullout of the grains, triggered by the hydrodynamic forces of the water, here called erosion and sometimes referred to as local or internal instability.

There are two criteria for the application of hydrodynamic forces; this chapter is only concerned with the first criterion:

– the seepage flow that causes internal erosion;

– the stream flow that can be found on the surface of the porous medium, which produces external erosion.

To get a more precise idea about the relative significance of each of these erosions, it is useful to refer to the statistics compiled by Foster *et al.* [FOS 00] regarding large dams. The following table shows that erosion (whether internal or external) represents the greatest danger, being far ahead of the problem of stability against sliding phenomena. Of the failures, 94% are caused by erosion. Over time, the failure probability for backfill dams has decreased, but the relative significance of erosion among the causes of failure has increased (Table 1.2).

Modes of failure in large dams	Number of failures to number of dams ratio in 1970–1979	Number of failures to number of dams ratio in 1980–1989
Internal erosion	20×10^{-4}	16×10^{-4}
External erosion	26×10^{-4}	19×10^{-4}
Sliding	4×10^{-4}	1×10^{-4}

Table 1.2. *Relative significance of failure modes (ICOLD 1995, excluding China)*

Although it is less known, internal erosion by no means is less dangerous, as is shown by the recent surveys on the failure of large dams.

1.2.4. *Internal erosion: one failure per year in France*

Internal erosion has been targeted as a recurrent maintenance problem. The first European symposium of ICOLD in 1993 confirmed that internal erosion is a major concern at international level, thus requiring a European research team to work on such issues. Soon after, the flooding of the Rhône in October 1993 and January 1994 caused 16 breaches in the dikes of the Camargue. The investigations carried out by Cemagref identified the internal erosion as the only cause. Following this incident, the French Committee for Large Dams (CFGB) appointed a team of researchers from the relevant field to collect feedback. The team submitted their report on the typology, detection, and repair of disorders caused by internal erosion, which was then distributed at the ICOLD Congress in Florence in 1997 [FRY 97]. The team identified that from 1971 to 1995, 71 incidents, of which 23 failures, had been reported on 550 large dams, several thousands of small dams, and 1,000 km of dikes. This study did not take into account the flood-protection dikes or the navigation canals.

Despite this publication, the frequency of failures remained high and constant during the following years, that is approximately one failure per year: the failure of the Arroux gutter in 2001, the Ouches dam in 2001, the Briare canal at Montambert in 2002, the North canal in 2003, the Rhône-Rhine canal in 2005, and the Roanne canal at Digoin in 2007. The most significant water damage took place at the time of the great floods in the Lower Rhône. The breaches of the Gard flood on 9 September 2002 cost the lives of five people in the village of Aramon and caused damages of € 1,200 million. The following year, when the Rhône flood took place in December 2003, more than 100 points of disorder were noted, of with half a dozen breaches, causing €845 billion in damages, spread over five *déparments*. More generally, in Europe, the yearly cost of floods between 1980 and 2003 was estimated at three billion dollars. In France, it reached a billion euros in 2002 and 2003. The vulnerability of these structures is such that a decrease of only 1% in these damages would be enough to justify the profitability of a research project on internal erosion, whose predicted annual gains should be clearly higher.

1.3. The impact of incidents on embankment dams and dikes

1.3.1. *Terminology*

In what follows, we will adopt the definitions given by the ICOLD as well as the most recent definitions given by French regulations.

An *embankment dam* is a facility that retains a permanent body of water (and is thus considered to be a hydroelectric canal or a navigation canal). A *dike* is a facility that retains the water only temporarily (i.e. flood-protection dike).

A *failure* is a collapse or a displacement of the entire backfill dam or of the foundation of the facility so that the water can no longer be retained by the facility.

A dam *accident* is a collapse or a displacement of the backfill dam that, in the absence of immediate intervention, can lead to failure.

A *failure mode* is a breaching process generated by a generic type of force. We can distinguish between two modes of failure: sliding, which happens when the resistance to mechanical shearing is overcome, and erosion, which is caused by hydrodynamic forces.

A *process of breaching* by erosion is a set of mechanisms that leads to the breaching of the facility. Before reaching the final stage of breaching, there are four mechanisms of erosion that occur successively: *initiation, filtration, continuation*, and *breaching*.

Finally, in order to distinguish between these various phenomena, a particle of cohesionless soil is called a *grain*, whereas an agglomerate of argillaceous particles is called a *floc*.

1.3.2. *Initiation areas*

The knowledge we have acquired so far on internal erosion failures rests on the public survey published by Foster *et al.* [FOS 00]. This study regrouped 71 cases of failure of large dams, and several cases of small dams. In 2011, Vogel updated the number of failures to 125 large dams and 340 small dams. In France, although internal erosion has not yet destroyed any large dams, it has damaged more than 50 smaller dams.

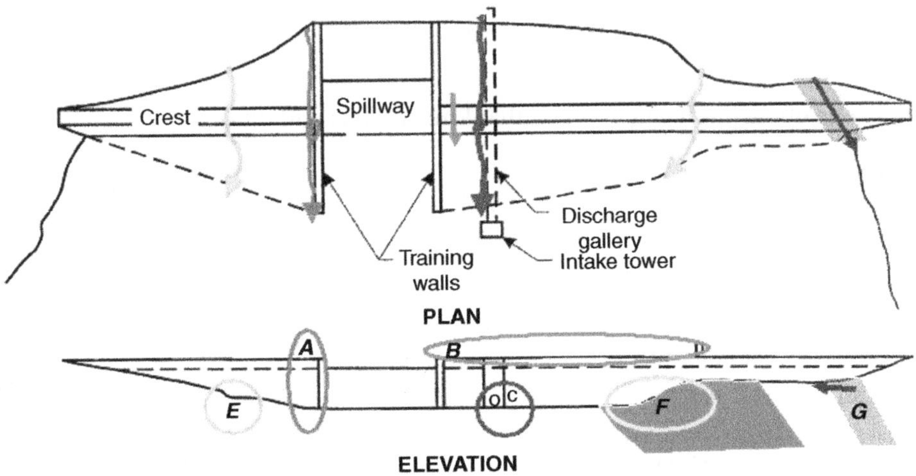

Figure 1.1. *Areas of initiation for internal erosion in an embankment*

Figure 1.1 shows the main points of initiation. Half of the failures start throughout the pipe that traverses the facility, or at its loose foundation (area *C* in Figure 1.1). Five important causes are responsible for this abnormality:

1) The lack of filters that surround the pipe at the exit on the face of the dam, or on the downstream foundation, is a widely claimed cause. The failure of the Sheep Creek dam took place in 1970 at its first priming. On the other hand, the failure of the Santa Rosa dam took place in 1975 after 46 years of exploitation.

2) When the circular pipe is not placed within a cradle made of grout, concrete, or when it is not placed at the bottom of a gallery, the concrete compaction is always incomplete. This was the case of the Saint-Julien des Landes dam where the first priming could never be completed.

3) The differential displacements imposed by the foundation that end up cracking the pipe. The outlet pipe of the small dam of La Prade was cracked in 1987, five years after it was first put into service, as a result of significant settlement of the foundation, i.e. around 25 cm. The crack was part of the faulty welding of one of the metallic anti-piping shields, thus questioning the viability of these shields. A deep seepage having a 30 cm diameter was discovered in 1981 on the face of the

Buget dam; it was the result of a leakage on one of its corroded pipes, which, being under the heavy pressure of water, had cracked as a result of settlement.

4) The deterioration of the pipe is a latent problem that presents itself in various shapes. The Houches dam broke down in 2001, after two centuries of exploitation, because its initial outlet pipe mad of wood, had rotted.

5) Hydraulic breakdown along a pipe that is placed inside a trench that is too narrow or that is placed between walls that are too steep is usually caused by the load transfer of the overlying weight of the layers, which, in turn, becomes insufficient to resist the Archimedean buoyancy force. This is how a breach developed within the pipe of the Zoeknog dam (1993), breaking it into two at its first priming.

The erosion that concerns only the peripheral foundation or its center (Figure 1.1, area *F*) is the second localization of failures, covering 20% of cases. This type of erosion differs according to the makeup of the ground. In the case of loose foundations, the retrogressive erosion of a layer of sand can often be found under the estuary dikes whose filling is made of clay. Contact erosion appears frequently at the interface of a layer of fine alluvial deposits whose particles of silt are carried away by the water flow that traverses the adjoining layers of coarse-grained alluvia. In the case of rocky foundations, the pads of spoiled rock, fractured by the decompression joints or by tectonic faults (with a thickness of over 10 cm), partly obstructed by an unconsolidated clay or sand filling, with a dip that is parallel to the surface, can cause extremely dangerous water circulation. The areas that became less cohesive due to the alteration or the areas that contain soluble materials, such as gypsum, are the most vulnerable areas. The rocky foundation of the Itiyuro dam has thus lost more than 1,000 m^3 of solid matter through the rockfill, as a consequence of leakage in its supporting gallery.

The contact between the filling and the structures that cross it, such as the spillway (Figure 1.1, zone *A*), is the third area where failure can happen as a result of load transfer, traction cracks, lack of compaction, and differential settlements. These problems account for 8% of failures. The concrete buttress, which is supposed to be anti-piping and against the spillway of the dam, did not help in the case of the Rowallan dam in avoiding the formation of subsidence only a year after its first priming (1968) because of the difficulty of compacting it all around.

The core of the backfill dam with 95% of the volume of the material well put into place represents only 8% of the failures out of the six surveyed cases. This small volume rate of failure tends to show that the failure initiation areas can be found rather more toward the periphery of the filling. The fact that small dams are under represented does not allow us to verify this interpretation with regard to homogeneous dams.

Internal erosion is likely to occur near the crest as well (Figure 1.1, areas *B* and *D*). Of failures, 95% took place when the water had reached its highest level. The main reasons are the presence of cracks in the dams (drying, shearing of the core, differential settlement, or earthquakes), the presence of excavation holes in dikes (13 to 16 dike failures in the Camargue in 1993 and 1994 come from excavation holes), or infiltration above the core (filling settlement, lack of filter above the core, a flooding more powerful than predicted, etc.). Four failures started at the crest, which represents 6% of failure cases.

The contact with the irregularities and heterogeneities of the foundation represents yet another source for disorders (Figure 1.1 zone *E*). Three out of 71 failures started upon the contact of the foundation. The hydraulic breakdown of the core in the canyon of the Mud Mountain dam in 1982 is one of the most spectacular examples. The quasi-vertical canyon on the right bank caused a heavy load transfer on the upper-core banks and a generalized fracture of the lower part, where leakages and filler transport were observed, after 41 years of exploitation.

The erosion that started out in the filling and continued toward the foundation, or conversely the erosion that started out in the foundation and continued toward the filling, has only caused two failures (Figure 1.1, zone *G*). However, one of them is more spectacular: the failure of the Téton dam (1976) during its first priming. The cutoff trench with steep walls caused a load transfer that presented – on the silt at the bottom of the trench – a total stress value that was too low to be able to oppose the pressure of the pondage. The hydraulic breakdown of the silt caused heavy flows in the widest joints of the cracked rock. The quickly washed-out joints, which opened into the downstream of the filter, allowed for the streams to break, in their turn, the outlet pipes of the downstream face. Retrogressive erosion ensued by creating a breach and thus emptying the container in a few hours.

The analysis of these accidents enables us to evaluate the reliability of the laws that governed the design of embankment dams and dikes during those times.

1.3.3. *The importance of design*

There have not yet been any reported cases of failure in large dams in the presence of a granular material called "filter". This material is not cohesive; it is carefully elaborated and placed downstream for water tightness. This confirms that the first defense barrier against internal erosion is by far the filtration carried out with an "appropriate" filter. Several failures have taken place in the presence of "inappropriate" filters.

However, although the filter is necessary almost everywhere, it does not always suffice. It can be bypassed by a faulty foundation, a faulty design, or by a fault in the execution. We should bear in mind the lesson of the Téton dam, where the filter had been bypassed. In fact, among the three possible tracks that erosion can take – through the filling, from the filling toward the foundation, and throughout the foundation – the filter helps to intervene particularly in the first track.

Foster *et al.* [FOS 00] have classified the problems encountered according to the type of design. Twelve profiles have been studied and are represented in Figure 1.2.

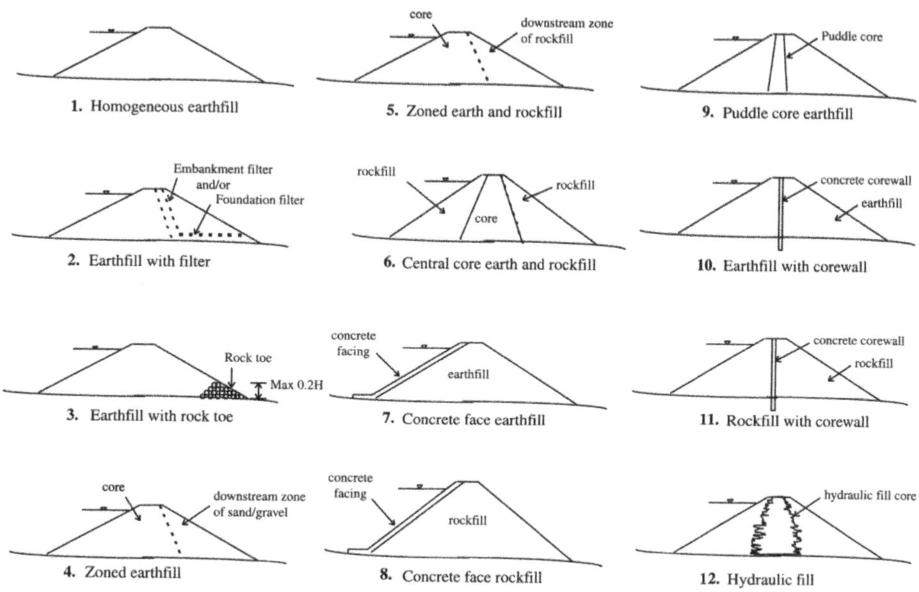

Figure 1.2. *Profile type classification after Foster et al. [FOS 00]*

Homogeneous earthfill: This zoning-free profile is very frequent in older dikes and has the highest failure rate, i.e. around five times higher than all the other types taken together. This figure is an underestimation for the totality of the dams, as a consequence of the underrepresentation of small dams. The failures have often been caused by leakages along the pipes that traversed the filling (nine cases), by retrogressive erosion or erosion of the cracks (16 cases), or by piping phenomena that took place in poorly compacted materials (12 cases) and piping in dispersive types of soil (four cases). In Europe, it is worth noting that the rate of failure cases involving homogeneous backfill dams built before 1900 is approximately 10 times higher than for those built after 1950. This general remark which applies to an old set of dams must be considered in connection with the frailty of the set of small French dams.

Earthfill with rock toe: This type of backfill dam also has one of the highest rates of failure by internal erosion. Nine failures have been inventoried, starting from internal erosion throughout the filling (five cases) or starting from internal erosion at the foundation level (three cases) and, lastly, starting from internal erosion in the filling and continuing toward the foundation (one case). The failure was generally the consequence of an erosion focused around the pipe, of backward erosion of homogeneous filling throughout the empty spaces between the rock ripraps, or, finally, of water streams flowing through the cracks that had formed within the filling as a result of several irregularities in the foundation or of excessively steep banks.

Zoned earthfill: This profile has a relatively low failure rate, mainly because most of them have a downstream shell of alluvia, which behaves like a filter. The majority of these embankments having been built after World War II have also benefited from a more careful approach. It is noteworthy that three out of four failures took place at the time of the first priming.

Earthfill with a thin core made of puddle core: This type of dam is the ancestor of the zoned dam, built almost two centuries ago in the United Kingdom. Its failure rate is relatively high. Out of the four failures of large dams, three were built before 1856 and each time they were the consequence of a hydraulic fracture of the thin core, where the core had not been protected by any transition material. More than half of the piping phenomena took place after more than 50 years of exploitation, whereas the failure of the small dam at Clandeboye that was only 5 m high took place after 80 years of exploitation and was attributed to a wooden pipe that traversed the dam.

Earthfill with upstream core and downstream zone of rockfill: There is only one case of failure regarding this type of design, i.e. the Avalon dam that broke in 1904, because of a backward erosion caused by the contact with the rockfill in the absence of a filter.

Earthfill with filters and drains: This type of dam has a relatively low rate of failure. The failures are usually associated with the presence of preferential streams and hole erosion, or backward erosion throughout the dam made of dispersive material at the contact of the traversing pipes (three cases) or happening at the contact of the spillway (one case). The failures started out in those places where the filter had been removed: the two failures of backfill dams with a vertical filter were caused by backward erosion where there was no filter around the pipe, whereas in the other two cases, the dam had only a horizontal filter.

Earthfill with upstream concrete facing: Five failures due to internal erosion have been reported. Two of them happened throughout the backfill, and three happened throughout the foundation. The main cause of failure has been identified as being the connection between the pipe and the revetment (two cases out of three). As regards the third case, the problem had been the settlement and the failure of the backfill in the pipe. If internal erosion causes a network of pressure that had not been predicted by the design, it compromises the stability of the downstream face of the dam, since it had not been designed to support such a hydrostatic load. This is how the Gouhou dam fell apart, after a significant decrease in water tightness.

Rockfill with upstream concrete facing: Internal erosion had little connection with this type of design, probably because of the good quality of the foundation as well as the stability and draining capacity of the bulk of the backfill.

Zoned dam with central core, filters, and rockfill: No failure has been reported on this up-to-date type of design, although 21 accidents have been recorded. Of those, 15 were associated with a core suffusion with widespread grain size distribution of glacial origin throughout coarse or segregable filters. This great number of accidents and the absence of failures lead us to think that this type of dam has a very low failure rate and that the erosion kinetics is very slow. The analysis of these accidents indicates that this is not only a result of the very slow kinetics of volume erosion at the heart of the backfill, through suffusion, but also a result of the strong influence that filters, the draining capacity, and the general stability of the downstream shell have on controlling the process of erosion.

1.3.4. *Four mechanisms of erosion, classified according to their boundary conditions*

Terzaghi had distinguished between two different failure mechanisms in hydraulic structures: "failures by subsurface erosion and failures by heave". He immediately discerned between the mechanisms used by hydrodynamic outflow forces, grouped them under the terms "internal erosion" and "subsurface erosion", i.e. the hydrostatic mechanisms that cause the so-called heave condition (also known as blowout). However, the heave condition is not considered here.

Subsequently, the studies carried out on the incidents of internal erosion have shown that there had existed different outflow kinetics and different erosion kinetics that corresponded to different levels of danger. To try and classify these levels of danger, it might be useful to distinguish between different initiation mechanisms, or rather to discern the boundary conditions of the outflow that lie at the bottom of these manifestations of internal erosion. There are four types of boundary conditions that cause internal soil erosion (Figure 1.3):

1) erosion in concentrated leak, i.e. the water flows through a crack, a hole, or a hollow, thus eroding the walls;

2) backward erosion, i.e. the water flow erodes the surface of the soil from where it comes, thus causing a backward erosion of this exit surface;

3) contact erosion, i.e. the water flows though a very permeable soil, thus eroding a fine soil at the interface; and

4) suffusion, i.e. the outflow erodes the small grains through the pores of the coarser grains.

Pipe or crack internal erosion		*Backward internal erosion (generalized or local, i.e. at the bottom of the pipe)*	
	Entrainment of the particles from the tips of the crack or from the surface of the hollow, under the action of the outflow.	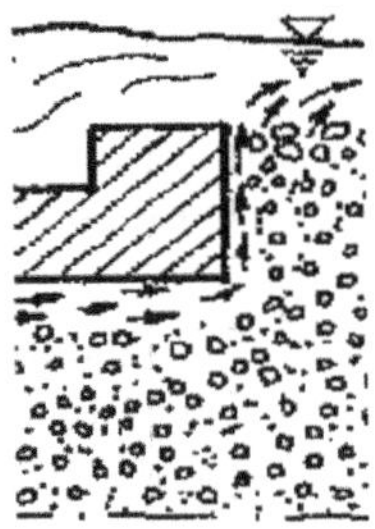	Entrainment of particles underneath a cohesive surface, or a rigid surface at the outlet pipe of the subterranean stream flow, in the beginning on the downstream wall, and then at the bottom of the hollowed pipe, starting downstream and going upstream.
Internal contact erosion (between gravel and silt)		*Suffusion, or the internal instability of a soil, or volume erosion*	
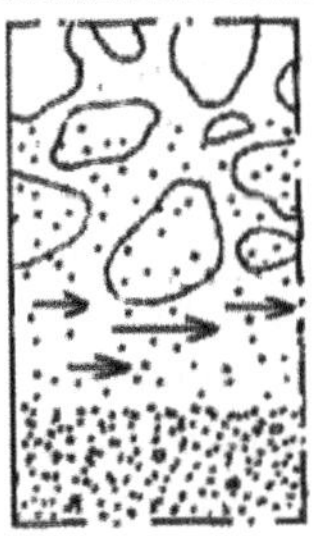	Entrainment of the particles of a fine soil throughout its contact with a coarse soil; this entrainment takes place as a result of the water stream that traverses the coarse soil.	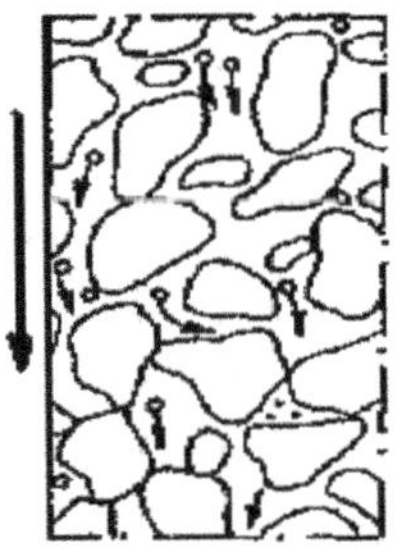	Entrainment of free moving fine particles that can move throughout the gaps between the coarse particles of a soil with discontinuous grain size distribution.

Figure 1.3. *The four types of erosion analyzed [FEL 07]*

The succession of internal erosion phenomena, observed alongside the great rivers descending from the Alps, the Rhône and the Rhine, illustrates quite well the four types of phenomena observed.

The water of the mountain stream traverses shingle, colluvia, or moraine, and it causes a suffusion in those soils whose grain size distribution level is too widespread; it transports the finer particles through the gaps between the coarser particles.

On the plain, the water stream recovers the gravel that had been deposited by the great floods as well as the fine alluvia deposited during the annual flood. This way, it fulfills the conditions for contact erosion between the two.

The gentle slope of the valley that reaches the estuary gives it enough energy to transport only the bedload sand and the clay (through suspension). The sand underneath the dikes is then subjected to retrogressive erosion: the exit gradients pull out the grains, digging little canals upstream and depositing them in the tumuluses at the base of the dike.

Finally, in the delta, the dikes made of argillaceous sediments can fall apart very quickly as a result of the erosion of the walls because of the holes dug by burrowing animals or because of the shrinkage cracks that inadvertently connect the upstream with the downstream.

1.3.5. *Triggering mechanisms*

There are two conditions that need to be met for internal erosion to take place: detachment and transport.

The equations governing each phenomenon are different. The detachment is the breaking of the balance between three types of forces (Figure 1.4):

– hydrodynamic forces (driving forces);

– the effective stress forces (stabilizing forces that were transmitted at the contact of neighboring grains, and are controlled by the geometrical assembling);

– the surface forces that emanate ions and floc atoms and the mass of the grain saturated with water (resistance or susceptibility to erosion).

The most obvious (and too often overseen) condition of the detachment is the quasi-null value of isotropic effective stress forces. Consequently, the initiation areas are naturally the areas where the effective stress is the weakest.

On the basis of the incidents that were recently recorded, it is possible to deduce some recurrent situations for each mechanism as well as the initiation sources of the erosion (also simply called initiators).

Concentrated leak erosion is caused by different agents when it takes place in the short term rather than in the long term. At the first filling, the main causes for initiation are:

– the absence or the deficiency of the compaction around the circular pipes, or at the edge of the structures that cross these pipes;

– the deficiency or absence of compaction that leaves significant gaps between the clods of clay at the heart of the backfill;

– the vault effect or the load transport that can lead to a null effective stress in the trenches of the pipe or the cutoff of the walls that are so rigid that the weight is taken over by the friction alongside the walls;

– the hydraulic fracturing following the postponement of the loading of the core on the banks of a near-vertical canyon in the valley;

– the detachment of the backfill along the spillways through the shrinkage of the backfill, or the settlement and the displacement of the backfill downstream through the undersizing of the training walls that swivel downstream;

– the scraping of the tectonic faults, fractures, or karsts in the foundation, which carry along a part of the backfill; and

– the presence of dispersive materials.

The inventory of the causes for long-term pipe erosion confirms that we have reason to believe that internal erosion is initiated by weathering:

– The cracks that appear at the level of the crest are a result of significant settlement of the core.

– The shearing cracks found in the core are because of the significant settlement of the upstream and downstream shoulders.

– The cracks that appear in the core as a result of differential settlement in the foundation.

– The differential settlement cracks caused by the irregularities of the contact between the backfill and the foundation.

– The tectonic movement of the faults.

– The post-earthquake shearing cracks.

– The pipe leaks caused by corrosion, differential settlements of the pipe, or earthquakes.

– The burrows dug by animals that have been seen in the crest of the narrower dikes.

– The holes caused by dead roots in argillaceous dikes (often where $h < 6$ m).

– The alluviation of the masonry joints, which can be found in the hydraulic structures completed before the 19th Century.

– The spalling cracks of the concrete slabs or the loss in watertightness of the joints that make up the upstream water tightness.

– The shrinkage of the crest of the clay backfills during the dry season (less than 250 mm of rain per six months).

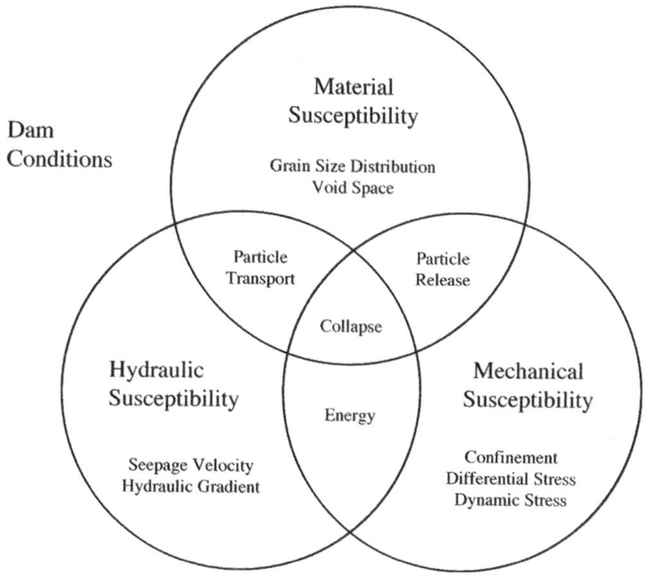

Figure 1.4. *Internal erosion conditions according to Garner and Fannin [GAR 10]*

– The porous areas, cracked by the ice or liquefied because of the thaw, or undergoing strong hydraulic gradients, as a consequence of the drain getting plugged by ice.

– The constant change of the nature of the water, i.e. heavy raining or large floods, which cause a progressive dispersion of the filler.

– The chemical degradation of the filter over time.

The initiators for backward erosion frequently are:

– the flood that increases the gradient beyond its normal level;

– the geological facies, which alternate with layers of sand, clay, or loess.

Contact erosion is frequently associated with:

– the stratification of the alluvia: silt-covered gravels (at the level of the foundation);

– the segregation of a layer of coarse material (in the backfill);

– the contact between the backfill and the foundation, as a result of a high permeability ($K > 10^{-3}$ m/s).

The suffusion is initiated in the following materials:

– cohesionless materials having less than 30%–40% filler;

– the cores of the moraines composed of blocks, gravels, sand and silt, or fairly plastic types of clay;

– through the segregation of filters or coarse transition materials, where more than 40% of these are higher than 5 mm and a maximum diameter is higher than 75 mm; and

– in the foundation, in the alluvial cones, the landslides, the alluvial replenishment.

1.4. Main results of erosion trials

1.4.1. *Which law of erosion?*

Although the mechanical behavior of a random type of soil is described by well-established patterns, its erosion law still makes for the subject of research. It is surely possible to state a law of erosion, starting from an experimental protocol, but its application to a random situation is not yet fully legitimate. What does this law turn into when the surface is no longer plain, when the flow changes its direction, when the turbulence rises in intensity, and when the soil is undergoing a condition of effective stress forces leading almost to failure? In the attempt to have a unified approach on the phenomenon of erosion, and in the aim of pinning it down, researchers agreed on having to analyze separately the four forms of erosion previously described.

1.4.2. *Concentrated leak erosion*

The erosion trial performed through a pinhole, the "Pinhole test", identifies, according to its author, the dispersion sensitivity of a type of clay used to ensure the watertightness of a dam [SHE 76]. In fact, it categorizes the sensitivity to internal erosion. Wan and Fell [WAN 04a, WAN 04b] have modified it, to quantify the erosion kinetics, in the HET, “Hole Erosion Test”. According to the results, the initiation of the erosion depends on the critical stress τ_c and the velocity of the coefficient of erosion Fell C_e (or, which is equivalent, the coefficient of erosion Hanson k_d [HAN 91). If v_{er} represents the eroding velocity and ρ_d the volume mass, then $\rho_d v_{er}$ is the variation of eroded mass per time unit and per surface unit. The law of erosion thus follows:

$$v_{er} = k_d(\tau - \tau_c) \quad [1.1]$$

$$\rho_d v_{er} = C_e(\tau - \tau_c) \quad [1.2]$$

The erosion threshold τ_c (Pa) is the resistance to the pullout force τ of the fluid, also called friction force or shearing force. It can be deduced from velocity measures or from the hydraulic gradient of the fluid. It depends on the product between the dimension of the grain and its buoyant density for cohesionless soil and on the product between the forces that unify particles and the corresponding surface for a cohesive soil. It is also very sensitive to the condition of the soil and thus to its compaction, as well as to the chemical nature of the water.

The volume coefficient of erosion k_d (m/Pa.s) or the mass coefficient of erosion C_e (s/m) is the coefficient of proportionality between the loss of thickness of the wall in a given period of time and the unbalance between the pullout force and the resistance. This coefficient does not depend so much on the condition of the soil but rather on the nature of the soil. Wan *et al.* [WAN 04a] define an erosion indicator I_{HET} as the decimal logarithm of $1/C_e$, increasing proportionally with the resistance.

Hanson, Fell, and Briaud [BRI 01] have each proposed different classes of erodibility (Figures 1.5–1.7). Regazzoni [REG 09], comparing the two authors, has shown that to keep the results and the classifications adequate, we should have as a priority the improvement of the interpretation models.

Underneath its apparent simplicity, the law of erosion needs usage instructions. In fact, it represents the force applied to the grain that varies greatly in time and space, through the constant value of entrainment stress. This means that the variations in the laboratory are identical to those on-site. Thus, the law is applicable to a site whose context is similar in terms of velocity and hydraulic regimen to the site measured in the laboratory. For instance, every study of crack erosion or pipe erosion in a cohesive soil can and should be conducted with the HET; the HET is a procedure whose boundary conditions are simple enough to be well interpreted and applied accurately to the possible velocities within the work.

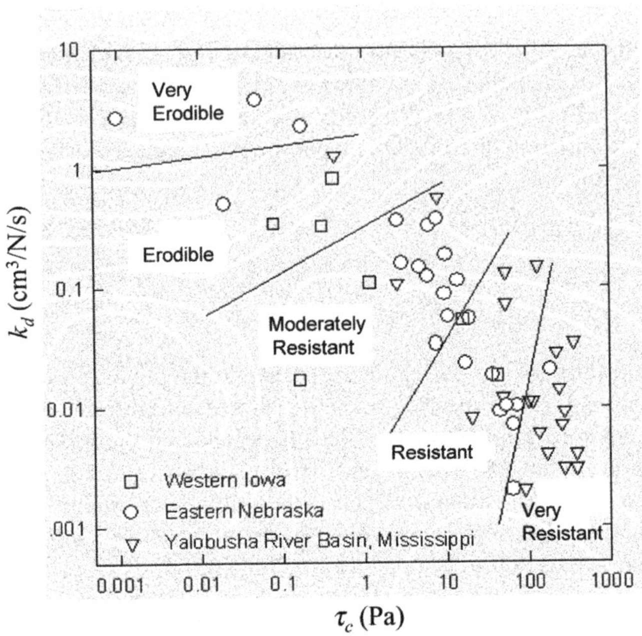

Figure 1.5. *Classes of soil erodibility according to Hanson [HAN 01]*

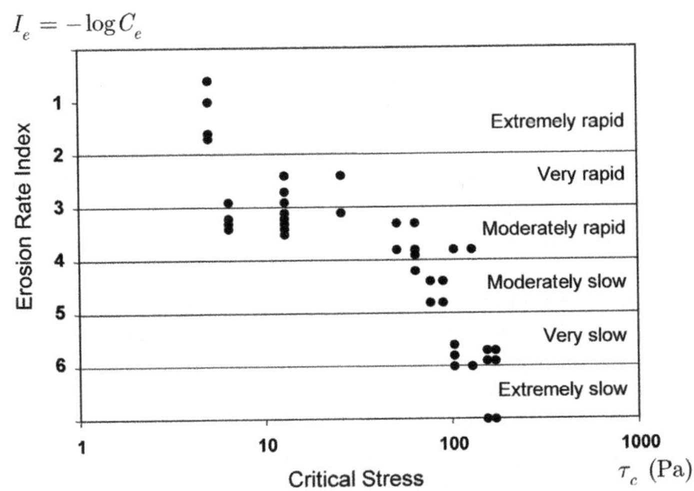

Figure 1.6. *Classes of soil erodibility according to Wan and Fell [WAN 04a, WAN 04b]*

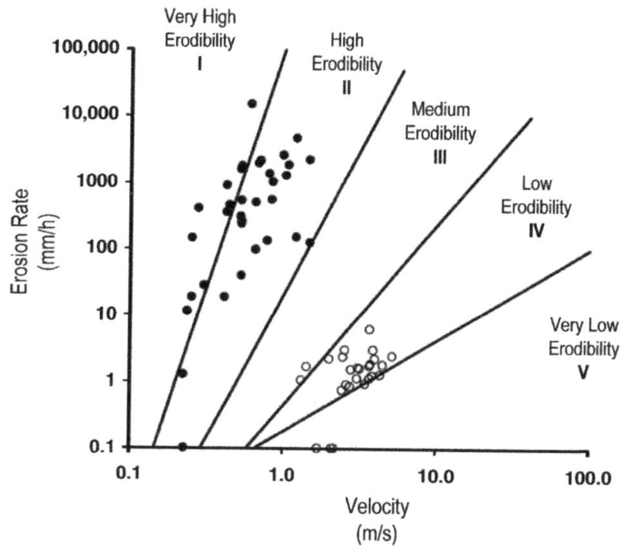

Figure 1.7. *Classes of soil erodibility according to Briaud [BRI 08]*

Is the HET adapted to every study of surface erosion? *A priori*, no! Other types of procedures have been developed in the United States and they measure the erosion in other configurations: the EFA (Erosion Function Apparatus) [BRI 01] and the JET (Jet Erosion Test) [HAN 91]). The EFA measures the erosion of a cylindrical hand specimen that is placed with the aid of a piston at the surface of an outflow with controlled velocity. The JET measures the deepening of a hole created by a vertical jet of water falling on a compact soil in a Proctor mold, and the way this varies with time. The JET has three advantages over the HET: it tests the erodible materials that collapse because of saturation in the HET; it measures the resistance in position; and, finally, it measures a value that is characteristic of overflow erodibility, through its jet that recreates the staircase pattern of retrogressive erosion. Its measures show a higher erodibility at the crest of the hydraulic structure where cohesive materials usually crack. However, the improvement of its interpretation remains a priority.

1.4.3. *Backward erosion*

There are two types of backward erosion. The first type is associated with the general erosion of the particles on the seepage surface, for example on the downstream toe of the dike. We can find out the erosion threshold by calculating the surface stability under gravitational forces (mass) and outflow forces (hydraulic gradient).

The second type is the local retrogressive erosion, which is initiated under a cohesive wall and under fine channels that have a maximum height of a dozen grains per sand layer. This phenomenon is located at the downstream toe of the dike through the presence of a sandboil in the middle of which flows a resurgence that carries fine particles. Quite frequent in river deltas, this phenomenon has been studied by Dutch researchers at Delft. One of these researchers, Sellmeijer, after studying the phenomenon, designed a nomogram in the 1980s. He reevaluated it, as shown in Figure 1.8, having tried more than 70 models, ranging anywhere from tens of centimeters to several tens of meters [VAN 11, SEL 06, SEL 11]. His rule replaces the rules drawn up by Lane and Bligh for fine sands.

The critical hydraulic gradient that takes place at the contact between the clay and the sand is expressed according to the length L of the upstream–downstream flow (m), of the geometrical or intrinsic permeability of the sand K (m^2) and of the significant diameter of the poorly graded sand (*d70*) according to the progression of the length of the pipe l (m). The pipe progresses very slowly according to floods and it increases suddenly when its length surpasses a third of the dike for loose sands, or half of the base of the dike on a dense sand.

The critical gradient of a hydraulic structure can decrease all the way to just several % (if the relative density of the sand is lower than 30%) and is about 0.3 for the coarser types of sand. Its values for gravel rises very quickly with the uniformity coefficient Cu, beyond $Cu = 6$, thus being close to the critical gradient of Terzaghi. This was indeed the case of the failure of the gravels in the foundation of the gravity dam at Puentes; heave occurred under a gradient close to one.

Erosion resistance according to Sellmeijer is the result of three factors:

– the resistance of the material, as a function of relative density, grain size distribution, and angularity;

– the scaling effect, which is the ratio between the size of the grain and the cubic root of the traversed poral volume; and

– the geometric effect, taking into account the impact of the thickness of the sand layer on the value of the driving hydraulic gradient at the bottom of the channel.

Sellmeijer's approach expresses the detachment condition starting from a mechanical model of the boundary balance of the grains at the edges of the hole, whereas Hoffman's approach interprets these erosion thresholds through the transport condition, with the Shields parameter [HOF 08], thus emphasizing the scaling effect in a different way.

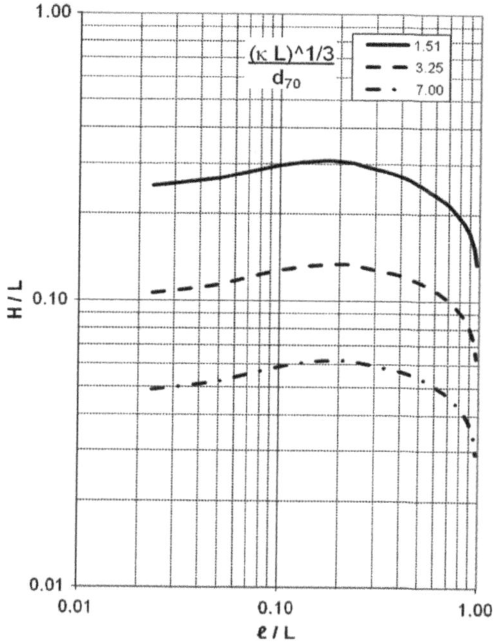

Figure 1.8. *Length of the erosion channels according to the width L of the base of the dike and the global hydraulic dike H/L, the geometric permeability K, and the sand layer d_{70} according to Sellmeijer [SEL 88]*

1.4.4. *Contact erosion*

The erosion threshold presented by Beguin [BEG 11] may be summarized as follows: as long as the Darcy velocity within the hydraulic work or its foundation does not surpass 1 cm/s, the contact erosion is not activated (Figure 1.9). This minimal filtering velocity value of 1 cm/s is an average for 0.2 mm sands (delta sands). When dealing with sands that are significantly finer, the critical velocity increases with the cohesion. Conversely, when dealing with coarser soils, the critical velocity increases with the diameter.

Other researchers emphasize critical velocities that are sometimes smaller. Several reasons account for this discrepancy. This chapter is mainly concerned with alluvia. The measurement of the critical velocity differs from author to author. Some authors take it into account starting with the first evidence for solid transport, while others define it starting with the clear acceleration of solid transport. Finally, the critical velocity varies with time: it decreases with each disturbance (gradient increase, changing of the direction) and increases throughout time. The reason for this increase is the emergence of the pavement, similar to that of the low-water channels composed of coarser particles that remain in place, thus forming a natural filter that protects the finest particles. This observation allows us to comprehend why the representative particle of the transported material is comprised, according to the grain size distribution, between the d_{70} for the soils with compact grain distribution and the d_{100} for the widely spread soils, or, on average, the d_{85}.

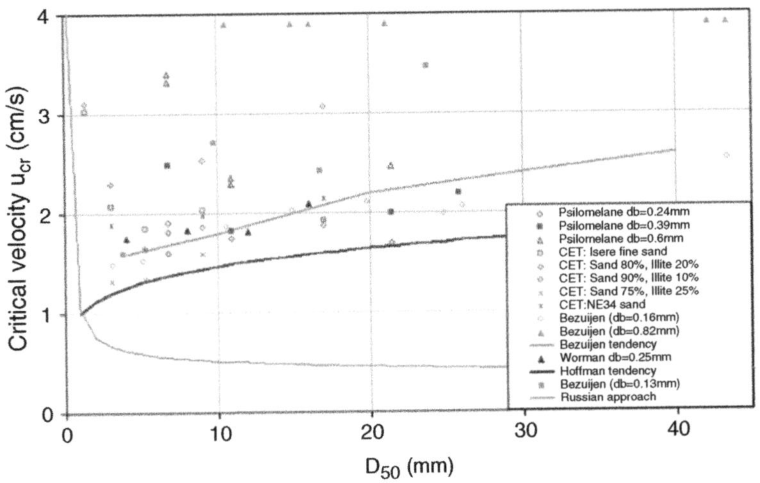

Figure 1.9. *Critical velocity of contact erosion according to Beguin [BEG 11]*

1.4.5. *Suffusion*

Suffusion affects cohesionless soils where a fraction of the fine particles can move between coarse particles because it does not receive the effective stress forces transmitted by the hydraulic structure. Its emergence needs two conditions to take place.

The first condition is a geometrical condition: the pores of the coarse particles allow for the smaller particles to pass between them. This condition was fulfilled by the alluvia composed of more than 65% gravel (of more than 5 mm) with less than 35% sand sable 0/0.5 mm (Figure 1.10). The self-filtering soils do not have suffusion. A self-filtering soil is a cohesionless soil that observes the filtering rule, whatever its fine fraction is F(%) as long as it is less than the diameter d and its fraction H is comprised between d and $4d$. According to Li and Fannin [LI 08], a soil is self-filtering when $H > F$ (Kenney and Lau) if its grain size distribution is continuous and $F < 15\%$ or when $H > 15\%$ (Kezdi test). A discussion and a comparison between these criteria and those of Wan and Fell, Burenkova, and Sherard are given in Chapter 2. In the case of the hydraulic work, the segregation triggers the geometrical condition of suffusion. Thus, previous tests should be applied to grain size distribution by taking into account "segregation".

Suffusion increases with the difference of size between fine and coarse fractions. If *H(n)* is the grain size fraction comprised between d and $n.d$, we can assume that no particle is mobile for $n < 4$, several particles are mobile but they finish by being stopped for $4 < n < 9$, whereas all the fine particles can be transported if $n > 20$. This approach was practical as long as the real filtering condition that compared the diameter of the eroded particles to the diameter of constraints was not established. However, Li and Fannin propose such an approach, which remains to be validated, by the ratio between d_{85} and the fraction of the eroded fine particle, d'_{85} and the average diameter of pores, and O_{50} as defined by Kovacs. When d'_{85}/O_{50} goes over 0.42, there is no suffusion. This criterion is comparable to the criterion used by Russian engineers who take the absence of suffusion to be the case where the maximum diameter of the fraction of fine particles is larger than half of the diameter of the pores.

The second condition of suffusion is the hydraulic criterion that triggers the phenomenon of entrainment [MUC 86]. This criterion considers critical velocity, whose value can be deduced from the relation of Shields [GOL 10]. If the fraction of fine mobile particles has a continuous grain size distribution, this value is not unique. According to Fannin and Li, a second threshold for the mobility of fine particles is more definite. There would have to be a critical hydraulic gradient i_c, which would cause a local quicksand condition of the mobile particles, due to their aggradation. This gradient i_c would be proportional in a ratio α to the critical gradient Terzaghi i_T and to the local actual vertical stress force σ'_{vm}. In fact, it annuls the actual stress force of the fraction of fine particles, fraction α of the vertical stress force applied on the soil, according to the hypothesis of Skempton and Brogan.

The values of α and i_c are deduced from the experiments carried out by Li [LI 08]:

$$i_c = \frac{\alpha}{1-0.5\alpha}\left(\sigma'_{Vm} + 0.5\gamma'/\gamma_w\right) \qquad [1.3]$$

$$\alpha = 1 - \frac{\left(0.42 - d'_{85}/O_{50}\right)}{0.26} \qquad [1.4]$$

Li and Fannin [LI 08] emphasize the fact that this gradient is local and non-intrinsic. Its value varies with the size of the test tube and, very likely, with its boundary conditions. Is this a scaling effect? Future research should confirm these results and explain the nature and value of this scaling effect.

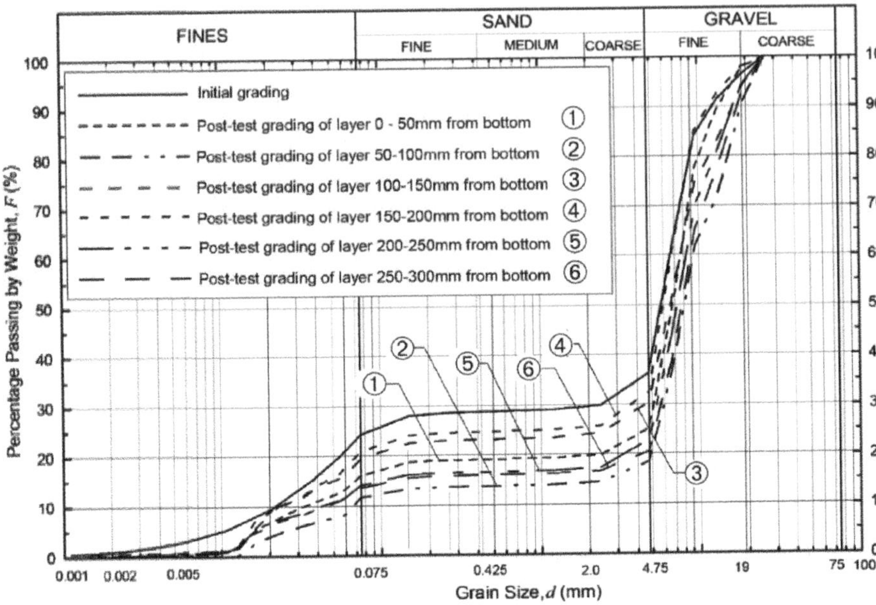

Figure 1.10. *Grain size distribution subject to suffusion, tested by Wan [GOL 10]*

1.5. Remarks on the applicability of erosion trials

1.5.1. *Problem of passing on the results from the laboratory to the hydraulic structure*

The application of erosion thresholds, measured in the laboratory with the aim of predicting on-site erosion, demands that the following three conditions be met in the laboratory:

– The mechanism of erosion that is being modeled needs to be the same as the one on-site.

– The velocities measured must be similar to the velocities on-site.

– Nature, the condition of the soil, and the boundary conditions of the soil within the hydraulic structure must be known and recreated.

The absence of any of these points causes a deviation from reality. For, on the one hand, not observing these conditions of similarity hampers the results of a study of outflow on a reduced scale to be transposed on to the real outflow and, on the other hand, the emergence of erosion does not depend on the average values of the medium, but on the maximum values. However, in nature, it often appears that the maximum length of heterogeneity (cracks in the rocks, segregation in the backfill, boundary effect throughout the structures) increases with the size of the volume; this disturbance is called the scaling effect. Is the laboratory enough? To know that, it is worth comparing the value of the initiation threshold to different scales for each mechanism. Sellmeijer has also quantified this scaling effect for retrogressive erosion. It seems to come from the porous outflow in the erosion channel, which varies along with the size of the aquifer.

Le *et al.* [LE 10] also pointed out a scaling effect on the argillaceous sands that are confined by several effective stress forces. In Chapter 2, Marot suggests, however, that the energetic analysis, by taking into account potential energy, allows us to overcome this scaling effect. However, comparing physical models of contact erosion from 70 cm to 7 m does not emphasize this scaling effect, but it rather highlights the extreme dependence that these results have on the outflow boundary conditions (filter opening, contact between the soils).

1.5.2. *Scaling effect of outflows in the absence of similarity*

The soil is a medium that is traversed by countless capillary openings, composed of a succession of pores separated by various constrictions. The flow in these capillary openings can, at a first approximation, be considered as a pipe flow. At a lower velocity, the energy exchange is made uniquely through viscosity. The stream lines are parallel, hence the name laminary flow. There is no exchange of matter between the stream lines. The transfer of the linear momentum, which causes the friction, comes from molecular agitation. Reynolds showed that the friction caused by the water flow on the walls of a capillary opening is transferable to another, if the ratio between inertial forces and viscosity forces remains unchanged. This adimensional ratio R_e, which bears his name, is the product between the average velocity U and the pipe diameter D, divided by the kinematic viscosity ν. To conclude, all the subterranean outflow recreated in the laboratory is transposable on site, if the Reynolds number stays unchanged.

However, when the velocity increases or the flow is no longer uniform, the nature of the friction changes. A transfer of matter between neighboring layers takes place. The variation of normal velocity to the wall creates a varying force in the wall. The linear momentum is transferred from one layer to the other via temporal variation and local variation of velocity: this is the regimen of turbulence. Under such conditions, the pullout force is the sum between a viscosity friction and a turbulent friction, the latter being defined by Reynolds as the product between local velocity fluctuations and the mass volume of the water. The friction is applied to the wall by a limited layer thickness, called boundary layer, beyond the turbulent movements that generate an average velocity practically constant.

In laminar condition, the friction force is proportional to the velocity and the diameter. When the velocity increases, the regime becomes turbulent, the friction becomes proportional to the velocity raised to a power of 7/4 to finish with a result that is proportional to the square of the velocity when the movement is purely turbulent (depending solely on the roughness of the surface). To conclude, the water friction that pulls out the grains of the walls depends not only on the velocity, but also on the regime of the outflow, hence the dimension, which changes the distribution of the energy: this is the hydraulic scaling effect.

The flow in a capillary opening remains laminary up to approximately $R_e = 2{,}000$. This represents 20 cm/s in a capillary opening having the diameter of 1 cm, which is extremely rare for a soil. To conclude, the initiation of internal erosion often takes place in a laminary regimen.

However, the initiation is only one of the two conditions of internal erosion. The second condition is transport. The transport of grains causes turbulences that will in turn alter the laminary regimen and the water friction on the walls. The impact of turbulent forces is measured by the difference between the theoretical velocity of a real fall of a grain due to viscous friction (Figure 1.11) and the measured velocity [MUC 89]. Stockes expresses the theoretical velocity of the fall V according to diameter d, the saturated mass volume δ, and the dynamic viscosity μ:

$$V = \frac{\gamma'}{18\mu} d^2 \qquad [1.5]$$

According to Figure 1.11, the turbulent forces are insignificant for a grain with a diameter lower than 100 microns and a velocity lower than 1 cm/s. Other than that, they increase strongly with the velocity and the Reynolds number R_e.

However, when a grain falls in a large volume of water, R_e is not yet defined, hence a Reynolds number at the level of the grain R_e* takes its place with the velocity U* at grain level and the grain diameter d. This adimensional number is the key that gives access to the friction force and to the scaling influence:

$$R_e^* = \frac{U^* d}{\nu} \qquad [1.6]$$

1.5.3. *Influence of the geostatic structure of the soil on the erosion threshold*

For the erosion to be initiated in a laminary regimen, there has to be a force, or a friction force per wall unit, that is large enough to enable a grain to roll over. This notion of a threshold stress force or a critical stress was introduced by a French researcher, Du Boys, in 1879: "a pebble placed at the bottom of a liquid stream may be displaced by the impulse of the stream lines that encounter it: the movement will take place if the velocity is higher than a certain limit (...) [that] depends on the density, volume and shape of the pebble; it also depends on the density of the liquid [DUB 79]". The threshold of the stress force is also called the critical friction force, the critical pullout force, or the critical entrainment force. Is this critical force the force applied on a unitary surface at any given time? No, it is rather a global variable, average in time and space, defined by a uniform outflow, whereas the instant force pressing on the grain has a statistic distribution that changes together with the conditions of non-uniformity of outflow. In what follows, we advance the hypothesis that statistical variation is sufficiently insignificant in a laminary regime for the hypothesis of uniform outflow to be very similar to reality.

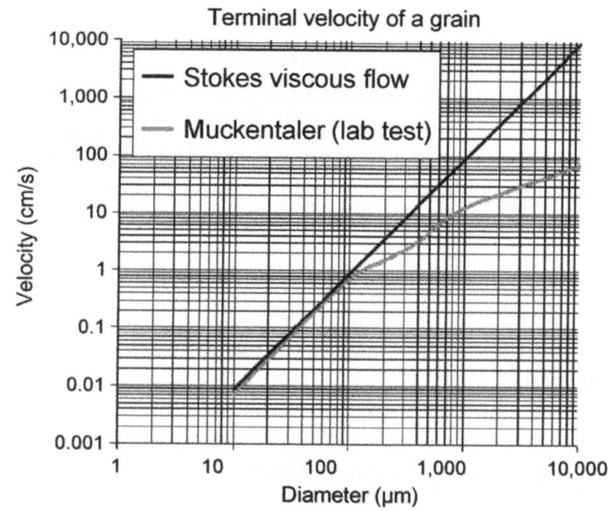

Figure 1.11. *Comparison between the real velocity of the fall of a grain having the diameter d and the theoretical velocity in viscous outflow [MUC 89]*

The friction force of a uniform and permanent outflow on the surface of the soil wet from the flow is assessed simply by writing the balance of the forces of water volume. It represents the friction on the walls that take over the load of that volume. It is thus the projection of the mass of the water on a unitary area of the wet surface. For a river that is ideally long and wide and has a constant slope, the stress is the product between the specific mass of the water γ_w, the height of water h, and the slope i (hydraulic load slope or hydraulic gradient). Shields correlated the critical value of this force τ_c with the diameter of the grains, d, and the immersed mass volume γ', which is set into motion by this force at the bottom of a river. The coefficient of proportionality is the number Shields Ψ:

$$\tau = \gamma_w h\, i \tag{1.7}$$

$$\tau_c = \Psi \gamma' d \tag{1.8}$$

This coefficient translates the average friction force into a fraction of the immersed mass, applied parallel to the bottom of the river, thus setting the grain into motion. Thus, Ψ depends on the flow regime (laminary or turbulent). Ψ is approximately 20% when its velocity is extremely low; it decreases to 3% when the velocity reaches several cm/s, and it increases and stabilizes around 4% and 6%. It depends on the particular Reynolds number Re^* (Figure 1.12). Its value depends on the definition of the erosion threshold. It varies with a 2–3 value, depending on whether the threshold is the occasional movement of a grain or the generalized movement of grains.

In the same manner, the value of the friction force in the hollow of a porous medium is deduced from the balance of the water volume in the hollow, slowed down by the friction alongside the wall of the hollow. If we generalize the formula for this force to the entire soil, it remains the same. Only the height of the water, a particular form of the hydraulic ray R_h, is thus replaced by its generalized expression:

$$\tau = \gamma_w R_h i \tag{1.9}$$

where R_h is the ratio of the surface wet by the outflow, per wet perimeter where the outflow causes friction.

The soil being traversed by several pores and constrictions, the difficulty of estimating the stress is transferred onto the estimation of the hydraulic ray.

Let us suppose that the hydraulic ray corresponds to its definition (the Kozeny hypothesis); it is the ratio of porosity per specific volume surface, and the outflow is considered as a whole. In this case, the outflow is no longer uniform. The velocity may have a normal component alongside the wall. Also, the friction force is increased by the linear momentum of that component. However, this approach is limited by the absence of a simple laboratory measurement of this volume surface.

Let us now suppose that the outflow is laminary, i.e. it has a low velocity that is supposedly uniform, throughout channels of a constant hydraulic ray R_h (the Poiseuille hypothesis), whose volume is called actual porosity n_e. The soil is thus likened to a medium of length L, surface S, crossed by a number m of rectilinear pipes having the length L/T, where T is the tortuosity, with the diameter 0_p, having the hydraulic ray $R_h = 0_p/4$, the outflow porosity $n_e = m\pi 0_p^2/(4S)$ facilitating the transit of the outflow Q under a gradient i with a permeability k and a dynamic viscosity μ. The hydraulic ray is determined by:

$$k = \frac{n_e T R_h^2}{2} \frac{\gamma_w}{\mu}, \quad R_h = \frac{0_p}{4} = \sqrt{2 \frac{k}{n_e T} \frac{\mu}{\gamma_w}} \qquad [1.10]$$

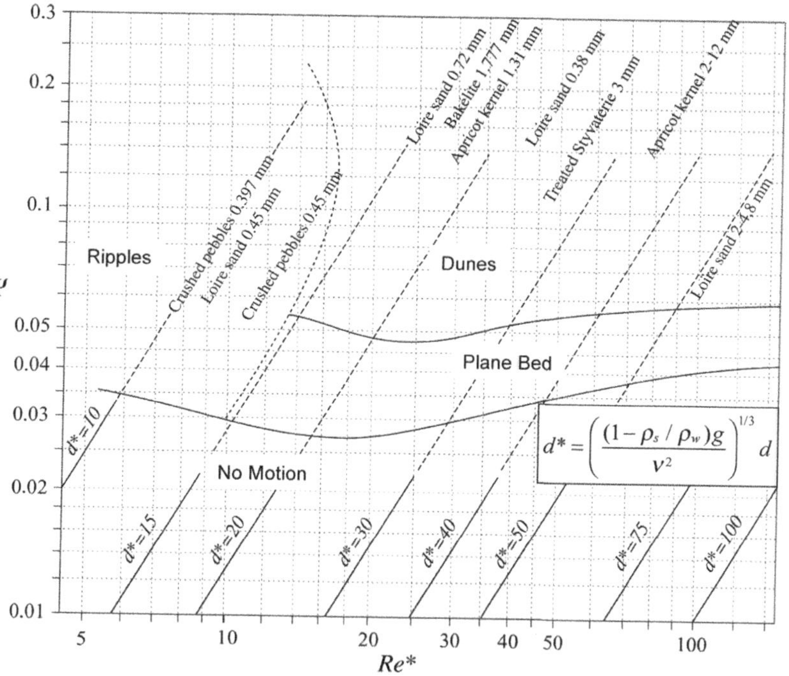

Figure 1.12. *Number of Shields varying according to the corresponding Reynolds number*

In practice, this currently deployed hypothesis has its limits. It does not recreate the heterogeneity that is present in real soil. In reality, porosity and permeability are the product of a quadratic series of the size of its constrictions. If the soil is composed of a series of n capillary openings, having an increasing ray $r_1, r_2, \ldots, r_{max}$, that traverses the surface S of a volume SL, nothing proves that the permeability (equation [1.11]) and the porosity (equation [1.12]) are independent of S and L, if the structure of the soil is not carefully analyzed. On the other hand, the erosion concerns only the capillary openings whose hydraulic ray r_m allows the eroded particle d ($2r_m > d$) to pass through, and the initiation starts with $r_{max} = r_m = 0_{max}/2$.

$$k = \frac{\pi \gamma_w T}{8 \mu S}\left[r_1^4 + r_2^4 + \cdots + r_n^4\right] \qquad [1.11]$$

$$n_e = \frac{\pi}{S}\left[r_1^2 + r_2^2 + \cdots + r_n^2\right] \quad [1.12]$$

$$r_m = \frac{i\gamma_w}{\Psi\gamma'}\sqrt{\frac{\left[r_1^4 + r_2^4 + \cdots + r_n^4\right]}{4\left[r_1^2 + r_2^2 + \cdots + r_n^2\right]}} \quad [1.13]$$

If there is an autocorrelation between r_n and the dimension of the model L, then the critical stress depends on L. To evaluate the critical stress of a site, not only do we have to know the permeability of that site, but we must also verify that the statistical distribution of the porous medium is very homogeneous and that it is not geostatically dependent. This is the geostatic scaling effect.

The influence of the similarity conditions emerges in equation [1.17], when the initiation condition is inscribed in the capillary opening where the erosion is initiated with equation [1.8] and from equations [1.12]–[1.16]:

$$\tau = \tau_c \Rightarrow \gamma_w \frac{r_{max}}{2} i_c = \Psi\gamma' d \quad [1.14]$$

$$i_c = \frac{2\Psi\gamma' d}{r_{max}\gamma_w} \Rightarrow i_c = \Psi\frac{4d}{0_{max}} i_T \quad [1.15]$$

The critical gradient is thus the product of the critical Terzaghi gradient i_T, of the Shields parameter Ψ, and the geometric filtration condition $d/0_{max}$. Istomina [IST 57] considers that filtration becomes effective as soon as $d/0_{max}>1/2$, which agrees with the results found by Li. Knowing that the Shields parameter depends on the friction velocity, and assuming the following hypotheses, it is possible to write the expression for the hydraulic gradient at the time when the piping was formed:

$$\Psi = aR_e^{*b} \quad [1.16]$$

$$U = cU^* \quad [1.17]$$

$$i_R = \frac{(H1 - H2)}{L} \quad [1.18]$$

$$i_R = \left[\frac{ac^b g^{1/2}}{\sqrt{2}(n_e T)^b (v)^{b+1/2}} \frac{d^{1+b}}{k^{1/2-b}} i_T\right]^{\frac{1}{1-b}} \quad [1.19]$$

This relation confirms a general relation that links the diameter d of the eroded particle, the permeability k of the supposedly homogeneous soil, with b varying from –3/5 to +1/5 following the value R_e* of the most frequent outflows in soils.

1.5.4. *Initiation of internal erosion in a cohesionless soil*

The friction force is also calculated through the loss of energy per volume unit, be it a fraction C_f of the kinetic energy, $C_f\rho U^2$, with C_f being the friction coefficient. In laminary outflow, the velocity is small enough so that the integration of the linear momentum leads to the Darcy law: the hydraulic gradient is proportional to the velocity, since the friction coefficient is inversely proportional to velocity and is equal to $16/R_e$. The velocity is null at the walls but the friction force is at its maximum. The velocity is maximal at the center of the capillary opening and the stress there is null. Between them, the stress varies linearly, whereas the profile of the velocity is parabolic.

The friction velocity U^* is the value of the velocity whose square multiplied by the volume mass of the water is equal to the friction force. Its value was determined by Shields [SHI 30] according to the size of the grains and the Reynolds number (Figure 1.13):

$$\tau = \rho U^{*2} = f(R_e^*) \qquad [1.20]$$

The friction velocity is linked to the percolation velocity U given by Von Karman, Prandtl, and Schlichlting, and the results of Nikuradsé [SCH 87], showing that the profile of velocities in a pipe follows a remarkable constant, adimensional profile, regardless of the regimen, so that the ratio of the difference between the maximal and the average velocity, divided by the friction velocity, equals 4.07. In laminary regimen, the maximal velocity is double the average percolation velocity. Therefore, friction velocity equals the percolation velocity divided by approximately 4. The percolation velocity is the filtration velocity divided by the efficient porosity and tortuosity. With an average porosity of 1/3 and a tortuosity between 1 and $\pi/2$, the friction velocity is very close to the filtration velocity.

Starting from the pullout force, which varies with the diameter of the pulled-out particle from a cohesionless soil, measured by Shields, the initiation of internal erosion can be evaluated as follows:

– evaluating the maximum filtering velocity in the work according to outflow laws, the boundary conditions, and the analysis of heterogeneities;

– moving from this filtration velocity to the maximum friction velocity; and

– evaluating the maximum erodible diameter according to Schields and the erodible grain size distribution (mobile particles).

This approach was successfully applied to:

– suffusion, by Goltz and Muckentaller [GOL 10];

– contact erosion, by Béguin [BEG 11]; and

– retrogressive erosion, by Hoffmans.

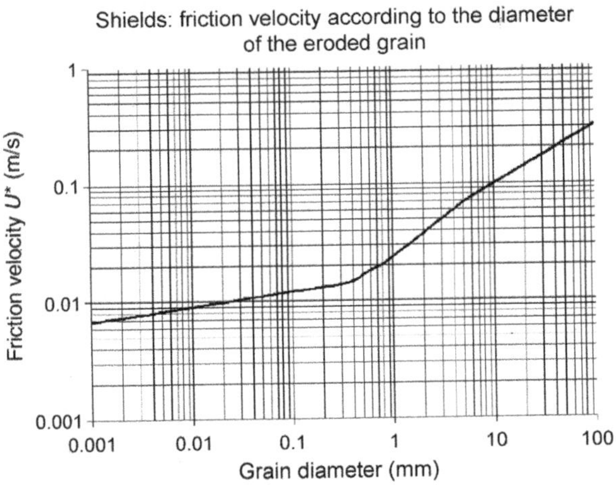

Figure 1.13. *Velocity U* pulling out a grain according to its diameter [SHI 30]*

1.5.5. *Erodibility and researching erosion laws*

To apply the erosion of cohesive soils studied in the laboratory to the hydraulic structures, we must enquire about the physical meaning of erosion and the nature of its resistance.

The balance sheet of dam failures shows that the absence or the weakness of cohesion causes the greatest sensitivity that backfills have against failure. The small value of hydrodynamic forces that initiate the erosion in comparison to the value of effective stress forces leads to a second trivial remark that is too often omitted: the areas where the effective stress forces approach the value zero are the areas where erosion initiates. The Terzaghi principle, when applied to the edges of argillaceous soils, states that the hydraulic forces represent the total stress, resumed by interstitial pressure and resistance. As the permeability increases, so does the interstitial pressure. The resistance is either the cohesion, if the shearing is being compressed, or the resistance to traction. In the ideal case of an impermeable soil that is being subjected to a purely laminary outflow, parallel to a plain surface, the resistance of the soil should be the cohesion. However, in reality, the rough surface disturbs the flow that becomes turbulent, the water pressure seeps through the cracks and it increases up to the point where the effective stress becomes negative and a traction effort adds to the shearing (given that the soil resists only through resistance to traction, resistance that is a lot lower than cohesion). Knowing that the turbulent water pressure is cyclical to variations amplified by the presence of air bubbles in the interstices, the soil resistance would be the resistance to traction weariness. The resistance to erosion is then proportional to the traction resistance, and inversely proportional to the logarithm of the number of interstitial pressure cycles. This resistance decreases when the depth where the water has penetrated the soil increases. The resistance to erosion strongly depends on the structure of the pores and in particular of the permeability to saturation, besides depending on surface forces [REG 09]. The soil drily compacted has a macroporosity, a strong permeability and a low resistance, whereas when it is compacted in a humid medium, it becomes more impermeable and erosion resistant. The erosion of certain soils is amplified by the disappearing of the capillarity along with saturation. If there is a dispersion phase among flocs under the changing of the curvature of the meniscus rays or as a result of the migration of air bubbles, the soil becomes very erodible and is considered to be dispersive, within the classification of the pinhole test or within the "crumble test". However, this phenomenon of disintegration by reduction of the capillarity needs to be distinguished from the phenomenon of dispersion, which is a repulsion of argillaceous particles as a result of a chemical process called cation substitution.

There is not yet any transducer that is sufficiently miniaturized so that it can evaluate the value of pressure variations applied to the surface of the soil. However, it is possible to measure the distribution of the velocity, as it was done by Béguin. The variation of the pressure is close to the inverse of the variation of the velocity squared if the loss of energy is small. Nikuratze [SCH 87] has shown that the water velocity is stronger, more homogeneous, and more uniform when the distance from the surface of the wall increases. The velocity of the water follows a logarithmic law according to distance.

Nezu [HOF 97] has shown that the standard deviation of the water is at its maximum when close to the wall and diminishes offshore following a logarithmic law that involves distance. Supposing that the maximum velocity is m times greater than the average velocity, the maximum hydraulic gradient should be m^2 times its average. Having access to the distribution of velocities, it should be possible to have access to the maximum and minimum hydraulic gradients as well, when at the erosion threshold. Hoffmans [HOF 97] proposes to take into account the turbulent variations starting from Pilarczyk's concept. He explains erosion through the concept of turbulent energy, the sum of the squares of the standard deviation of each component of the velocity on every outflow surface, expressed by $(r_oU)^2$, where r_o is the average variation coefficient for every surface of the kinetic energy of an average velocity U. He does not use turbulence at the level of soil surface, but rather uses the global turbulence of the outflow, because the boundary conditions of the outflow generate a global turbulence that has a strong impact on the walls of the soil. By using the turbulence ratio r_o, generated by the on-site boundary conditions, it would be possible to deduce from them the size of the detached floc d, according to the tensile fatigue (N) with equation [1.21]:

$$\frac{\text{Resistance}}{\text{Maximum action}} = \frac{\gamma' d + R_t'}{\dfrac{\rho_w (r_0 U)^2}{\Psi}} = 1 \qquad [1.21]$$

$$R_t' = 0 \Rightarrow \frac{\gamma'}{\dfrac{\rho_w (r_0 U)^2}{d}} = \frac{1}{\Psi} = \frac{\gamma'}{i_{moy}\gamma_w} = \frac{i_T}{i_{moy}} \qquad [1.22]$$

The interpretation of the previous formula is that the minimal resistance of cohesive soils is the sum of the traction weariness resistance and the stress exerted by the saturated mass of the block d^3 with γ' being the saturated mass volume. The destabilizing maximum driving action is $1/\psi$ times the average turbulent action $\rho_w(r_o U)^2$. Resistance can be accessible only through experiment. We are dealing with a resistance to effective stress. The minimum resistance is obtained when the hydrostatic pressure is at its lowest and the interstitial pressure reaches a maximum value. Analogous to corrosion, erosion should increase with the ratio of the local pressure per dynamic pressure, which is the cavitation number.

For cohesionless soils, the interpretation of equation [1.22] is that γ' represents volume resistance, whereas $\rho_w(r_o U)^2/d$ represents the average destabilizing driving action. The destabilizing force is the maximum hydraulic gradient on the scale of the floc, which would be around $1/\Psi$ times the average gradient. In this interpretation, Ψ, the Shields coefficient, appears as inversely proportional to the ratio between the maximum gradient and the average gradient.

The turbulence ratio r_o mentioned by Hoffmans [HOF 97], which reflects the fluctuation of the kinetic energy, increases the size of the eroded flocs d if the average velocity U is constant, according to the relation $\Psi d = r_o^2$, supposing that the erosion threshold is set by the maximum value of the increasing hydraulic gradient that, multiplied by the volume mass, overcomes the tensile strength. At a constant average velocity, size d of pulled-out flocs varies according to $d = r_o^2$. Given that the flocs have variable sizes, the threshold would be defined by the intersection of the histogram of these actions with the histogram of resistances, as proposed by Hoffmans. This is perhaps the explanation for the deviation between the coefficient of erosion and the energy spent in the trials of the HET and the JET, otherwise negligible for the more erodible soils [REG 09], but persistent for certain cohesive soils with a crumbly aspect. In fact, the law regarding the stress underestimates the effects of the turbulence. The global law of erosion regarding stress or energy does not distinguish the effects of the traction stress from those of compression; however, it is only the former that needs traction resistance.

While waiting to find a general law of erosion, it is possible to use the HET on a site, if the conditions of the HET are representative for the on-site conditions. If the soil is homogeneous and porous, the on-site friction stress is estimated by equation [1.20]. The friction velocity is estimated at 1/4 of the average velocity in the largest capillary opening. The average velocity is deduced from the permeability and from the gradient, knowing that in equation [1.10] it is best to assign a security coefficient of 10 to model the size of the largest capillary. The effective porosity is the difference between the porosity and the content of retained water. If the soil is cracked or pierced, it is best to take into account the influence that the roughness has on the friction force. For example, if the size of the nibs goes from ks_1 to ks_2 and if the hole ray goes from R_1 to R_2 with $ks_1U_1/\nu > 70$, the ratio of friction forces is set by equation [1.23]:

$$\frac{\tau_2}{\tau_1} = \left[\frac{2\log\dfrac{R_1}{ks_1} + 1.74}{2\log\dfrac{R_2}{ks_2} + 1.74} \right]^2 \qquad [1.23]$$

Finally, the erosion trials allow us to characterize the level of erodibility of the ground. Briaud [BRI 08] has shown that after hurricane Katrina, the embankment dikes of New Orleans that resisted the submersion had a low erodibility, as opposed to those that were carried away by the floods, which had a high erodibility. Erodibility is now a property that completes the identity card of every piece of material used to build hydraulic structures, by knowing the coefficient of erosion and the critical stress.

1.6. Conclusion

This chapter has introduced the work of the National Project ERINOH. Having started with an analysis of the pathology of erosion, we proposed a classification of the relevant erosion mechanisms. Laboratory trials were conceived and carried out to quantify the initiation thresholds of these mechanisms. An application of the results of these laboratory tests was then sketched out. Thus, this chapter emphasized the importance of similarity laws, of boundary conditions, and of the structure of the soil.

1.7. Bibliography

[BEG 11] BEGUIN R., Etude multi-échelle d'érosion de contact au sein des ouvrages hydrauliques en terre, Doctoral Thesis, University of Grenoble, p. 320, 2011.

[BRI 01] BRIAUD J.-L., TING F.C.K., CHEN H.C., CAO Y., HAN S.W., KWAK K.W., "Erosion function apparatus for scour rate predictions", *Journal of Geotechnical and Geoenvironmental Engineering*, vol. 127, no. 2, pp. 105–113, 2001.

[BRI 08] BRIAUD J.-L., CHEN H.-C., GOVINDASAMY A.V., STORESUND R., "Levee erosion by overtopping in New Orleans during the Katrina Hurricane", *Journal of Geotechnical and Geoenvironmental Engineering*, vol. 134, no. 5, pp. 618–632, 2008.

[DUB 79] DU BOYS M.P., "Etude du régime du Rhône et de l'action exercée par les eaux sur un lit à fond de graviers indéfiniment affouillable", *Annales des Ponts et Chaussées*, vol. 5, no. 18, pp. 141–195, 1879.

[FEL 07] FELL R., FRY J.J. (eds), *Internal Erosion of Dams and Their Foundations*, Taylor & Francis, p. 245, 2007.

[FOS 00] FOSTER M., FELL R., SPANNAGLE M., "The statistics of embankment dam failures and accidents", *Canadian Geotechnical Journal*, vol. 37, pp. 1000–1024, 2000.

[FRY 97] FRY J.J., DEGOUTTE G., GOUBET A., "L'érosion interne: typologie, détection et réparation", *Barrages & Réservoirs*, no. 6, p. 126, 1997.

[GAR 10] GARNER S., FANNIN J., "Understanding internal erosion: a decade of research following event", *Hydropower and Dam*, vol. 15, no. 3, 2010.

[GOL 10] GOLTZ M., AUFLEGER M., PERZLMAIER S., MUCKENTHALER P., "Assessing the critical seepage velocity causing transport of fine particles – the approach of Muckenthaler", *8th ICOLD European Club Symposium*, Innsbruck, Austria, 22–23 September 2010.

[HAN 91] HANSON G.J., "Development of a jet index to characterize erosion resistance of soils in earthen spillways", *Transactions of the ASAE*, vol. 34, no. 5, pp. 2015–2020, 1991.

[HAN 01] HANSON G.J., SIMON A., "Erodibility of cohesive streambeds in the loess area of the Midwestern USA", *Hydrological Processes*, vol. 15, pp. 23–28, 2001.

[HOF 97] HOFFMANS G.J., C.M., VERHEIJ H.J., *Scour Manual*, Balkema, 1997.

[IRE 03] IREX, Revue bibliographique sur l'érosion interne, Ed. IREX, 2003.

[IST 57] ISTOMINA V.S., *Soil Stability to Seepage* (in Russian), Moscow, 1957.

[LI 08] LI M., Seepage induced instability in widely graded soils, PhD Thesis, University of British Columbia, 2008.

[MON 98] MONNET A., "Boulance, érosion interne, renard: les instabilités sous écoulement", *Revue française de Géotechnique*, vol. 82, pp. 3–10, 1998.

[MUC 89] MUCKENTHALER P., "Hydraulische Sicherheit von Staudämmen", *Institut für Wasserbau und Wassermengenwirtschaft, Technische Universität München*, Bericht Nr. 61. 1989.

[REG 09] REGAZZONI P.L., Confrontation et analyse d'érodimètres et caractérisation de la sensibilité à l'érosion d'interface, Doctoral Thesis, University of Nantes, 2009.

[SCH 87] SCHLICHTING H., *Boundary Layer Theory*, 7th ed., McGraw-Hill, New York, 1987.

[SEL 88] SELLMEIJER J.B., On the mechanism of piping under impervious structures, PhD Thesis, Delft University of Technology, LGM-Mededelingen, no. 96, 1988.

[SEL 06] SELLMEIJER J.B., "Numerical computation of seepage erosion below dams (piping)", *Proceedings of the 3rd International Conference on Scour and Erosion*, Amsterdam, November 2006.

[SEL 11] SELLMEIJER J.B., LÓPEZ DE LA CRUZ J., VAN BEEK V.M., KNOEFF J.G., "Fine-tuning of the backward erosion piping model through small-scale, medium-scale and IJkdijk experiments", *European Journal of Environmental and Civil Engineering*, vol. 15, no. 8, pp. 1139–1154, 2011.

[SHE 76] SHERARD J.-L., DUNNIGAN L.P., DECKER R.S., STEELE E.F., "Pinhole test for identifying dispersive soils", *Journal of the Geotechnical Engineering Division*, ASCE, vol. 102, no. GT1, pp. 69–84, 1976.

[VAN 11] VAN BEEK V.M., KNOEFF J.G., SELLMEIJER J.B., "Observations on the process of piping by underseepage in cohesionless soils in small, medium and large-scale experiments", *European Journal of Environmental and Civil Engineering*, vol. 15, no. 8, pp. 1115–1138, 2011.

[WAN 04a] WAN C.F., FELL R., "Investigation of rate of erosion of soils in embankment dams", *Journal of Geotechnical and Geoenvironmental Engineering*, vol. 30, no. 4, pp. 373–380, 2004.

[WAN 04b] WAN C.F., FELL R., "Laboratory tests on the rate of piping erosion of soils in embankment dams", *Geotechnical Testing Journal*, vol. 27, no. 3, pp. 295–303, 2004.

Chapter 2

Suffusion, Transport and Filtration of Fine Particles in Granular Soil

2.1. Introduction

2.1.1. *Chapter objectives*

Flows within the interstitial pores of a soil may cause suffusion. This complex phenomenon corresponds to the process of detachment and then transport of the finest particles within the porous network. Suffusion may cause changes in porosity and can also lead to important modifications in the hydraulic and mechanical characteristics of the soil. Such modifications of the porous medium can also be the catalyst for significant instabilities at real scale of hydraulic structures.

Suffusion, i.e. the interaction between water and soil, is conditioned by the evolving and interdependent characteristics of the fluid and the solid phases. Therefore, its study is a complex matter as the phenomenon of suffusion depends on numerous geometrical, hydraulic, mineralogical, and physicochemical parameters, all in connection with solids and fluids.

The complexity of the mechanisms and phenomena at stake may cause discrepancies in the interpretation of the results of the same study [SKE 94, MON 98]. Besides, suffusion is governed by the principles of soil mechanics as well as by hydraulic principles.

Chapter written by Didier MAROT and Ahmed BENAMAR.

This multidisciplinary characteristic explains the duality of the main approaches, which are detailed in the relevant literature, with the aim of drawing up criteria for evaluating the initiation of this phenomenon. When suffusion takes place, a fraction of the detached particles can resettle or be filtered at the center of the porous network. This process can then locally induce a clogging. The processes of suffusion and filtration are thus indissoluble and their coupling must give a realistic account of the mechanisms that lead to the instabilities of hydraulic earth structures.

The role of a filter is to retain the particles that were suspended during a flow. The design of the filters in hydraulic works is particularly important given its role in stopping or at least slowing down erosion. The criteria used for designing these filters (especially [SHE 89]) are founded mainly on the characteristics of grain size distribution of the materials. The operating of a filter depends on the geometrical conditions (size and distribution of pores, size of particles and their concentration), hydraulic conditions (velocity, gradient), and physicochemical conditions. This process of filtration takes place through the deposition of particles on the surface of the filter or by trapping the particles in the porous medium. Experiments have enabled us to measure the indirect indicators of clogging (i.e. pressure and concentration) and also they helped us to adapt an appropriate modeling.

The first objective of this chapter is to identify the predominant suffusion parameters and present the main criteria used for evaluating the initiation of this phenomenon. By confronting several criteria, we can elaborate new criteria for the initiation and development of suffusion. Moreover, suffusion is a pathology of the filter that hydraulic works must be prevented from. The second objective is, therefore, to characterize the way filters work in order to set out the clogging conditions that can result from the properties of the filter or from the properties of the particles that need filtering. The classic criteria used for the filters are compared with the experimental results and they are discussed in this chapter.

2.1.2. *Terminology*

In relevant literature, several terms are used to describe the process of detachment and transport of fine particles within the porous network of a soil, under the action of an internal water flow.

Researcher Pavlov coined the term "suffusion" in 1898. Kovacs [KOV 81] used the term "suffusion" to describe the movement of fine particles. Undoubtedly influenced by the distinction advanced by Istomina, [KOV 81] distinguished between two types of suffusion: internal and external. The first type is characterized by the redistribution of fine particles that alters the local hydraulic conductivity of the material without causing a significant modification in the overall solid volume.

The second type of suffusion corresponds to the pulling out and then evacuation of fine particles from the solid matrix considered; during this time, the solid volume of the matrix decreases and its hydraulic conductivity increases. However, according to Kovacs [KOV 81], the stability of the granular skeleton composed by coarser fraction of the grains remains unaffected.

To describe the transport of small particles of a type of soil, Ken and Lau [KEN 85] use the term “suffosion”, and in specialized literature both “suffusion” and “suffosion” are frequently used interchangeably. The term “suffusion” is equally used in medicine to designate the blood suffusion or any other liquid that seeps through a surface. To avoid ambiguity, Chapuis and Aubertin [CHA 04] suggest that we should use the term “suffossion”. Li [LI 08a] considers that “suffossion” does not describe the migrating process and shows that the term “internal stability” (or “internal instability”) is generally accepted without any confusion. However, the terminology “internal stability” or “internal instability” does not mention the origin of the process and allows us to suppose a mechanical consequence of the detachment and transport of fine particles.

Following the workshop on the topic of internal erosion in April 2005 at Aussois (France), several researchers and experts of various countries have adopted a terminology that includes the term “suffusion” [FEL 07]. Consequently, this is the term that we will use in this chapter. Several types of filtration are described; however, the description of the results of the experiments regards the deep filtration that concerns hydraulic works.

2.2. Dominant parameters that influence suffusion

Suffusion is the result of a seepage flow in the pores of a soil. Consequently, its initiation and development depend, on the one hand, on the parameters that influence the characteristics of the fluid and solid phases and, on the other hand, on the interstitial geometry, i.e. on the exchange interface between the two phases. Garner and Fannin [GAR 10] describe the main initiation conditions for suffusion using a diagram comprising three components: material susceptibility, critical hydraulic load, and critical stress condition.

2.2.1. *Parameters that modify the geometry of the porous medium*

The detachment of fine particles and their subsequent transport throughout the porous network of the soil requires that the pore space is sufficient. This space is conditioned by the granular distribution, which, in turn, depends on the size of the grains, their form, and the fabric. Given that the flow of the water and of the detached particles depends on the tracking conditions of their entire run, the opening of the pores at the downstream of the considered soil volume can equally influence suffusion.

2.2.1.1. *Grain size distribution*

There are three main gradation curves that can be distinguished [LAF 89]: linear distribution (curves 1 and 2 in Figure 2.1), discontinuous distribution (curve 3), and, finally, upwardly concave distribution (curve 4).

In the case of the curve 2, the coarser fraction is a minority (<40%) and "floats" within the fine fraction. In the case of the discontinuous distribution, a series of intermediary-size grains is missing. The concave distribution consists of a poorly graded coarser fraction associated with a highly graded fine fraction. In this type of soil, the volume between the grains of the coarser fraction is higher than the volume of fine particles, thus enabling the fine particles to migrate.

The soils whose grain-size distribution curve corresponds to curves 1 or 2 are stable to suffusion [LAF 89]. The soils that are likely to suffer from suffusion are, according to [FEL 07], "internally unstable", i.e. their grain-size distribution curve is either discontinuous (curve 3) or upwardly concave (curve 4).

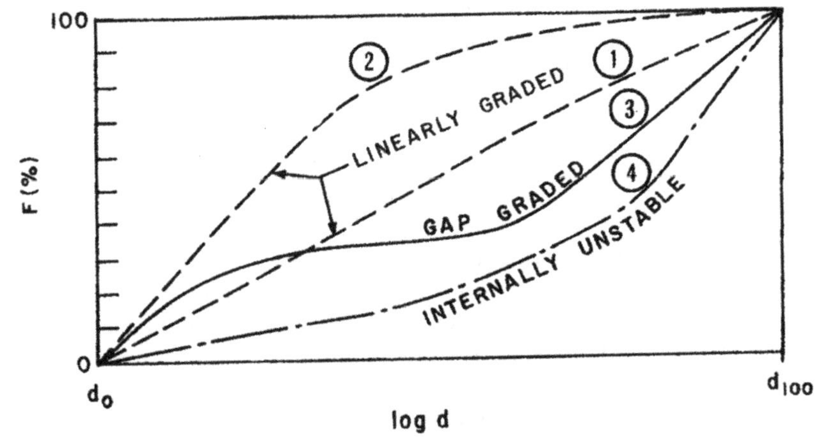

Figure 2.1. *Classification of the grain size distribution of soils [LAF 89]*

2.2.1.2. *Effective stress*

The Terzaghi principle consists of postulating the existence of effective stress that governs the behavior of the granular skeleton. For the same granular distribution, the modification of the effective stress can induce grain rearrangements.

Several tests performed in oedometric conditions on unstable soils show that a rise in the effective stress causes an increase in the soils' resistance to suffusion [MOF 05]. In the same manner, when tests were carried out under isotropic confinement [BEN 08], the increase in the confinement pressure allowed for a decrease in the suffusion rate (Figure 2.2).

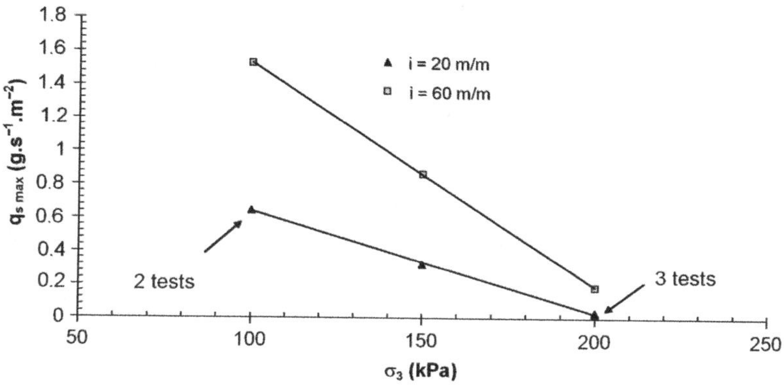

Figure 2.2. *Influence of the confinement pressure on the rate of suffusion for two values of an applied hydraulic gradient (clayey sand samples) [BEN 08]*

2.2.1.3. *Grain angularity*

Little data regarding the influence of the shape of the grains on suffusion were published. However, let us turn toward the model of the capillary tube proposed by Kovacs [KOV 81], which represents the poral space by a set of cylindrical, parallel tubes. The average diameter of the pore d_0 is calculated by using the following equation:

$$d_0 = 4\frac{n_c}{1-n_c}\frac{D_h^c}{\alpha_D}, \quad D_h^c = \left(\sum_i \frac{\Delta F_i^C}{D_i^C}\right)^{-1} \qquad [2.1]$$

where n_c is the porosity of the coarser fraction, α_D is the shape coefficient, D_h^c is the average diameter in the interval i of the granular distribution of the coarser fraction, and ΔF_i^C is the corresponding mass percentage. The porosity of the coarser fraction is expressed by $n_c = n + F_n(1-n)$, where n is the porosity of the soil and F_n is the mass percentage of the fine fraction.

One of the criteria for evaluating the initiation of suffusion is confronting the average value of the pore diameter with the size of the fine fraction (see section 2.3.1). It is worth noting that in equation [2.1] there emerges the shape coefficient whose values are comprised between 6 for the spherical particles and from 7 to 9 for the more angular particles [KOV 81]. Thus, grain angularity probably plays a significant role in suffusion.

2.2.1.4. *Downstream pore opening*

Suffusion can be described as the result of the process of surface erosion, the surface being here the surface of the pores, together with the fate of eroded particles [RED 00]. Furthermore, the distinction between internal suffusion and external suffusion [KOV 81] demonstrates the need to take into account the conditions for the evacuation of the detached particles in relation to the soil volume. This problematic is at the center of the study of filter efficiency, whether they are made up of a geotextile material or of a layer of draining soil. For cohesionless soils, [LAF 99] identifies the ideal diameter d_i of the opening of the filter, by starting from the uniformity coefficient and from the aspect of the grain-size distribution curve. Depending on the real opening O_F of the pores of a filter, several phenomena can be observed: suffusion, if $O_F >> d_i$, clogging if $O_F << d_i$, and self-filtration if $O_F \approx d_i$.

Even when filters that allow the migration of all the fine particles are being used, the size of the opening of the filter's pores constitutes a parameter that can modify significantly the initiation and the development of suffusion. When testing suffusion on clayey sands, replacing a filter with a 4 mm opening by another filter of a 0.08 mm opening has enabled the critical hydraulic gradient to rise by 6 for the initiation of clay suffusion [MAR 09].

2.2.2. *Parameters that modify the physicochemical characteristics of the medium*

The physicochemical characteristics of the fluid and solid phases are crucial, particularly in the case of cohesive soils. According to Arulanandan and Perry [ARU 83], the size of the eroded aggregates is a function of either the state of flocculation or of de-flocculation. The state of flocculation or de-flocculation depends particularly on the chemical composition of the interstitial fluid, as well as on the quantity and nature of the clay.

2.2.2.1. *The content of sodium chlorate in the interstitial fluid*

Reddi *et al.* [RED 00] have carried out a series of suffusion tests on clayey sand, injecting fluids that had different concentrations of sodium chlorate. The rate of erosion by suffusion decreases when the concentration of sodium chlorate increases. There are two distinct processes that can be responsible for this decrease in the quantity of eroded particles: the decrease in the detachment of particles and the increase in the filtration of detached particles. This reduction in suffusion could be attributed, as is the case of surface erosion [ARU 83], to the influence of sodium chlorate on the flocculation of the soil. If the rate of sodium absorption is high, the forces of physicochemical interaction decrease, which, in turn, causes a deflocculation of the clay and a decrease in the resistance of clay particle detachment.

2.2.2.2. *The content and mineralogical nature of clay*

Clay particles are made up of flakes, which, in turn, are made of atoms that are arranged in tetrahedral or octahedral meshes. By studying these different packings of atoms and flakes, we can classify the different types of clay into eight categories. Among these, we should mention in particular the kaolinite group, the smectite group (mainly comprising montmorillonite), and the illite group. Their structure and their differing chemical composition give them different particle sizes and a different sensitivity to water.

For the same type of clay, several tests performed on clayey sands have shown that suffusion decreases with the increase in the clay percentage [BEN 08]. Arulanandan and Perry [ARU 83] show that the nature of the clay influences the indicator of dielectric dispersion. This indicator is obtained by measuring the electric properties of the soil and of the interstitial fluid. By taking into consideration the concentration of the interstitial fluid and the rate of sodium absorption, the dielectric dispersion indicator enables us to characterize the initiation of the detachment of clay particles [ARU 83]. Thus, as with any other type of erosion, suffusion, which depends on the conditions of detachment of particles or aggregates of particles, appears to depend on the nature of the clay, and for a given type of clay, on the clay percentage.

2.3. Main initiation criteria for suffusion

In what follows, we do not intend to offer a treatise on the initiation criteria, or to make up a list of all the criteria proposed in the relevant literature. For a more in-depth study of these matters, the reader can refer to [WAN 04] and [LI 08a]. The main objective here is to present the criteria that are most widely used or most suitable to characterize the initiation of suffusion. We distinguish two types of criteria: grain-size distribution and hydraulic.

2.3.1. *Grain-size distribution criteria*

The capacity for the geometric retention of a filter was initially characterized by a criterion that was generally attributed to Terzaghi and that can be expressed by the following inequality:

$$\frac{D_{15}}{d_{85}} < 4 \qquad [2.2]$$

where D_{15} is the sieve size for which 15% of the weighed filter material is finer and d_{85} is the sieve size for which 85% of the weighed base material is finer.

In the aim of applying this filtering criterion to the case of suffusion, Kézdi [KÉZ 79] proposed the distinction between the coarser fraction and the fine fraction in a point chosen randomly in the distribution of the grains. He considers that the coarser fraction will act as a filter for the fine fraction. The sieve size for which 15% of the weighed coarser fraction is finer (called D'_{15}) characterizes the size of the constrictions of this fraction, and the sieve size for which 85% of the weighed fine fraction is finer (d'_{85}) is used to characterize the size of the particles in this fraction. Thus, [KÉZ 79] advances the hypothesis that the soil will be stable to suffusion if the ratio D'_{15} / d'_{85} is well below 4 for every random division point throughout the grain size distribution. By using the same method, Sherard (quoted by [WAN 04]) proposed different numeric values for the inequality [2.2] depending on the percentage of fine particles in non-cohesive soils. Starting from the model of capillary tubes, Kovacs [KOV 81] compared the average diameter of pores d_0 (equation [2.1]) with the representative diameter of the fine fraction d'_{85}. If $d_0 < d'_{85}$, then the soil is considered to be stable to suffusion. However, this criterion was not tested experimentally.

One of the most currently used criteria is the criterion proposed by Kenney and Lau [KEN 85]. This criterion was established based on results of tests that were carried out on samples composed of sand and gravel. It consists of determining, for each size of the particle d, the mass percentage F of the grains, having a size that is lower than d, as well as the mass percentage H of the grains having a size between d and $4d$ ($H = F_{4d} - F_d$). These researchers consider that the grains smaller than d can detach themselves if the soil does not have enough grains within the interval between d and $4d$. They represent the values of H depending on the values of F (where the obtained curve is called shape curve). Within the interval $0 < F < X$, they propose that the soil be deemed unstable if its shape curve is situated completely or partially under the line represented by $H = 1.3\,F$ (the equation once modified in 1986 becomes $H = 1.0\,F$). In the case of soils whose coefficient of uniformity $C_U = d_{60}/d_{10} = 3$, X is chosen to be equal to 0.3 and if $C_U > 3$, then $X = 0.2$.

As a consequence of dry mixing tests carried out on soils composed of sand and gravel (C_U up to 200), Burenkova [BUR 93] assumes that a fraction of the small-size grains is not part of the granular skeleton of the soil if its insertion in the mixture does not entail any increase in the volume of the mixture. Conversely, if the volume of the sample rises with the addition of the fine fraction, this fraction is supposed to belong to the granular skeleton. The author suggests that the stability to suffusion depends on the ratios $h' = d_{90} / d_{10}$ and $h'' = d_{90} / d_{15}$. [BUR 93] defines the domain of the stable soils by:

$$0.76\log(h'') + 1 < h' < 1.86\log(h'') + 1$$

Furthermore, the size of the largest particle d_{dv}, representing the limit between the granular skeleton and the grains sensitive to suffusion, can be estimated by:

$$0.55\ h''^{-1.5} < \frac{d_{dv}}{d_{100}} < 1.87\ h''^{-1.5}$$

Following the research works of [BUR 93] and in order to distinguish more clearly between the unstable soils and the soils stable to suffusion, Wan and Fell [WAN 08] define the probability P for suffusion by:

$$P = \frac{\exp Z}{1 - \exp Z} \quad [2.3]$$

For the mixtures of silt–sand–gravel and clay–silt–sand–gravel:

$$Z = 2.378 \log h'' - 3.648\ h' + 3.701 \quad [2.4]$$

For the mixtures sand–gravel with less than 10% of non-plastic fine particles ($d < 0.075$ mm):

$$Z = 3.875 \log h'' - 3.591\ h' + 2.436 \quad [2.5]$$

For the soils whose granular distribution is upwardly concave (type curve 4 in Figure 2.1) Wan and Fell [WAN 08] propose a different method as an alternative for the criteria proposed by Burenkova [BUR 93], based on the ratios d_{90}/d_{60} (used to characterize the slope of the coarser fraction) and d_{20}/d_5 (characterizing the range of the fine fractions). Figure 2.3 shows $15/\log(d_{20}/d_5)$ depending on $30/\log(d_{90}/d_{60})$, [WAN 08] discerns an area where the soils are stable to suffusion, an area where they are unstable and, finally, a transition area.

2.3.2. *Confronting granular criteria*

Starting from the tests that dealt with sand–gravel mixtures, Li and Fannin [LI 08b] have carried out a comparative study of the criteria offered by Kézdi [KÉZ 79] and Kenney and Lau [KEN 85]. These two criteria evaluate the risk of suffusion based on the slope of the granular distribution.

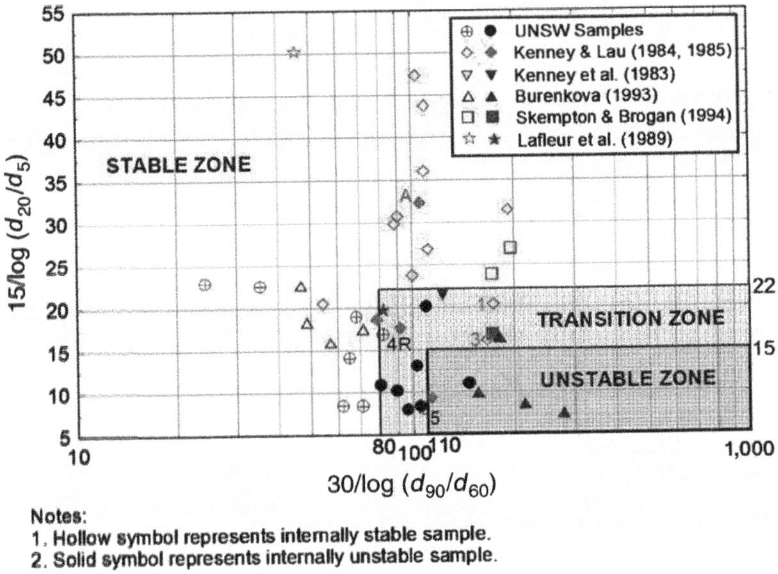

Figure 2.3. *Alternative method for determining the stability to suffusion [WAN 08]*

According to Li and Fannin [LI 08b], the risk evaluation for suffusion made with these two criteria can be reformulated by: within an interval that corresponds to four grain diameters, the slope is lower than 15% in the case of Kézdi's criterion and lower than F% in the case of the criteria proposed by Kenney and Lau [KEN 85]. The authors summarize the comparison between the two criteria by representing the increase in mass H depending on the variable F (Figure 2.4) for the minimal value of the ratio H/F.

Li and Fannin [LI 08b] draw the following conclusions: the two criteria converge for $F = 15\%$. In the case of soils with discontinuous grain size distribution, Kézdi's criterion seems to be more suitable for distinguishing the stable soils from the unstable soils. For the soils with continuous grain size distribution, the best results are obtained using the Kenney and Lau criterion. Kézdi's criterion is more conservative for $F < 15\%$ and, conversely, Kenney and Lau's criterion is more conservative for $F > 15\%$.

Wan and Fell [WAN 08] have carried out 20 suffusion tests on mixtures of clay–silt–sand–gravel and silt–sand–gravel mixtures. A downward flow is generated by a hydraulic gradient of 8. Three criteria are used for predicting the initiation of suffusion: the Sherard's criterion, the Kenney and Lau's criterion, and, finally, Burenkova's criterion.

The comparison between the experimental results and the predictions made with these three criteria shows that the Sherard's and the Kenney and Lau's criteria are quite definitely conservative. The best predictions are made using the Burenkova's criterion, and Wan and Fell [WAN 08] have proposed the calculation of probability (equations [2.3]–[2.5]) in order to improve these predictions.

We can deduce that the limits of the different grain-size distribution criteria come from the fact that several important parameters, such as the mineralogy of the material, the angularity of the grains, the effective stress force, and the viscosity of the fluid, were not taken into account. Furthermore, most of the tests were performed on cohesionless materials and did not concern cohesive soils. Finally, even if the transport of particles is geometrically feasible, the action of the hydraulic flow must be sufficient for detaching the soil particles. Thus, Kovacs [KOV 81] acknowledges that not observing a grain-size distribution criterion does not necessarily entail a strong sensitivity to suffusion, due to the contact forces between the grains that resist their detachment.

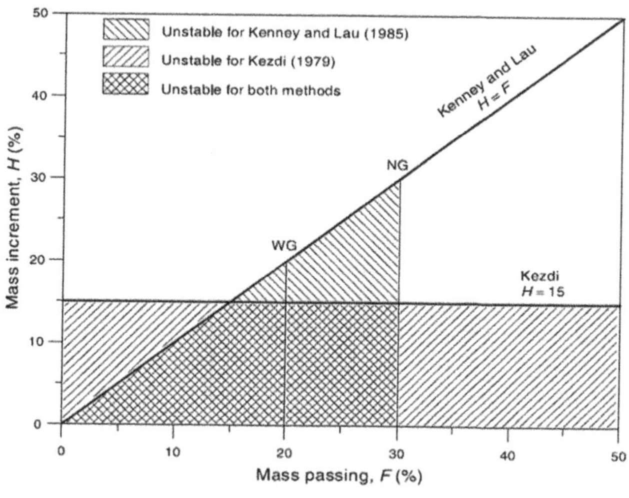

Figure 2.4. *Comparison between Kézdi's criteria and Kenney and Lau's criteria [LI 08b]*

2.3.3. *Hydraulic criteria*

The action of the hydraulic flow on the grains is often described following three distinct approaches: the hydraulic gradient, the hydraulic shear stress, and the pore velocity. The critical values of these three quantities can be used to characterize the initiation of suffusion.

2.3.3.1. *Critical hydraulic gradient*

Under the effect of an upward water flow, the effective stress of a volume of soil can annul itself when the hydraulic gradient reaches the critical value defined by Terzaghi. This is called critical gradient of heave. This heave or blow out gradient can be expressed as the ratio between the submerged density of soil and the water density. The gradient is generally of the order of one unit.

By carrying out tests on the mixtures of sand and gravel with an upward flow, Skempton and Brogan [SKE 94] concluded that the value of the critical gradient of the initiation of suffusion, marked as i_{ch}, can be approximately a third or a fifth of the value of the critical heave gradient i_c. As the authors explain, the deviation between the values of the critical gradient is caused by the fact that the major part of the effective stress is supported by the granular skeleton made of gravel. They propose to express the critical suffusion gradient of the sand i_{ch} by:

$$i_c - i_{ch} = \alpha \frac{\gamma'}{\gamma_w}$$

where α is a reduction factor for the effective stress, which is lower than 0.1 [SKE 94].

Monnet [MON 98] used the results of the same tests to draw up a method of evaluation for the suffusion gradient. This method aims at comparing the hydraulic conductivity k of the material with the value $0.01(d_{15})^2$. If $k < 0.01(d_{15})^2$, then the instability will be obtained by heave, and if $k > 0.01(d_{15})^2$, the instability will be obtained by suffusion with a critical gradient estimated by:

$$i_{ch} = 0.01 \frac{(d_{15})^2}{k} i_c$$

where d_{15} is expressed in millimeters and k is expressed in meters per second.

By taking into account the hydraulic conductivity, the criterion proposed by [MON 98] is equivalent to a critical Darcy velocity criterion ($v_{ch} = k i_{ch}$).

2.3.3.2. *Critical hydraulic shear stress*

Hillel (quoted by [RED 00]) expresses the hydraulic shear stress generated by a hydraulic flow in a tube of length L and of the radius r through the equation:

$$\tau = \frac{\Delta P}{L}\frac{r}{2}$$

where ΔP is the pressure drop throughout the tube. To generalize this approach to the study of soils, Khilar *et al.* [KHI 85] model the soils by using a network of capillary tubes that are parallel to the constant radius. The intrinsic permeability K of the system of tubes may be expressed as follows:

$$K = n\frac{r^2}{8}$$

where n is the porosity of the soil. By replacing r with K in the expression of the hydraulic shear stress, Red *et al.* [RED 00] came up with the expression:

$$\tau = 1.414\ \frac{\Delta P}{L}\sqrt{\frac{K}{n}} \qquad [2.6]$$

The hypothesis that the hydraulic loading responsible for the suffusion is essentially a shear allows comparing the rate of suffusion with the hydraulic shear stress. With the aim of helping the representation of erosion rates versus the hydraulic shear stress, Shields (quoted by [ARU 83]) defines, by making a linear extrapolation, the critical hydraulic shear stress as the maximum value of the hydraulic shear stress corresponding to the absence of particle detachment.

Using equation [2.6] and by extrapolating from the curve of the maximum erosion rate versus the hydraulic shear stress, the critical value of the hydraulic shear stress was evaluated for mixtures of sand and kaolinite, consolidated under isotropic confinement of 100 kPa [BEN 08]. Depending on the percentage of the clay, the value of the critical stress is 0.13, 0.23, and 0.32 Pa for 10%, 20%, and 30% of kaolinite, respectively. Red *et al.* [RED 00] have compared the values of the critical hydraulic shear stress of suffusion with the values of critical hydraulic shear stress of surface erosion, measured by hole erosion test. This confrontation emphasized the fact that the critical hydraulic shear stress value of suffusion can be several orders of magnitude higher than the critical hydraulic shear stress of surface erosion. These deviations can be the result of the extremely significant difference between the exchange surfaces of the solid and the fluid, for these two types of erosion. According to [RED 00], the potential surface erosion is of the order of 5.5×10^{-4} m^2

in the case of the hole erosion tests, whereas the porosity tests using mercury reveal a potential surface erosion in the case of suffusion of 7.065 m^2. It is worth noting that this very important discrepancy is, perhaps, due to the measuring method used by [RED 00] (which can be used *a priori* on a rigid structure). These deviations also emphasize the great difficulty in defining and measuring the critical hydraulic shear stress.

2.3.3.3. *Critical pore velocity*

To consider the spatial variability and the anisotropic characteristic of the hydraulic conductivity, Perzlmaier [PER 07] predicts the use of the average pore velocity for characterizing precisely the hydraulic solicitation. The average pore velocity is expressed as the ratio between the Darcy velocity and the porosity multiplied by the tortuosity.

The tortuosity is the squared ratio of the distance between two points and the length of the real course of the flow between these two points. Wittmann (quoted by [PER 07]) has evaluated the tortuosity to $2/\pi = 0.6366$. Starting from several theoretical approaches that were developed in the relevant literature, [PER 07] has calculated the critical value of the average velocity of a pore according to the size of the particles of cohesionless soils. He concludes that for particles of a size higher than 0.1 mm, the critical pore velocity increases with the size of the particles, and for particles of a diameter lower than 0.1 mm, the critical velocity is about 10^{-3} m/s.

For certain tests used to determine the critical velocity, it is worth noting that the initiation of the suffusion was visually detected. In the case of the detachment of silt or clayey materials, we should not exclude the possibility that this observation is not accurate enough and consequently that the value of 10^{-3} m/s is an overestimation.

2.3.4. *Summary and final remarks*

The initiation of suffusion is governed by:

– the geometry of the porous network, which is influenced by the granular distribution, porosity, and the shape of grains;

– the hydraulic loading that was mainly modeled by the hydraulic gradient, the hydraulic shear stress, or the pore velocity.

The different confrontations carried out between the predictions of these criteria and observations emphasize some significant differences. These deviations may come from the absence of coupling between the two types of initiation criteria.

Furthermore, the critical hydraulic shear stress values for suffusion seem very different from that of the surface erosion. These differences can be caused by the intrinsic characteristics of the different erosion processes, but the models of interpretation that were deployed can also cause these differences.

2.4. An initiation criterion formulated using a geohydromechanical approach

The approach proposed by Li [LI 08a] can be divided into two stages: the first stage aims at evaluating the potentiality for the initiation of suffusion by using a geometric criterion. The objective of the second stage is to evaluate the capacity of hydraulic loading to trigger the suffusion by taking into account the effective stress.

2.4.1. *Geometric criterion*

To identify the fine fraction within the granular distribution, [LI 08a] recommends choosing a point of division at the level of the discontinuity in the case of a discontinuous grain-size distribution curve, and at the level of the minimal value of the ratio H/F (emerged from the criteria of Kenney and Lau) in the case of a soil with continuous grain size distribution. The fine fraction having been identified (the value of F cannot be higher than 35% or else the coarser fraction floats within the fine fraction [SKE 94]), [LI 08a] calculates the average diameter d_0 of the model of the capillary tube proposed by Kovacs (equation [2.1]). Starting from the results of his tests and from several tests in relevant literature, [LI 08a] suggests that the soils whose $d_0 < d'_{85}/0.42$ could be considered potentially stable.

For a soil that is potentially unstable to suffusion, [LI 08a] advances a hydromechanical criterion to evaluate the hydraulic strain required for the initiation of suffusion.

2.4.2. *The hydromechanical criterion*

From the concept of stress reduction supported by the fine fraction [SKE 94], Li [LI 08a] advances the notion of the hydromechanical envelope that depends on the effective stress. This concept is expressed by:

$$i_c = \frac{\alpha}{1-0.5\alpha}\left(\bar{\sigma}'_{\text{vm}} + 0.5\frac{\gamma'}{\gamma_w}\right), \quad \bar{\sigma}'_{\text{vm}} = \frac{\sigma'_{\text{vm}}}{\gamma_w \Delta z}, \quad \sigma'_{\text{vm}} = \frac{1}{2}(\sigma'_{t0} + \sigma'_{b0}) \qquad [2.7]$$

where α is the effective stress reduction factor, $\bar{\sigma}'_{\text{vm}}$ is the normalized average vertical effective stress value, Δz is the height of the layer of soil, σ'_{t0} is the effective stress at the top of the soil layer, and σ'_{b0} is the effective stress at the base of the soil layer.

The correlation research made by [LI 08a] between α and three geometric indicators ($(H/F)_{\min}$, D'_{15}/d'_{85}, d'_{85}/d_0) leads to a linear correlation between α and the ratio d'_{85}/d_0. Thus, equation [2.7] enables us to determine the critical hydraulic gradient and, in comparison with the applied hydraulic gradient, to predict the initiation of suffusion.

2.4.3. *Summary and final remarks*

The approach proposed by Li [LI 08a] for estimating the initiation of suffusion combines a geometric criterion (founded, on the one hand, on the Kovacs model for the capillary tube diameter and, on the other hand, on the criteria proposed by Kenney and Lau [KEN 85]) and a hydromechanical envelope (mainly based on Skempton and Brogan's concept of effective stress). The consecutive stages of this framework are shown in Figure 2.5.

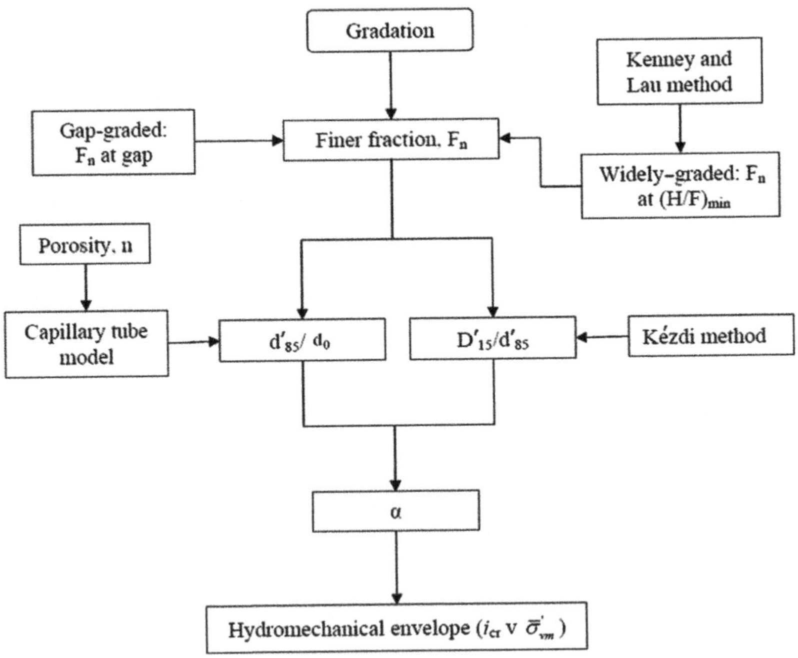

Figure 2.5. *Framework for evaluating suffusion process [LI 08a]*

The critical gradient formed by Li [LI 08a] depends on the normalized average vertical effective stress, which is itself inversely proportional to the height of the layer of soil taken into consideration. The same remark can be made regarding the criterion for surface erosion proposed by Sellmeijer [SEL 88], where the critical gradient decreases depending on the length of the seepage flow [MAR 12].

This remark prompts us to further investigate the relevance of an approach based on the expression of a hydraulic gradient or on the relevance of all the laboratory tests that were carried out on a smaller scale, in relation to the large scale of the real structure. The following section discusses this scaling effect as well as the method proposed for avoiding this effect.

2.5. The scaling effect and the energetic approach

2.5.1. *Identifying the scaling effect*

Li [LI 08a] has carried out suffusion tests on mixtures of glass beads with discontinuous granular distribution, using two permeameters of different sizes. For the same type of tested specimen and the same applied effective stress, the values of the critical gradient to initiate suffusion can be eight times higher with the small permeameter than with the large permeameter. For a more in-depth study of this scaling effect, Mar *et al.* [MAR 12] have carried out a set of suffusion tests on clayey sand with different sample lengths. With the aim of controlling the condition of effective stress, the experimental device was placed on the swinging basket of a geotechnical centrifuge, thus enabling the application of different gravitational forces. Figure 2.6 represents the erosion rate that varies with the length of the tested samples and with the applied hydraulic gradient. For the same value of the hydraulic gradient and the same effective stress, the values of the suffusion rate increase linearly, depending on the length of the tested sample.

For this type of clayey sand, the values of the critical suffusion gradient obtained according to the length H_m of the samples are $i_c = 44$ m/m for $H_m = 60$ mm, $i_c = 34$ m/m for $H_m = 90$ mm, and $i_c = 28$ m/m for $H_m = 120$ mm [MAR 12]. The characterization of both initiation and development of suffusion based on the hydraulic gradient is thus perturbed by the length of the tested sample.

2.5.2. *Energetic approach*

This approach aims at determining the energy dissipated by the fluid, when it flows within the porous medium. Let M be the mass of the fluid, V its volume, ρ_w

its density, e_{int} its internal energy, and $\vec{U}$ its velocity (components u, v, w). Let S be the contact surface between the fluid and its environment, having the normal value $\vec{n}$. Then [REG 09] and [MAR 11] express the energy E dissipated by the fluid as:

$$\frac{dE}{dt}=\frac{\partial}{\partial t}\iiint_{\text{Volume}}\left(e_{\text{int}}+\frac{u^2}{2}+\vec{g}\,\vec{z}\right)\cdot\rho_w\,dV+\oint\!\!\oint_S\left(e_{\text{int}}+\frac{u^2}{2}+\vec{g}\,\vec{z}\right)\cdot\rho_w\,(\vec{U}\,\vec{n})\cdot dS$$

where $\vec{g}$ is the vector of the gravitational force and z is the coordinates. The total energy E is the sum of the mechanical energy and the energy exchanged with the environment.

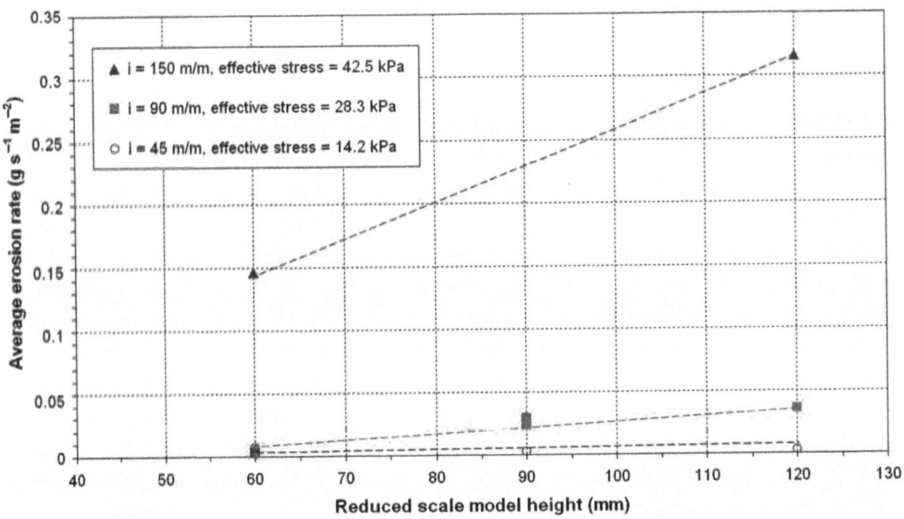

Figure 2.6. *Erosion rate varying according to the height of the sample, according to the applied hydraulic gradient, as well as according to the value of effective stress [MAR 12]*

In the case of suffusion, Mar *et al.* [MAR 12] use four hypotheses: the system is supposed to be adiabatic and isothermal, the state is supposed to be steady, and the dissipations of energy through viscosity and turbulence within the fluid are considered to be negligible in comparison with the dissipation through the contact with the soil that generates erosion. The erosion power P_{er} can thus be determined by $P_{\text{er}} = Q\,\Delta P + \rho_w\,g\,L\,Q$, where Q is the flow rate and ΔP is the pressure drop between the upstream and downstream of a sample of length L.

The average erosion rate increases linearly, depending on the average erosion power, and independently of the length of the sample (Figure 2.7(a)). For every test, the total mass of the eroded clay can be plotted according to the energy dissipated by erosion, which, in turn, is calculated by time integrating the erosion power (Figure 2.7(b)). The total mass of the eroded clay is linearly correlated to the dissipated energy and this correlation is unperturbed by the scaling effect.

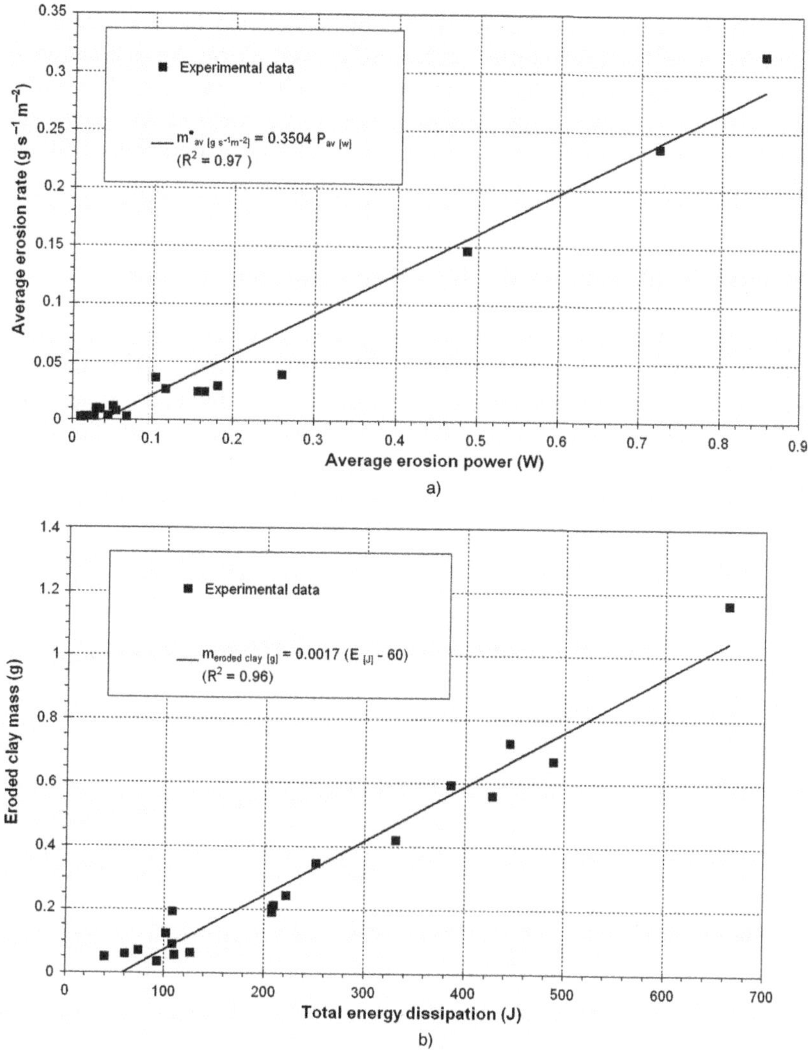

Figure 2.7. *(a) Erosion rate, varying with the average erosion power and (b) total mass of eroded clay, varying with the dissipated energy [MAR 12]*

2.5.3. *Summary*

Several experimental campaigns of suffusion [LI 08a, MAR 12] and of surface erosion experiments [SEL 88] emphasize the fact that the value of the critical hydraulic gradient to initiate erosion decreases with the size of the tested sample. The energetic approach that consists of determining the power of erosion and the energy dissipated by the fluid allows us to avoid this scaling effect.

The different approaches that were previously detailed have a bearing on the characterization of the initiation of suffusion as well as on the estimation of the average erosion rate and on the estimation of the total mass eroded by suffusion. Recent research has centered on the simulation of the coupling between suffusion and filtration; this relationship was emphasized through experiments that we will present in more detail below.

2.6. Coupling the phenomena of suffusion and filtration-clogging

Several tests that were carried out on mixtures of sand and clay [RED 00, BEN 08, MAR 09] show that hydraulic conductivity decreases throughout the phenomenon of suffusion (Figure 2.8). This decrease in hydraulic conductivity is attributed to a clogging of some pores caused by the clay particles that were detached and then redeposited, or that were detached and then filtered in the soil.

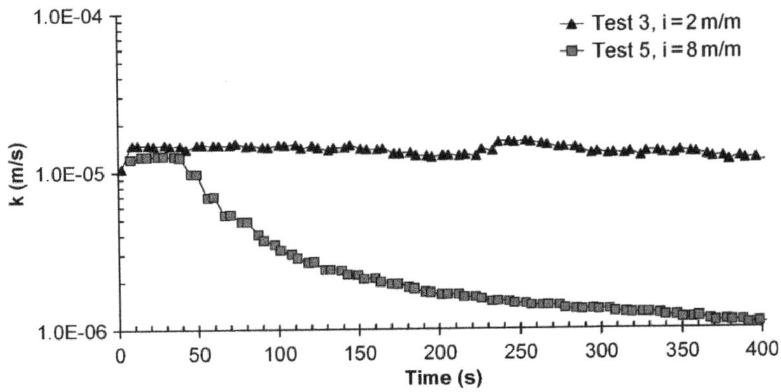

Figure 2.8. *Evolution throughout time of the hydraulic conductivity without erosion (test 3) and in the presence of clay suffusion (test 5) [MAR 09]*

Furthermore, in the suffusion tests involving a constant global hydraulic gradient, which were carried out on cohesionless materials with discontinuous grain size distribution [SAI 11], an increase in the local hydraulic head was measured before the massive detachment of fine particles. This local increase in the pressure is attributed to a filtration of certain particles that were detached in the upstream of the area in question. However, in the case of non-cohesive materials, the filtration is localized in very restricted areas on the tested sample. The global hydraulic conductivity is not significantly modified.

Suffusion, i.e. the process of detachment and then transport of fine particles, is completely dependent on the fate of eroded particles. A fraction of the fine particles will be able to migrate toward a much more porous medium, while other particles will be filtered in the soil. The amplitude of the incidence of this type of filtration on hydraulic conductivity varies with the nature of the fine particles and the porosity of the granular skeleton. In the case of cohesive fine particles, a decrease in hydraulic conductivity can be measured, and in the case of non-cohesive fine particles, the effect of the filtration can be limited to an increase in the interstitial pressures, increase that is much localized.

The next section deals with a more in-depth analysis of filtration.

2.7. Processes causing filtration

The clogging of pores in a permeable granular filter by eroded fine particles can drastically reduce its drainage capacity and lead to the instability of the hydraulic earth structures. The granular filters are subjected to a long accumulation of fine particles, and must remain permeable in order to guarantee the safety of the hydraulic structure. The efficiency of the process of filtration depends on various parameters, which can be classified into two categories: geometrical parameters (porosity, shape and size of grains and particles, concentration of suspension, and density) and hydraulic parameters (flow velocity, Reynolds number, and Péclet number). We can distinguish between four types of filtration: the bed filtration (retention of particles whose dimensions are higher than a fixed size), tangential filtration (the flow of the suspension running parallel to the surface of the filtering medium), the filtration with cake formation (i.e. very dense suspensions formed through autofiltration), and deep filtration (diluted suspensions; the particles of a size lower than that of the pores that infiltrate deep into the porous medium and that deposit at different depths). Predicting the importance of the transport of particles is difficult because of the complexity of the processes that govern the detachment, the transport, and the deposition of the particles in a porous medium. We can discern four types of retention sites: the surface sites, the crack sites, the constriction sites (in the upstream of the shrinkage of pores whose size is lower than that of the particles), and the cavern sites (Figure 2.9).

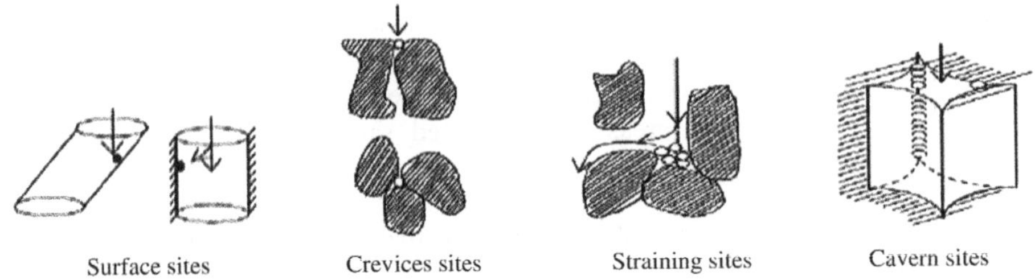

Figure 2.9. *Different types of retention sites [HER 70]*

2.7.1. *Background knowledge*

The absence of a filter in an earthen hydraulic structure can cause its failure by erosion, and the use of a poorly adapted filter can cause either clogging or a massive migration of particles [FOS 00]. If the initiation of internal erosion is the first condition of its existence, the absence or the unsuitability of the filter to stop the transport of particles is the second condition. The criterion based on the ratio between the opening of the sieve that allows for 15% of the material of the filter to pass through it (D_{15F}) and the opening of the sieve that allows 85% of the soil particles to be filtered (d_{85B}) is often used by engineers when they design the filters [SHE 89]. However, the porosity and the permeability are also significant parameters and the geometrical criteria are complemented by a hydraulic criterion. Although the deep filtration is a well-known technology, the mechanisms responsible for the retention of particles are not yet fully comprehended.

The deep filtration is a complex process in view of the different mechanisms at stake [HER 70]. We can discern two major types of forces: hydromechanical forces and physicochemical forces. The experiments carried out in two columns on porous media with latex spheres have shown that the diameter of particles plays a major part in the transport of particles [CUM 99]. Significant research was carried out regarding the movement of particles in suspension as well as the mechanical blockage (or straining), interception, sedimentation, or diffusion [MCD 86]. For coarser particles, those that are larger than 10 μm, the hydrodynamic, gravitational, and inertial effects are predominant, whereas for the small particles whose dimension varies between 0.1 and 1 μm [GOH 99], all the forces and mechanisms are accounted for.

The results obtained by Khilar and Fogler [KHI 84] on the colloidal materials explain the interaction between the chemical conditions and the hydrodynamic conditions in the processes of retention and release of fine particles. The remobilization arises below a critical concentration of salt and above a critical flow velocity. The studies carried out on the detachment and filtration of clayey particles in a porous medium have shown that the trapping of particles does not take place below a certain threshold value of the concentration of particles.

The trapping of fine particles in the constrictions is partly due to the direct interception and partly due to the size exclusion [MAY 05]. The effect of the physicochemistry on the filtration of colloidal particles in a saturated porous medium was studied by [TUF 04], where the trapping efficiency of the collector is modeled as the sum of the contributions of different transport mechanisms.

The research works of [FRE 99] on the physicochemical effects have shown that they determine not only the distribution of the deposition in the porous medium but also the clogging process. The importance of the relationship of pore solution in the medium was largely studied for the colloidal particles that present the adequate surface properties [GRO 01, TOS 09], and was less studied for the soil particles [KHI 84, BLU 05].

The first criteria for deep filtration [HER 70] are founded on the ratio between the dimensions of the suspended particles (d_p) and the grains of the porous medium (d_g), and they predict the blockage of particles if the diameter of these particles is 0.05 times higher than the average diameter of the medium grain d_{50} ($d_p/d_{50} > 0.05$). Bradford *et al.* [BRA 04] have shown that even for a ratio of d_p/d_{50} equal to 0.002, the trapping of particles can well take place. Other authors [SIL 95] stress the importance of not only the effect of the size of particles but also the importance of their shape and nature. Angular particles cause their deposition and a fall in the permeability that is more important than in the case of spherical particles. According to [GHI 94], the (spherical) particles are systematically blocked starting from a threshold of $d_p/d_g = 0.155$. On the contrary, when this ratio is below 0.065, the retention of particles is negligible. The studies carried out on the pore scale advance a capturing model with two kinetics: the transport of particles up to the interface with the grain of the porous medium, and then the attachment of the particles on the surface of the collector. Probabilistic models [RED 05] founded on a capillary flow were developed to quantify the capturing of particles in a porous medium. Several factors, such as the velocity and the pressure, can lead to a detachment of particles that were previously deposited in the medium.

2.7.2. *Theoretical background*

The basis for the understanding of the process of flow and particle transport in the porous media was largely developed from experimental and theoretical studies on the macroscopic models, where the coupling at the pore scale is not often measured or observed directly. The experimentation was focused on the measuring of the variables and average properties of the medium, such as the porosity, the permeability, the fluid pressure, and the concentration of particles in the laboratory reactors. There are two types of forces that act on the retention of particles in a porous medium: hydromechanical forces and physicochemical forces.

The hydrodynamic forces are the following:

– *The gravitational pull*: it causes the sedimentation of the particle and it is expressed as:

$$F_G = \frac{1}{6}\pi d_p^3(\rho_p - \rho)g$$

where ρ_p is the volume mass of the particle, ρ is the volume mass of the water, and g is the gravitational constant.

– *The drag force*: in the case of a Newtonian fluid and a laminary flow that acts upon a spherical particle, this force is expressed by:

$$F_D = 32\mu a_p U_{\text{ap}}$$

where U_{ap} is the velocity of the fluid on a radial distance that is equal to the radius of the particle a_p, and μ is the viscosity of the water.

– *The inertial force*: this centrifugal force is approached by:

$$F_G = \frac{1}{3}\pi d_p^3(\rho_p - \rho)\frac{u^2}{d_g}$$

where u is the average interstitial velocity.

The physicochemical forces are the following:

– *The Van der Walls* forces: these are intermolecular forces of attraction of an electromagnetic origin, which act on very short distances (10 nm); for a sphere that interacts with a plane, these forces are expressed by:

$$F_{\text{vdw}} = \frac{\pi H d_p}{12h^2}$$

where H is the Hamaker constant that characterizes every surface (10^{-19} to 10^{-20} J for every condensed phase) and h is the distance between the particle and the solid grain.

– *The electrostatic forces* of the double electric layer: in the specialized literature, there have been several formulations of the electrostatic force, starting from the resolution of the Poisson–Boltzman equation, with well-defined boundary conditions [KHI 84]; in the case of the interaction between a plain surface and a sphere, the appropriate expression is the following:

$$F_{dl} = 2\pi\varepsilon a_p \kappa \exp(-\kappa h)\left[2\psi_p\psi_g - \left(\psi_p^2 + \psi_g^2\right)\frac{\exp(-\kappa h)}{(1-\exp(-2\kappa h))}\right]$$

where ε is the dielectric constant of the fluid, κ is the Debye–Huckel parameter that characterizes the extension of the load of counter-ions in the solution, ψ_p and ψ_g, respectively, are the surface potential of the particle and of the wall of the porous medium.

These different forces will compete against one another in the porous medium and they contribute to the retention or remobilization of the particles. Several retention mechanisms depend on the nature of these suspended particles (mineralogy, shape, size, etc.) of the properties of the fluid (pH, ionic force, etc.) and of the properties of the medium (the nature of its components, size and distribution of the pores, shape of the grains, etc.). These mechanisms, which are illustrated in Figure 2.10, are the following:

– *The mechanical blockage* (or straining): this mechanism depends directly on the relation between the diameter of the grains d_g and the diameter of the suspended particles d_p; the blockage is the most dominant mechanism for the ratio $d_g/d_p \leq 10$ [MCD 86] and leads to clogging; it can be ignored if $d_g/d_p > 20$. More recent experiments involving well-calibrated porous media have reached a ratio of $d_{50}/d_p > 200$ from which the mechanical blockage can be neglected [BRU 05]. However, this ratio evolves according to the number of particles that arrive at the entrance of a constriction. The results of these recent experiments carried out on micromodels show that the only particles that are not blocked are the particles with a diameter lower than 1.5 times the diameter of pores [KEL 04];

– *Interception*: in view of its size, a particle may collide with the collector and can reside inside the convergent walls of the pore surface; this trapping is linked to the ratio of the dimensions of the grain and the particle called interception parameter.

– *Sedimentation*: the sedimentation velocity of these particles can be calculated using Stokes' law; sedimentation can contribute to the deposit of particles in the flows of low velocity.

– *Diffusion*: in the absence of electrostatic forces, diffusion causes the retention of particles on the collectors; this mechanism for transport and deposition can be quantified using the Brownian diffusion coefficient $D = K_B T / (3\pi\mu d_p)$, where K_B is the Boltzman constant and T is the temperature .

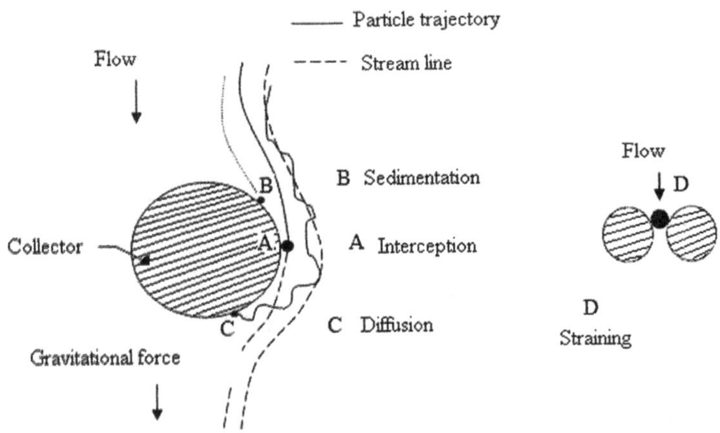

Figure 2.10. *Mechanisms of transport and trapping of a particle using a spherical collector [MCD 86]*

2.8. Filtration modeling

2.8.1. *Modeling in a continuous medium*

2.8.1.1. *Homogeneous medium*

There are two approaches that can describe the transport of particles in a porous medium. The first is the macroscopic approach, which is based on a mass balance at the level of each volume element of the porous medium and consists of following the average concentration in each point. The deposition of particles is described by a filtration coefficient. The second is the microscopic approach, which is founded on the individual study of different processes of interaction that could develop in the medium. In this approach, the capture and the detachment efficiency (or the adsorption/desorption) is described not only in terms of empirical transfer coefficient but also in terms of the contributions brought by the different transport mechanisms (section 2.7.2). In steady-state flow, by ignoring the dispersion and the diffusion mechanisms (by considering a clean porous medium), the empirical model describing the proportion of the deposited particles as a function of the first order of concentration is expressed by the so-called filtration model:

$$\frac{\partial C}{\partial x} = -\lambda C \tag{2.8}$$

where C is the concentration of particles in the fluid phase and λ is the filtration coefficient (L^{-1}). The solution of equation [2.8] describes an exponential decrease in concentration. Starting from the restitution curve obtained through filtration tests (step-input injection of the suspension of a concentration C_0 in a column of length L and the measuring of the outlet concentration C_L), we obtain $\lambda = -\ln(C_L/C_0)/L$.

The efficiency model for capturing a simple collector emerges from the microscopic considerations and it expresses the relation between the deposition kinetics of the particle and its velocity u of approaching a collector:

$$\eta = \frac{I_d}{uC\pi a_c^2}$$

where I_d is the global mass retention flux, obtained by adding the local fluxes of particles on the entire surface of the collector, and a_c is the collector radius.

The particular deposition can be described by two processes: the transport of particles in suspension toward the surface of the grain defined by the frequency of collisions with the collector and the attachment of the particles or the efficiency of the collision. In the presence of the repulsive forces of the double electric layer, the capture efficiency is multiplied by an efficiency collision factor α:

$$\eta = \alpha \eta_0$$

According to [TUF 04], the kinematics of the deposition of particles calculated by the sum of the individual contributions of the different transport mechanisms (diffusion, interception, and sedimentation) agrees very well with the one calculated using the solution of the equation of convection diffusion. The capture efficiency η_0 can be expressed by:

$$\eta_0 = \eta_D + \eta_G + \eta_I$$

where η_D is the capture efficiency by Brownian diffusion, η_G is the contribution of the gravity, and η_I describes the retention by interception.

The collision efficiency α represents the fraction of the particles attached to the grains following the collisions. Given that the classical theories are unable to predict the efficiency of the collision or of attachment, the experiments carried out in the reactors are often used to determine α by:

$$\alpha = -\frac{2}{3}\frac{d_g}{(1-n)L\eta_0}\ln\left(\frac{C}{C_0}\right)$$

where L is the length of the porous medium, n is its porosity, and C is the concentration of the particles at the outlet of the medium. The relation between the deposition kinetics coefficient K_{dep} and the capture efficiency of a collector η_0 is [RYA 96]:

$$K_{dep} = \frac{3}{2}\frac{(1-n)}{d_g n}U\alpha\eta_0$$

where U is the Darcy velocity. K_{dep} $[T^{-1}]$ is thus a coefficient that links the variation of the particle concentration to the concentration of the medium.

In the case of deep filtration, the filtration coefficient λ is linked to η_0 through the relation:

$$\lambda = \frac{3}{2}\frac{(1-n)}{d_g}\alpha\eta_0$$

2.8.1.2. *Heterogeneous medium*

To give a more realistic description of the heterogeneity effect on the flow and the transport in the porous media, a model of a medium with double porosity (macropores and micropores) is described by [BAI 95]:

$$\begin{cases} D\dfrac{\partial^2 C_1}{\partial x^2} - u\dfrac{\partial C_1}{\partial x} = f\dfrac{\partial C_1}{\partial t} + K(C_1 - C_2) \\ -\dfrac{u}{b}\dfrac{\partial C_2}{\partial x} = (1-f)\dfrac{\partial C_2}{\partial t} - K(C_1 - C_2) \end{cases}$$

where C_1 and C_2 are the concentrations of particles in the macropores and micropores, D is the dispersion coefficient in the macropores, u is the average velocity of the interstitial fluid in the region of the macropores, f is the fraction of porosity contained in the macropores, K is the celerity of the mass transfer between the stagnant fluid and the flowing fluid, x is the distance starting from the source, and b is the inverse of the factor of convection intensity, used to account for the difference in the flow velocity between the two regions of porosity.

The convection-dispersion model with the two regions mobile-immobile model (MIM) is founded on a hypothesis of fractioning the water in mobile area θ_m and immobile area θ_{im}. The particles are then transported by convection-dispersion within the mobile water, and they diffuse at the interface mobile water/immobile water according to a kinetics of the first order. The non-reactive transport is thus described by the equations:

$$\begin{cases} \theta_m \dfrac{\partial C_m}{\partial t} + \theta_{\text{im}} \dfrac{\partial C_{\text{im}}}{\partial t} = \theta_m D_m \dfrac{\partial^2 C_m}{\partial x^2} - q \dfrac{\partial C_m}{\partial x} \\ \theta_{\text{im}} \dfrac{\partial C_{\text{im}}}{\partial t} = \alpha (C_m - C_{\text{im}}) \end{cases}$$

where θ_m and θ_{im}, respectively, are the mobile and immobile water contents, and C_m and C_{im} [ML^{-3}], respectively, are the concentrations in the fractions of mobile and immobile water; D_m is the dispersion coefficient in the mobile fraction [L^2T^{-1}], q is the exchange kinetics coefficient, and α is the exchange coefficient between the two fractions [T^{-1}]. This model has three parameters: θ_m, D_m, and q.

2.8.2. *Convection model – dispersion with deposition kinetics*

We can distinguish between three main migration mechanisms of the elements of the solution in a porous medium: convection (advection), molecular diffusion, and kinematic dispersion. Convection is described by the principle of mass conservation:

$$\frac{\partial C}{\partial t} = -u \nabla C$$

where u is the average pore velocity, also called effective velocity.

The diffusion coefficient D_0 [$L^2\ T^{-1}$] of a concentration C of particles is given by Fick's second law:

$$\frac{\partial C}{\partial t} = D_0 \frac{\partial^2 C}{\partial x^2} \qquad [2.9]$$

Kinematic dispersion is a mixture of phenomena essentially linked to the heterogeneities of the microstructure of the medium that creates fluctuations in the flow velocities, regardless of the observation scale. This phenomenon results in a dilution in the concentration plume. The kinematic dispersion D and the molecular diffusion D_0 are generally described by the same equation [2.9], and the global coefficient is called hydrodynamic dispersion coefficient D.

Taking into account the stock variation of the matter (deposit or release), the molecular transport throughout granular porous media and under conditions of a saturated and uniform flow can be described, in the one-dimensional case, by:

$$\frac{\partial C}{\partial t} = D_L \frac{\partial^2 C}{\partial x^2} - u \frac{\partial C}{\partial x} - \frac{\rho}{n_0} \frac{\partial S}{\partial t}$$

In the hypothesis of first-order kinetics, the source term is expressed as follows:

$$\frac{\rho}{n_0} \frac{\partial S}{\partial t} = K_{\text{dep}} C - \frac{\rho}{n_0} K_r S \qquad [2.10]$$

where D_L is the longitudinal hydrodynamic dispersion coefficient of particles $[L^2 \cdot T^{-1}]$, u is the average velocity of the pores of the particles $[L \cdot T^{-1}]$, ρ is the apparent density of the solid matrix $[M \cdot L^{-3}]$, n_0 is the porosity $[L^3 \cdot L^{-3}]$, and K_{dep} and K_r are the deposition and release kinetics coefficients of the particles $[T^{-1}]$, respectively. These equations describe the evolution of the concentration of the particles in suspension and the mass of the deposited particles per mass unit of the porous medium S $[M \cdot M^{-1}]$, which varies according to the transport distance and to the period of time. The initial deposition and release kinetics of the particles are assumed to follow the first-order kinetics law.

If we neglect the release (small quantity of particles available), equation [2.10] becomes:

$$\frac{\partial C}{\partial t} = D_L \frac{\partial^2 C}{\partial x^2} - u \frac{\partial C}{\partial x} - K_{\text{dep}} C \qquad [2.11]$$

Equation [2.11] is no longer valid if the previously deposited particles influence the deposit kinetics of the newly injected particles [KRE 99]. We then obtain a deposition kinetics that depends on the concentration of particles in the aqueous phase and thus on the fraction of the porous medium available for the deposition Ψ. The latter is linked to the retention capacity through the porous medium, $X_{\max}$, by:

$$\psi = \frac{X_{\max} - S}{X_{\max}}$$

The differential equation with the second-order deposit kinetics is then expressed as follows:

$$\frac{\rho}{\omega_0}\frac{\partial S}{\partial t} = K_{\text{dep}}\psi C - \frac{\rho}{\omega_0} K_r S$$

The modeling through a macroscopic approach [COM 01] has led us to define the convection-dispersion equations by considering two first-order deposition kinetics sites (reversible–irreversible):

$$\frac{\partial S_s}{\partial t} = \frac{\partial S_{\text{rev}}}{\partial t} + \frac{\partial S_{\text{irr}}}{\partial t}, \quad \frac{\partial S_{\text{rev}}}{\partial t} = K_{\text{dep}} C - K_r C_{\text{rev}}, \quad \frac{\partial S_{\text{irr}}}{\partial t} = K_{\text{dep}} C$$

where S_{rev} is the mass of reversibly deposited particles per unit pore volume and S_{irr} is the mass of irreversibly deposited particles per unit pore volume.

2.9. Confrontation between the laboratory filtration tests and the modeling

To investigate the dispersion and deposition parameters of particles, the most frequent laboratory tests are carried out in columns that recreate the granular medium where the identified suspensions are injected. The tracing tests are usually carried out with a step-input suspension injection, but other experiments with a pulse injection are also carried out with different particles [COM 01, MAS 02, BEN 05]. This type of test, where we inject a small suspension volume, allows us to avoid a likely deposition of particles that would modify the porosity of the medium. In the case of a pulse injection ($C_0 dt = m / Q$), where m is the mass of particles, adequate boundary conditions provide an analytical solution of equation [2.11]:

$$C(t,x) = \frac{mx}{Q\sqrt{4\pi D_L t^3}} \exp(-K_{\text{dep}} t) \exp\left[-\frac{(x-ut)^2}{4D_L t}\right] \qquad [2.12]$$

2.9.1. *Validation of the model in the case of suspended particles*

The convection-dispersion model was used to simulate the transport and deposition of silt particles (having a modal dimension of 14 µm) in a column (Figure 2.11) filled with coarse sand or with glass beads [BEN 05, AHF 09]. The sand has an angular shape with d_{50} = 2.35 mm and the glass beads have a uniform size of 3 mm. We will first use the device with pulse injections of suspended materials, and then we determine, by using the convection-dispersion model (transfer function of equation [2.47]), the residence time and the dispersion coefficient of the medium as well as the deposition kinetics. Having introduced the tracer in the porous medium, these tracing experiments (dissolved tracer or particular tracer) consist of measuring the time variation of the concentration (breakthrough curve or restitution curve). Figure 2.12(a) shows the measured signal (dimensionless parameters C_r and t_r) with a pulse injection of silt particles and the value calculated with equation [2.12]. The shape of this type of curve characterizes the transport and the deposition in a granular medium. A homogeneous medium is characterized by a symmetrical curve, and the spread of the curve indicates a strong dispersion of the medium. A tail at the end of the curve and the maximum record (curve "peak") arising earlier than the average residence time indicates the presence of stagnant areas that are in a slow exchange with the part of the mobile fluid (heterogeneous medium). The curve shows a slight tail in view of the heterogeneity of the sand (angular grains) that may contain sites of low porosity.

2.9.2. *Results and preferential flow*

2.9.2.1. *Influence of the geometrical and hydrodynamic parameters*

The restitution of fine particles at the outlet of the reactor is better for the glass beads when this medium has a permeability that is lower than that of the gravel. Thus, this result shows that the overall parameters of a porous medium do not suffice for characterizing the process of mass transfer and that the organization of the poral space is a defining element. The dispersion in the porous medium is a parameter that characterizes the slowing down of the transport of particles in relation to the carrying fluid. The values obtained for the two media are very similar and they increase with velocity. The deposition kinetics of the sand is more important in view of the more abundant constrictions than in the medium made of glass beads. It increases with the flow velocity up to a "critical" value of the velocity, and then it starts decreasing. From this behavior, we can induce that there are two different flow regimes where the geometric effects and the hydrodynamic effects do not have the same importance.

2.9.2.2. *Size exclusion*

The comparison of the filtration between the dissolved tracer (fluorescein) and the suspended particles shows that the latter is retained shorter in the porous medium. The restitution of particles takes place before the restitution of the tracer (Figure 2.12(b)); the particles are thus transported faster than the dissolved tracer (velocity ratio close to 1.1). This transport behavior was associated with the phenomenon of size exclusion or the phenomenon of hydrodynamic chromatography. In fact, the coarser particles "sampled" the highest velocities in the column because they cannot approach the walls due to their size. This behavior shows the importance of the shape of the grains in the filter and its efficiency, and that the average values of the grain-size distribution parameters as well as the hydraulic parameters do not suffice for characterizing the transport and the filtration.

2.10. Filtration and clogging

2.10.1. *Phenomenological laws*

The phenomenological models take into account the equations for mass conservation, a kinetic equation that describes the rate of particle retention and an equation that characterizes the evolution of the pressures. These models are not founded on a physical modeling of the particle retention mechanisms. In the case where the diffusion is negligible (pure convection), the equation of mass conservation is written as follows:

$$\frac{\partial \sigma}{\partial t} + u\frac{\partial C}{\partial x} = 0$$

where C is the mass concentration of suspended particles, u is the pore velocity, and σ is the retention of particles (volume of captured particles per unit filter volume).

Using the kinetic law described in equation [2.8], we can express retention as follows:

$$\frac{\partial \sigma}{\partial t} = \lambda u C$$

The variation in permeability can be determined by using the Kozeny–Carman model. The modeling of pressure drop is expressed from the nonlinear variation of the hydraulic gradient according to the retention of particles [LEC 93]:

$$\frac{i}{i_0} = \frac{1}{(1 - j\sigma)^m}$$

where j and m are constants in the model, which vary according to the type of filter.

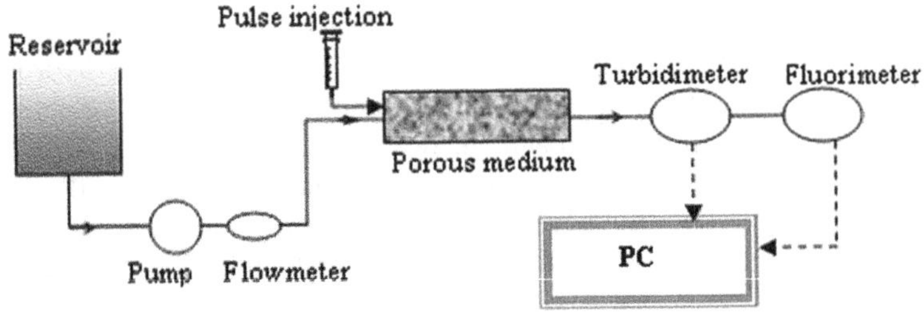

Figure 2.11. *Layout of the tracing device in a laboratory reactor*

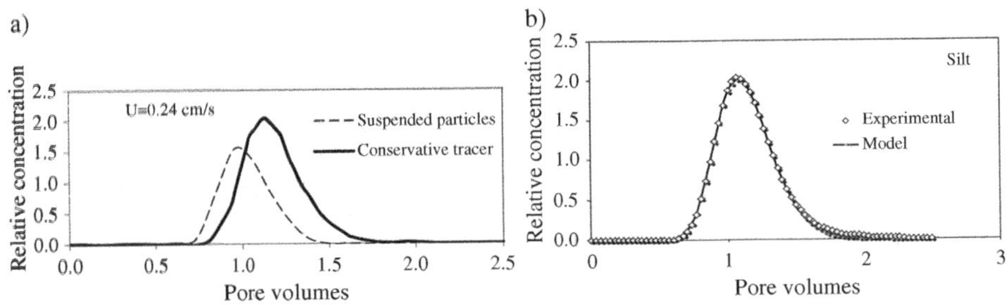

Figure 2.12. *(a) Modeling of the breakthrough curve using the convection-dispersion model and (b) facilitated transport of particles in relation to the solute*

In the case of a step-input injection of suspended particles in a porous medium, under certain boundary conditions, the convection-dispersion equation with deposition kinetics accepts an analytical solution given by:

$$C(t,x) = \frac{C_o}{2}\left[\exp\left(\frac{(u-v)x}{2D_L}\right)\text{erfc}\left(\frac{(x-ut)}{2\sqrt{D_L t}}\right) + \exp\left(\frac{(u+v)x}{2D_L}\right)\text{erfc}\left(\frac{(x+vt)}{2\sqrt{D_L t}}\right)\right]$$

where $v = u(1 + 4K_{\text{dep}} D_L / u^2)$.

2.10.2. *Physicochemical aspects*

The macroscopic and microscopic modelings of filtration have shown that it is difficult to separate the capture mechanisms from the physicochemical processes. The modeling of the transport and filtration of particles must integrate the coupling of the concentration of suspended particles into the ionic force of the solution. The study of the influence of the pH of the fluid on the transport of particles

(Figure 2.13(a)) shows that the retention of particles decreases when the pH increases. The acid medium thus favors the retention of particles (deposition kinetics). This property is emphasized by the decrease in the thickness of the double electric layer (Derjaguin, Landau, Verwey, Overbeek – DLVO theory) when the medium becomes acid. However, this effect of the pH is more important for the clay particles than for the silt particles. The deposition kinetics coefficient increases with the ionic force (Figure 2.13(b)) up to a certain value, and then it stagnates or decreases slightly. This behavior shows the existence of a critical salt concentration (CSC) starting from which the attraction effect of the suspended particles by the grains of the porous medium is inversed. The silt particles show a more important deposition kinetics because of their size (the effect of the gravitational pull). The influence of the salinity on the two types of particles is somewhat similar.

2.10.3. *Filtration and clogging*

The analysis of the breakthrough curve, in the case of step-input injection of particles, allows us to deduce the retention parameter of the granular medium (filter). Figure 2.14(a) shows a typical breakthrough curve, expressed according to the number of pore volumes of injected suspension. This typical curve shows a first phase that corresponds to a transient filtration regime, before reaching the permanent regimen where the proportion of filtered particles remains constant (constant kinetics). The increasing part characterizes the dispersion and the plateau is characteristic of the deposition kinetics, obtained by $K_d = -(1/t_c)\ln(C/C_0)$, where t_c is the convection time measured on the breakthrough curve (Figure 2.14(b)) and corresponding to a value of C/C_0 equal to 0.5. The results obtained for the clogging of filters quantify the variation of indirect parameters such as the pressure gradient and the permeability. The tests carried out with a constant flow rate or a constant hydraulic load show an exponential decrease in permeability with time (Figure 2.14(a)). The reduction in permeability is more drastic in the sand (stronger retention of particles) than in the glass beads, starting from a certain quantity of particles inserted into the filter. The reduction in permeability is important for a certain length of the filter. In the case of filtration with a constant hydraulic load, this reduction in permeability is the most important (clogging) and it concerns a less important length of the filter. The volume of the injected suspension influences the clogging kinetics and the distribution of the retention of particles. The decrease in permeability of the filter is not linear and so it is approached by a power law.

The application of the usual criteria has shown that the filter (sand 160–315 μm), for which $D_{15F}/d_{85B} = 11$, seems insufficiently permeable for avoiding the overpressures (the gradient reaches the value of 38 and the reduced permeability reaches the value of 90%), even if it filters all the eroded particles. Since the filter must retain the eroded particles while remaining permeable to water, this second property is not maintained for this type of filter. *A priori*, for all the filters whose $D_{15F}/d_{85B} < 11$, the permeability criterion is no longer satisfied and the Sherard's criterion [SHE 89] is no longer verified. The results show that this classical criterion ($D_{15F}/d_{85B} < 9$) of the filters' design is not appropriate while maintaining the permeability for the used soils. The Terzaghi filtration rule is not verified but filtration occurs. The geometric criterion that is defined according to the solid flow cannot be dissociated from the potential mass of the erodible particles (the concentration introduced in the filter). The evolution of the permeability of the filter according to the ratio D_{15F}/d_{85B} shows threshold values for this ratio (between 15 and 20) below which permeability falls and clogging quickly occurs.

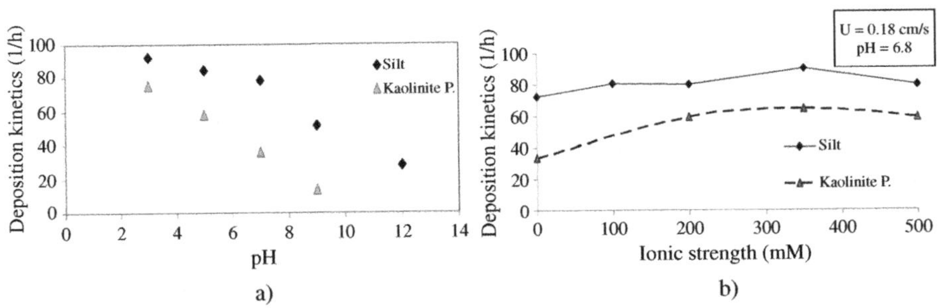

Figure 2.13. *Influence of (a) the pH and (b) the ionic force on the deposition kinetics of particles in a granular medium*

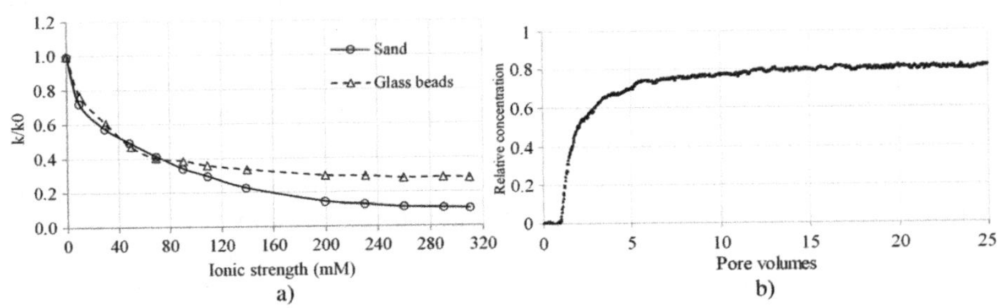

Figure 2.14. *(a) Breakthrough curve of the continuous injection of particles and (b) reduction of the measured permeability in a filtration test*

2.11. Conclusion

Suffusion and filtration are two coupled processes that are governed by the geometry of the porous network, the physicochemical interactions between the solid and the fluid phases, as well as by the hydrodynamic conditions. Several evaluation criteria for suffusion have been developed, starting from the study of the grain size distribution of the soil. The confrontation of the estimations obtained using these grain-size distribution criteria has emphasized the limits of these criteria and the necessity to associate the analysis of the grain size distribution with the evaluation of hydraulic loading. The hydraulic approach consists of evaluating the generated load by the fluid flow in order to initiate the suffusion. This evaluation is mainly carried out by expressing the critical value of the hydraulic gradient, the shearing stress, or the pore velocity. However, different experimental campaigns have pointed out the presence of a scaling effect that perturbs the hydraulic approach based on the expression of a global hydraulic gradient. The energetic approach that consists of determining the erosion power and the energy dissipated by the fluid is not affected by this scaling effect and allows us to estimate the average erosion rate as well as the total mass of eroded particles. Throughout the development of suffusion, we can measure a localized increase in interstitial pressures, as well as a decrease in permeability. These modifications of the hydraulic characteristics are attributed to a filtration at the center of the soil of a fraction of the detached particles. The main results of the study of filtration with a pulse injection concern the evolution of the deposition function of the particles with the geometric and hydrodynamic parameters. The introduction through a step-input injection of a significant quantity of fine particles throughout a filter, under controlled conditions, allows us to quantify the indirect indicators of clogging. The use of different grain-size distribution filters has allowed us to emphasize the influence of various parameters on the kinetics of filtration. The comparison of the results with the classical criteria for the design of filters has shown the insufficiency of these criteria and the importance of the hydraulic criterion. The filters must be sufficiently fine and also the pores between the grains must be small in order to retain the eroded particles that flow through the filters. However, these filters must also be made up of grains that are coarse enough to allow the setup of flow through the filter and prevent the formation of hydraulic overpressures. The results obtained show that the filtration and the clogging depend on the concentration of the suspension in particles and of the flow conditions. The results indicate an exponential distribution of particle retention with the depth of the filter and a clogging localized mainly in a thin layer at the inlet of the filter.

2.12. Bibliography

[AHF 09] AHFIR N.-D., BENAMAR A., WANG H.Q., ALEM A., "Influence of internal structure and medium length on transport and deposition of suspended particles: a laboratory study", *Transport in Porous Media*, vol. 76, pp. 659–668, 2009.

[ARU 83] ARULANANDAN K., PERRY E.B., "Erosion in relation to filter design criteria in earth dams", *Journal of Geotechnical Engineering*, vol. 109, no. 5, pp. 682–696, 1983.

[BAI 95] BAI M., BOUHROUM F., CIVAN F., ROEGIERS J.C., "Improved model for solute transport in heterogeneous media", *Journal of Petroleum Science and Engineering*, vol. 14, pp. 65–78, 1995.

[BEN 05] BENAMAR A., WANG H.Q., AHFIR N.-D., ALEM A., MASSEI N., DUPONT J.-P., "Effets de la vitesse d'écoulement sur le transport et la cinétique de dépôt de particules en suspension en milieu poreux saturé", *C. R. Geoscience*, vol. 337, pp. 497–504, 2005.

[BEN 08] BENDAHMANE F., MAROT D., ALEXIS A., "Parametric study of suffusion and backward erosion", *Journal of Geotechnical and Geoenvironmental Engineering*, vol. 134, no. 1, pp. 57–67, 2008.

[BLU 05] BLUME T., WEISBROD N., SELKER J.S., "On the critical salt concentrations for particle detachment in homogeneous sand and heterogeneous Hanford sediments", *Geoderma*, vol. 124, pp. 121–132, 2005.

[BRA 04] BRADFORD S.A., BETTAHAR M., SIMUNEK J., VAN GENUCHTEN M.T., "Straining and attachment of colloids in physically heterogeneous porous media", *Vadose Zone Journal*, vol. 3, pp. 384–394, 2004.

[BRU 05] BRUSSEAU M.L., OLEEN J.K., SANTAMARIA J., CHENG L., OROSZ-COGHLAN P., CHETOCHINE A.S., BLANFORD W.J., RYKWALDER P., GERBA C.P., "Transport of microsporidium Encephalitozoon intestinales Spores in sandy porous media", *Water Research*, vol. 39, pp. 3636–3642, 2005.

[BUR 93] BURENKOVA V.V., "Assessement of suffusion in noncohesive and graded soils", *Proceedings of the 1st Conference on Geo-Filters*, Karlsruhe, Germany, Balkema, Rotterdam, The Netherlands, pp. 357–360, 1993.

[CHA 04] CHAPUIS R.P., AUBERTIN M., "Reply to the discussion by D. Hansen on the use of the Kozeny-Carman equation to predict the hydraulic conductivity of soils", *Canadian Geotechnical Journal*, vol. 41, pp. 994–996, 2004.

[COM 01] COMPÈRE F. POREL G., DELAY F., "Transport and retention of clay particles in saturated porous media. Influence of ionic strength and pore velocity", *Journal of Contaminant Hydrology*, vol. 49, nos. 1–2, pp. 1–21, 2001.

[CUM 99] CUMBIE D.H., MCKAY L.D., "Influence of diameter on particle transport in a fractured shale saprolite", *Journal of Contaminant Hydrology*, vol. 37, pp. 139–157, 1999.

[FEL 07] FELL R., FRY J.J. (eds), *Internal Erosion of Dams and Their Foundations*, Taylor & Francis, 2007.

[FOS 00] FOSTER M., FELL R., SPANNAGLE M., "The statistics of embankment dam failures and accidents", *Canadian Geotechnical Journal*, vol. 37, pp. 1000–1024, 2000.

[FRE 99] FREY J.M., SCHMITZ P., DUFRECHE I., GOHR PINHEIRO I., "Particle deposition in porous media: analysis of hydrodynamic and weak inertial effects", *Transport in Porous Media*, vol. 37, pp. 25–54, 1999.

[GAR 10] GARNER S.J., FANNIN R.J., "Understanding internal erosion: a decade of research following a sinkhole event", *The International Journal on Hydropower & Dams*, vol. 17, pp. 93–98, 2010.

[GHI 94] GHIDAGLIA C., Filtration en profondeur de particules, Doctoral Thesis, University of Paris VI, 1994.

[GOH 99] GOHR PINHEIRO I., SCHMITZ P., HOUI D., "Particle capture in porous media when physico-chemical effects dominate", *Chemical Engineering Science*, vol. 54, pp. 3801–3813, 1999.

[GRO 01] GROLIMUND D., ELIMELICH M., BORCOVEC M., "Aggregation and deposition kinetics of mobile colloidal particles in natural porous media", *Colloids and Surfaces A: Physicochemical and Engineering Aspects*, vol. 191, pp. 179–188, 2001.

[HER 70] HERZIG J.P., LECLERC D.M., LE GOFF P., "Flow of suspension through porous media, application to deep bed filtration", *Industrial and Engineering Chemistry*, vol. 62, pp. 8–35, 1970.

[KEL 04] KELLER A.A., SIRIVITHAYAPAKORN S., CHRYSIKOPOULOS C., "Early breakthrough of colloids and bacteriophage MS2 in a water saturated sand column", *Water Resources Research*, vol. 40, no. W08304, p. 11, 2004.

[KEN 85] KENNEY T.C., LAU D., "Internal stability of granular filters", *Canadian Geotechnical Journal*, vol. 22, pp. 215–225, 1985.

[KÉZ 79] KÉZDI, *Soil Physics: Selected Topics*, Elsevier Scientific Publishing Co, Amsterdam, 1979.

[KHI 84] KHILAR K.C., FOGLER H.S., "The existence of a critical salt concentration for particle release", *Journal of Colloid Interface Science*, vol. 101, pp. 214–224, 1984.

[KHI 85] KHILAR K.C., FOGLER H.S., GRAY D.H., "Model for piping-plugging in earthen structures", *Journal of Geotechnical Engineering*, vol. 111, no. 7, pp. 833–846, 1985.

[KOV 81] KOVACS G., *Seepage Hydraulic*, Elsevier Scientific Publishing Co, Amsterdam, 1981.

[KRE 99] KRETZSCHMAR R., BORKOVEC M., GROLIMUND D., ELIMELECH M., "Mobile subsurface colloids and their role in contaminant transport" (review paper), *Advances in Agronomy*, vol. 66, pp. 121–194, 1999.

[LAF 89] LAFLEUR J., MLYNAREK J., ROLLIN A.L., "Filtration of broadly graded cohesionless soils", *Journal of Geotechnical Engineering*, vol. 115, no. 12, pp. 1747–1768, 1989.

[LAF 99] LAFLEUR J., "Selection of geotextiles to filter broadly graded cohesionless soils", *Geotextiles and Geomembranes*, vol. 17, pp. 299–312, 1999.

[LEC 93] LECLERC D., DODDS J., "De la filtration des colloïdes à la modélisation des milieux poreux", *Entropie*, vol. 29, no. 180, pp. 21–28, 1993.

[LI 08a] LI M., Seepage induced failure of widely graded cohesion-less soils, PhD Thesis, Department of Civil Engineering, University of British Columbia, Vancouver, Canada, 2008.

[LI 08b] LI M., FANNIN J., "Comparison of two criteria for internal stability of granular soil", *Canadian Geotechnical Journal*, vol. 45, pp. 1303–1309, 2008.

[MAR 09] MAROT D., BENDAHMANE F., ROSQUOET F., ALEXIS A., "Internal flow effects on isotropic confined sand-clay mixtures", *Soil & Sediment Contamination, an International Journal*, vol. 18, no. 3, pp. 294–306, 2009.

[MAR 11] MAROT D., REGAZZONI P.L., WAHL T., "An energy based method for providing soil surface erodibility rankings", *Journal of Geotechnical and Geoenvironmental Engineering*, vol. 137, no. 12, pp. 1290–1294, 2011.

[MAR 12] MAROT D., LE V.D., GARNIER J., THOREL L., AUDRAIN P., "Study of scale effect in an internal erosion mechanism", *European Journal of Environmental and Civil Engineering*, vol. 16, no. 1, pp. 1–19, 2012.

[MAS 02] MASSEI N., LACROIX M., WANG H.Q., DUPONT J.-P., "Transport of particulate material and dissolved tracer in a highly permeable porous medium: comparison of the transfer parameters", *Journal of Contaminant Hydrology*, vol. 57, pp. 21–39, 2002.

[MAY 05] MAYS D.C., HUNT J.R., "Hydrodynamic aspects of particle clogging in porous media", *Environmetal Science and Technology*, vol. 39, pp. 577–584, 2005.

[MCD 86] MCDOWELL-BOYER L.M., HUNT J.R., SITAR N., "Particle transport through porous media", *Water Resources Research*, vol. 22, no. 13, pp. 1901–1921, 1986.

[MOF 05] MOFFAT R., Experiments on internal stability of widely graded cohesionless soils, PhD Thesis, Department of Civil Engineering, University of British Columbia, Vancouver, Canada, 2005.

[MON 98] MONNET A., "Boulance, érosion interne, renard. Les instabilités sous écoulement", *Revue Française de Géotechnique*, vol. 82, pp. 3–10, 1998.

[PER 07] PERZLMAIER S., "Hydraulic criteria for internal erosion in cohesionless soil", in FELL R., FRY J.J. (eds), *Internal Erosion of Dams and Their Foundations*, Taylor & Francis, pp. 179–190, 2007.

[RED 00] REDDI L.N., LEE I., BONALA M.V.S., "Comparison of internal and surface erosion using flow pump test on a sand-kaolinite mixture", *Geotechnical Testing Journal*, vol. 23, no. 1, pp. 116–122, 2000.

[RED 05] REDDI L.N., XIAO M., HAJRA M.G., LEE I.M., "Physical clogging of soil filters under constant flow rate versus constant head", *Canadian Geotechnical Journal*, vol. 42, pp. 804–811, 2005.

[REG 09] REGAZZONI P.L., Confrontation et analyse d'érodimètres et caractérisation de la sensibilité à l'érosion d'interface, Doctoral Thesis, University of Nantes, 2009.

[RYA 96] RYAN J.N., ELIMELECH M., "Colloid mobilization and transport in groundwater", *Colloids and Surfaces A*, vol. 107, pp. 1–56, 1996.

[SAI 11] SAIL Y., MAROT D., SIBILLE L., ALEXIS A., "Suffusion tests on cohesionless granular matter", *European Journal of Environmental and Civil Engineering*, vol. 15, pp. 799–817, 2011.

[SEL 88] SELLMEIJER J.B., On the mechanism of piping under impervious structures, PhD Thesis, Delft University of Technology, The Netherlands, 1988.

[SHE 89] SHERARD J.L., DUNNIGAN L.P., "Critical filters for impervious soils", *Journal of Geotechnical Engineering*, July 1989.

[SIL 95] SILLIMAN S.E., "Particle transport through two-dimensional, saturated porous media: influence of physical structure of the medium", *Journal of Hydrology*, vol. 167, pp. 79–98, 1995.

[SKE 94] SKEMPTON A.W., BROGAN J.M., "Experiments on piping in sandy gravels", *Géotechnique*, vol. 44, no. 3, pp. 440–460, 1994.

[TOS 09] TOSCO T., TIRAFERRI A., SETHI R., "Ionic strength dependent transport of microparticles in saturated porous media: modeling mobilization and immobilization phenomena under transient chemical conditions", *Environmental Science and Technology*, vol. 43, pp. 4425–4431, 2009.

[TUF 04] TUFENKJI N., ELIMELECH M., "Correlation equation for predicting single-collector efficiency in physicochemical filtration in saturated porous media", *Environmental Science and Technology*, vol. 38, pp. 529–536, 2004.

[WAN 04] WAN C.F., FELL R., Experimental investigation of internal instability of soils in embankment dams and their foundations, UNICIV report no. 429, University of New South Wales, School of Civil and Environmental Engineering, 2004.

[WAN 08] WAN C.F., FELL R., "Assessing the potential of internal instability and suffusion in embakment dams and their foundations", *Journal of Geotechnical and Geoenvironmental Engineering*, vol. 134, no. 3, pp. 401–407, 2008.

Chapter 3

The Process of Filtration in Granular Materials

3.1. Introduction

What is at stake in preserving the cultural heritage of dikes and fluvial levees largely exceeds the scope of its exploitation since these developments, besides the contribution they bring to the economical development of a region, also direct the stream of water and ensure the safety of people and goods in case of flooding. The statistical comparison carried out by ICOLD [FRY 07] regarding the failure of levees show that internal erosion is the most common cause of failure among hydraulic structures, excepting the situations where failure is caused by overflow. Internal erosion is a consequence of fine particles being transported over long distances, i.e. generally alongside the cracks, where the flow velocity of the water is higher. These fine particles constitute the main body or the foundation of the hydraulic structure, and the displacement of these particles may cause piping in the downstream of the structure, as well as the thinning of downstream areas, which may have irreversible consequences on the stability of the structure. As early as the 1920s, filters made up of granular materials were built especially in the downstream side of the core of the levees to retain these fine particles. The diameter of these filters was generally lower than 50 mm. This was not the case for fluvial dikes (or certain levees that had been built before 1920), which were made up of occasionally poorly compacted materials with a non-controlled grain size distribution. Thus the phenomenon of filtration, which consists of retaining a part of or all of the fine particles through a coarser granular matrix, does not concern solely the filters but also, more broadly, all of the porous materials that are being traversed by a fluid.

Chapter written by Eric VINCENS, Nadège REBOUL and Bernard CAMBOU.

This chapter will not be concerned with the situations where the fine particles and the coarser particles are mixed (i.e. the case of fluvial dikes), but its main concern will be the situation where these two types of materials make up two very distinct media within the actual structure. Put more simply, we are dealing with filters in earth dams. The issue of filtration is thus reduced to an issue regarding transportation throughout the network of voids whose complex geometry depends on the shape of the coarser particles and on their arrangement, which is assumed to be mechanically stable. Since the stopping of the particles is conditioned by a geometrical property that arises from the arrangement of the coarse granular medium, we will assume that water only acts as a support for the transportation.

3.2. Fundamental characteristics of the filtering granular media

As a general rule, we separate the space of the voids that can be found between the solid grains in two distinct subsets: the set of pores, which are entities of a higher volume, and the set of constrictions, i.e. narrower passages between the pores. Suppose there is a fine, non-cohesive particle that is transported throughout a non-cohesive granular matrix. The stopping of this particle will only work if this particle encounters a bottleneck (a constriction) that is smaller than its own size. The stopping of this particle will then be associated with a geometrical property of size exclusion, which is strictly linked to the size of the constrictions.

3.2.1. *Pores*

First and foremost, the space between the pores does not seem to be of interest for the process of filtration. However, it is a fact that it conditions the filtration process indirectly because the porosity of the medium, which is an average characteristic, accounts for the compactness of the coarse material. The more important this compactness is, the more restricted becomes the space available for transportation. As we will see in what follows, this characteristic must be considered if we are to explain and quantify the properties linked to the space of constrictions.

The complexity of the actual materials (angular, non-spherical, polydispersed, etc.) and the ensuing space of the voids requires the study of the more fundamental materials to discern certain primary characteristics of this poral space. Thus, in a regular close packing of uniform spheres, the porosity of the medium is widely known as being comprised between two extreme values: 0.4764 and 0.2595. The higher value can be found when we consider a cubic arrangement (Figure 3.1(a)), whereas the lower value can be found in a hexagonal pattern (Figure 3.1(b)). The latter can also be found in a cubic, face-centered pattern. However, these values, which characterize a crystallized organization, can never be reached experimentally, because an inadequate dispersion, as thin as it may be, will cause a disorganization of the medium.

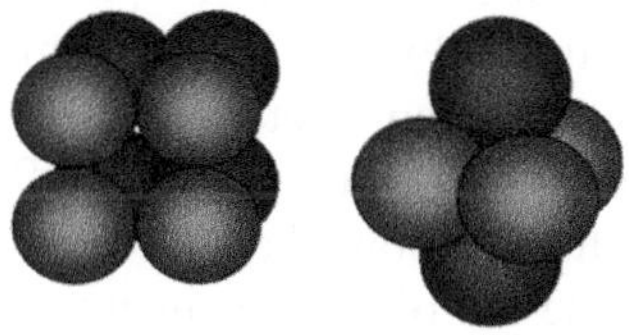

a) b)

Figure 3.1. *Orderly pattern of monodispersed spheres: (a) in a loose configuration; (b) a compact configuration*

The maximal porosities that were found experimentally are then quite dispersed: 0.399 for Scott and Kilgour [SCO 69], 0.445 for Onoda and Liger [ONO 90], and 0.491 for Nolan and Kavanagh [NOL 92]. On the contrary, the minimal porosities are agreed to be around 0.36 (Scott and Kilgour [SCO 69], Finney [FIN 70], Nolan and Kavanagh [NOL 92], and Roux and Chevoir [ROU 05]).

In the case of polydispersed materials, it is difficult to give values for an organization that depends on the proportion of elements regardless of their size, and on the grain size distribution. Certain results can be found for the materials with normal and lognormal distributions (Sohn and Moreland [SOH 68]).

3.2.2. *Constrictions*

Similarly, two extreme sizes of the constrictions can be defined by a pattern of orderly, uniform spheres. The constriction that has a maximal size for a regular network can be found by considering the same fundamental patterns that we have defined above. The diameter inscribed in the largest constriction equals $D(\sqrt{2}-1)$, given $D/2.4$ with D being the diameter of the solid sphere (Figure 3.2(a)) and the diameter inscribed in the smallest constriction being $D(2/\sqrt{3}-1)$ given $D/6.5$ (Figure 3.2(b)).

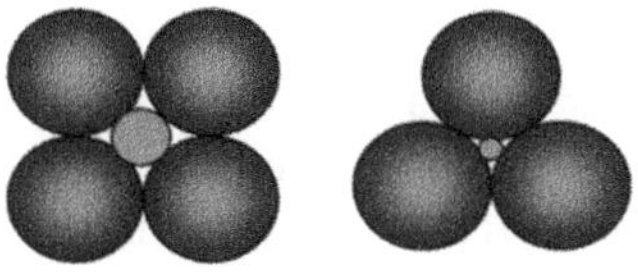

a) b)

Figure 3.2. *Constrictions for an orderly pattern of monodispersed spheres: (a) in a loose configuration; (b) in a compact configuration*

When the media are disorderly or polydispersed, although no general result can be given, we can at least maintain that the smallest size of the constriction is larger than the constriction formed by the three smallest particles of the medium, arranged according to a pyramidal pattern (Figure 3.2(b)). This is confirmed by numeric simulations (using the discrete element method (DEM)), which are carried out on sets of spheres that were deposited in a box under the effect of the gravitational pull. In fact, these simulations tend to show that the constrictions do not correspond necessarily to entities that came out of spheres interacting in a mutual contact, as is represented in Figure 3.1 (Reboul [REB 08]).

These average properties, i.e. the average porosity and the average constriction size of the filter, do not account for all the mechanisms used by the coarse material to retain the fine particles. In fact, a particle that traverses a polydispersed and disorderly medium will find, on its way, a sequence of confrontations (constrictions) of different sizes that are likely to retain it every time. This capacity of finding in its way a constriction of too small a size is linked to the proportion of the different sizes of the constrictions of the filter. Thus, it is important to establish this distribution for a given material.

Furthermore, it is worth noting that if the trajectory that needs to be covered by the fine particle is large, the probability of encountering on its way a constriction of a size smaller than its own size will increase. The process of filtration is thus governed by two geometrical characteristics that are different in nature: the distribution of the sizes of these constrictions and the thickness of the filter.

3.3. The distribution of constriction size

There are different experimental techniques that can be used to achieve the distribution of constriction sizes. Witt [WIT 86] injected epoxide resin and has studied the size of the constrictions associated with each pore. Soria *et al.* [SOR 93] have proposed an indirect method that is based on a probabilistic model of transportation. They rebuilt the size distribution of the constrictions by studying the grain size distribution of the base polydispersed material that was able to traverse the coarse material[1]. A distribution point is associated with a laboratory trial that implies a specific thickness of the filter.

Microtomography, together with a computerized processing of the images, also allows us to obtain local information (Schwartz *et al.* [SCH 94], Auzerais *et al.* [AUZ 96], Thovert *et al.* [THO 01], and Mourzenko *et al.* [MOU 08]). The quoted studies mainly cover porosity analyses or conductivity analyses, but this tool can be

1 We call "base material" a fine material that, because of the size of its particles, is likely to penetrate the filter.

used to recreate the distribution of constriction sizes. Some of the questions on this topic have been answered by Mehlhorn *et al.* [MEH 08].

Finally, in a purely digital framework, we can recreate a modeled pattern of material by using the DEM. By averaging the pavement of the space with the help of a generalized Delaunay approach (Richard *et al.* [RIC 99]), we can deduce local properties (i.e. pores and constrictions) that were obtained statistically (Mellor [MEL 89], Bryant *et al.* [BRY 93], Balhoff and Thompson [BAL 04], and Reboul *et al.* [REB 08]). This approach, which we will develop in the rest of this chapter, will be called DEMDT[2].

3.3.1. *The definition of pores and constrictions using a DEMDT approach*

There are different techniques that can be used to create or recreate a digital sample of spheres. Basically, they can be regrouped in two classes: the collective rearrangement or the random sequential addition for each of the variants. A synthesis of these methods can be found in Coelho [COE 97] or Bagi [BAG 05]. These techniques use geometrical or kinetic criteria for placing the particles in space without the dynamic balance equations ever having to be written and resolved. They have the advantage of being fast, as well as allowing recreating a sample of 10,000 polydisperse particles within hours. However, Al Raoush and Alsaleh [ALR 07] have noted that some of these techniques could lead to a non-realistic contact network.

Therefore, we can build samples by trying to entirely reproduce the experimental process of deposition under the force of the gravitational pull through the means of a DEM code. Although time consuming, this technique enables us to consider, simultaneously, the influence of the potential energy (height of the fall) and the interaction between particles within the final organization of the medium. In this last approach, the digital samples are carried out in two steps (Figure 3.3). First and foremost, the set of spheres with fixed grain size distribution is randomly generated in a space otherwise large enough so that each sphere does not overlap any other sphere. Subsequently, they are simultaneously released under the gravitational pull in the box of the same base, thus forming a sample, after achieving a balance criterion. The game of mechanical parameters, which is associated with the previously set out experimental protocol, leads to obtaining a sample that is close to the looser state for the chosen grain size distribution. At this point, the material is considered to be in its loosest state.

---

2 Discrete element method (DEM) coupled to a Delaunay tetrahedralization (DT) of the medium.

Several materials, corresponding to different grain size distributions, were studied: a quasi uniformly-graded (UG) material with a uniformity coefficient[3] U_c of 1.5 (Figure 3.4(a)), a well-graded (WG) material of U_c that is equal to 3.9 (Figure 3.4(a)), and a gap-graded (GG) material of U_c equal to 3.6 (Figure 3.4(b)), built on the foundation of the material WG.

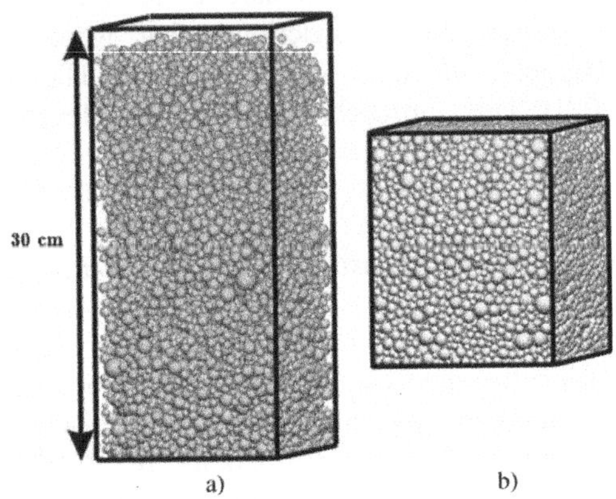

Figure 3.3. *Creating numerical samples: (a) generating the set of spheres; (b) sample created after the release under gravity*

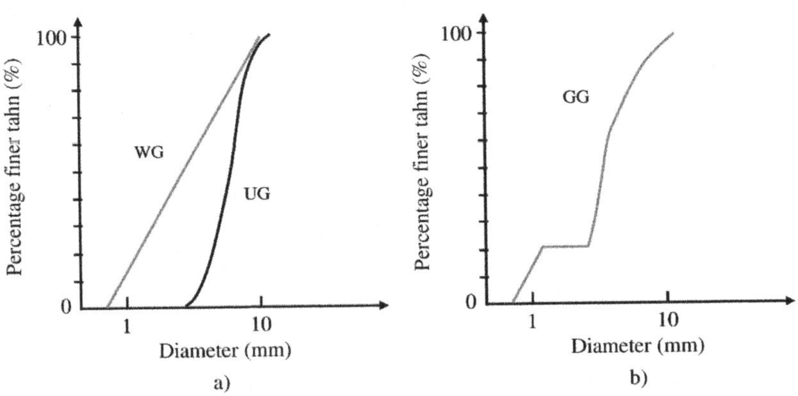

Figure 3.4. *Grain size distribution of the samples analyzed using the (a) DEMDT, UG, and WG materials; (b) GG material*

Once the sample has been created, the study of the morphology of the sample (the space between the pores) and the study of its topology (the space of constrictions)

---

3 $U_c = d_{60} / d_{10}$

are made by paving the sample with the help of a radical Delaunay schema. The space thus constituted is made up of a set of tetrahedrons[4] whose peaks cannot be distinguished from the centers of the spheres (Figure 3.5(a)). Thus, each center of a sphere belongs at least to one tetrahedron. In this representation, the volumes of the most important voids will be found mainly at the center of the tetrahedrons and the narrowest constrictions – on the surface of these tetrahedrons (Bryant *et al.* [BRY 93]). A constriction is hereby defined as the largest disk that is included in one of the faces of a tetrahedron, and which can be inscribed between the three peak-particles on the surface of a tetrahedron (Figure 3.5(b)). This disk must also be completely included in the space of the voids. Thus, in Figure 3.5(b), the disk indicated by the dotted line is not a valid candidate for characterizing the size of the constriction associated with the surface studied, because it overlaps with a neighboring solid particle.

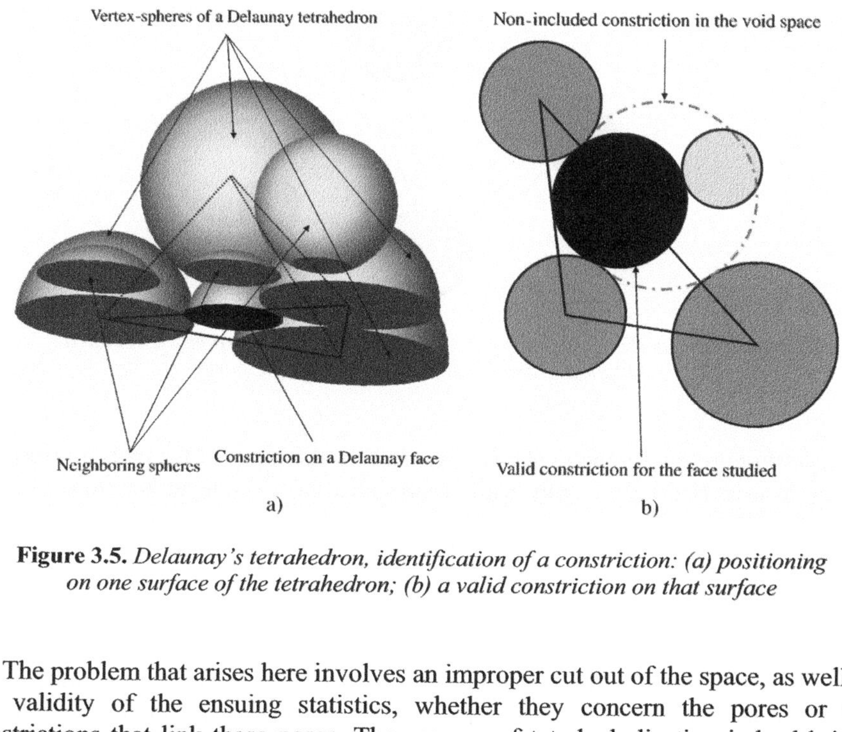

Figure 3.5. *Delaunay's tetrahedron, identification of a constriction: (a) positioning on one surface of the tetrahedron; (b) a valid constriction on that surface*

The problem that arises here involves an improper cut out of the space, as well as the validity of the ensuing statistics, whether they concern the pores or the constrictions that link these pores. The process of tetrahedralization indeed brings about an artificial partitioning of the elementary volumes (Al-Raoush *et al.* [ALR 03]) and implicitly a partitioning of the surfaces where the constrictions are defined. Certain adjacent tetrahedrons are thus likely to belong to a single poral entity and the constrictions identified between the tetrahedrons would then be

4 That is the tetraedralization of the space.

incompatible with their initial definition, given the narrow pathways within the void spaces. However, the localization of a pore whose contours are not very precise is not an easy task. Reboul [REB 08] gave different definitions of the pore: starting from the sphere inscribed within the void associated with a given tetrahedron (level qualified as 0) and continuing to a set of tetrahedrons whose largest inscribed spheres must overlap to be associated (see illustration for a 2D medium, Figure 3.6). If, for a given tetrahedron, the association implies only one of the direct neighboring tetrahedrons, the level of association is called level 1. If the neighbor of the neighbor enters this association, we may talk of a level 2. Let us take note of the level 1bis where the only neighbors that are selected (levels 1 or 2) are the tetrahedrons whose diameter of the inscribed spheres is smaller than the tetrahedron that is pivotal within the association.

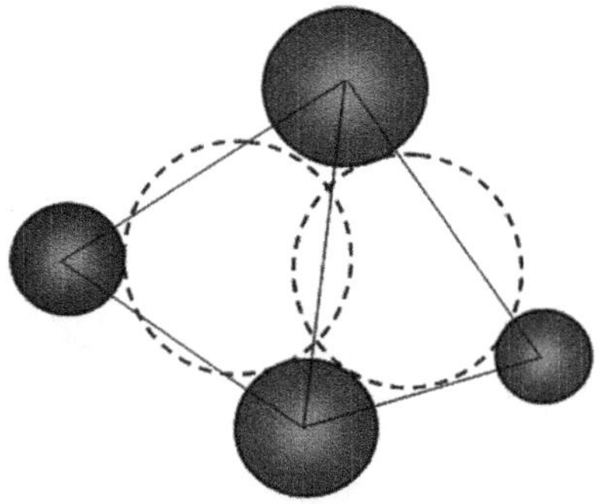

Figure 3.6. *In 2D, two neighboring Delaunay triangles strongly correlated that induce their fusion to define a single pore*

According to this geometrical criterion, 55% of the pores of the quasi uniformly-graded material (UG) can be physically represented by a single tetrahedron, whereas 95% will be constituted of more than three tetrahedrons. We may note in Figure 3.7 that the "distribution" for the approaches laid out in level 0 or 1bis is relatively similar, whereas the same cannot be said of how widespread their grain size distribution is. On the contrary, the mode obtained for the inscribed sphere starts shifting toward the smaller diameters. This is explained by the fact that, in this case, a certain volume of void is not accounted for in this calculation, to guarantee the total inscription of the sphere in the space of the voids. Whereas the impact that the partitioning has on the statistics regarding the volume of pores could well be established (Figure 3.7), the impact regarding the constrictions did not regard Reboul [REB 08].

What conclusions can we draw from this study? The first would be that it is difficult to outline the contour of a pore, since it is an object whose geometry is complex, which organizes itself largely around a tetrahedron, and whose impact on the quality of the results obtained from this study will be limited. Indeed, whatever the definition of the considered pore (whether it is a set of tetrahedrons), the transport can be viewed as a sequence of physical confrontations that a fine particle will have to successfully overcome. By taking as the definition of the pore one single tetrahedron and not a set of neighboring tetrahedrons, we impose only a reduced average distance between two confrontations on a linear trajectory. This aspect will thus be taken into account in the analysis of the process of filtration.

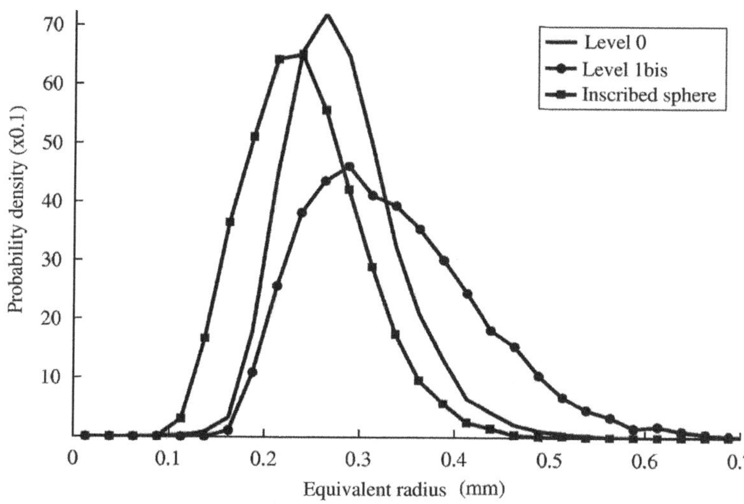

Figure 3.7. *Definition of a pore size and the associated distribution for the looser state of the material UG*

3.3.2. *The influence that the density of the medium has on the size of the constrictions*

For the convenience of the reader, in what follows we will call "constriction curve" what in reality is a cumulated distribution of the different sizes of the constrictions. The constrictions arising from the organization of the coarse medium, the constriction curve seems typical of a given grain size distribution. Furthermore, the representative constrictions of a granular medium are the constrictions that imply the most numerous particles in that medium (Silveira [SIL 65], Ziems [ZIE 68]). As regards the fine grain size distributions, whether linear or bimodal, we are dealing with "smaller" diameters of the grain size curve whose mass impact on this curve is mitigated in relation to larger diameters. This can be easily observed using the DEMDT approach.

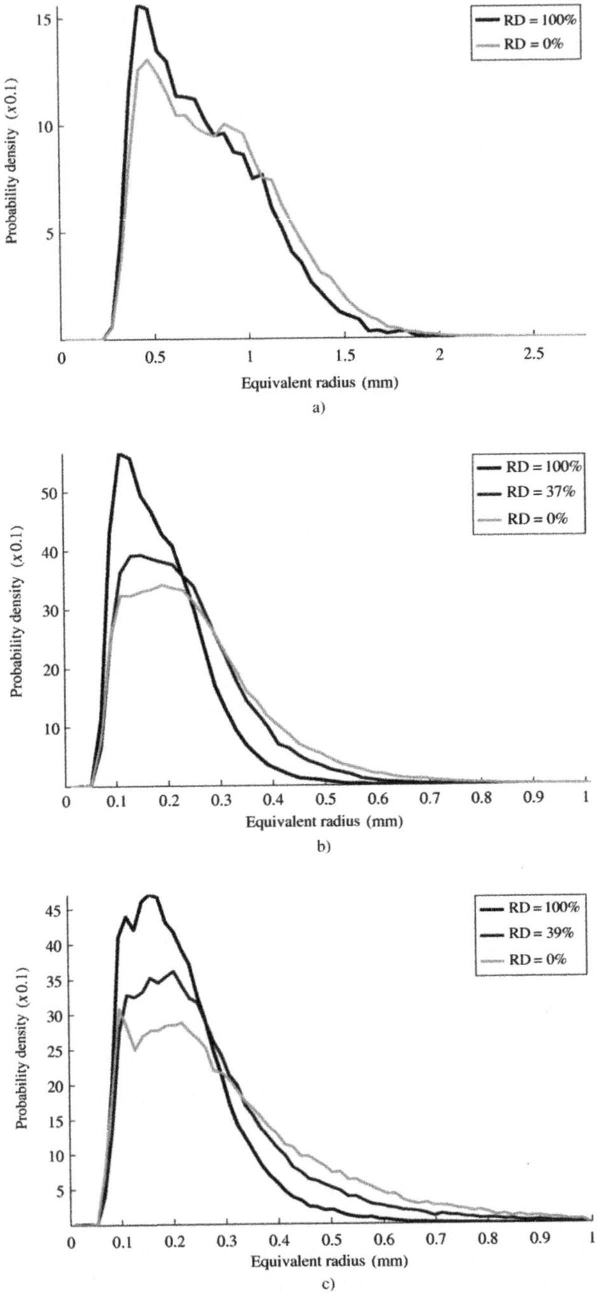

Figure 3.8. *Distribution of constriction sizes according to the relative density of the material, for a UG material (a), a WG material (b) and a GG material (c)*

Reboul *et al.* [REB 10] have studied the three materials UG, WG, and GG using the DEM associated with a Delaunay-type pavement of the space (DEMDT). The samples, created under normal gravity conditions, were adjusted into having different densities by decreasing the local friction stress between particles and by tri-axial cyclical, alternated straining (which contributes to an additional densification). In the looser states, we may note the presence of very heterogeneous media with densities of a constant probability whose diameters may range from single to triple (Figure 3.8). Along with the process of densification, the distribution, which can be found under a lognormal form, is clearly organized around a mode (Figure 3.8(a)–(b)) for continuous grain size distributions. In these figures, the relative density (RD) is defined by $RD = (e - e_{\max})/(e_{\min} - e_{\max})$.

In the case of the GG material, we have found two neighboring modes (Figure 3.8(c)). The mode of a smaller size corresponds to constrictions that were caused by the contact between small particles, whereas the mode of a larger size implies at least two particles whose diameter is larger than the gap. For this material, throughout the densification process, the large size mode tends to predominate. The sizes associated with these modes shift only very slightly during the densification process, thus being characteristic of a particular grain size distribution; the impact of the densification, which is especially visible in the larger constriction sizes or in intermediary sizes, manifests itself through a recompacting of distributions (Figure 3.8(a)–(c)).

Near the granular curve of the three materials: UG, WG, and GG (Figure 3.9), we will find the constriction curves of the different materials in their loosest state as well as the most compact state. The effect of the density is inconsequential on the material UG, but this was expected since such materials are very poorly compaction-sensitive. We may also note that for the GG material (Figure 3.9(c)), this discontinuity no longer appears on the constriction curve.

Finally, regardless of the material studied, less than 7% of the constrictions in the most compact state appeared to be formed by three particles in mutual contact (Figure 3.2(b)). The representation of a dense granular medium with the aid of this geometrical configuration is not therefore compatible with a state of mechanical balance. On the other hand, we may observe that approaching the filtration process assuming such a pattern for the compact states is not at all a conservative approach. Even if it constitutes an approximation of the reality, it underestimates the opening of a real network of constrictions.

Should we wish to quantify the influence that the density has on the properties of these materials, we may look at the 0.5 fractile of the constriction curves. Whereas the effect of the density proves to be limited for the materials with compact grain size distribution (5% decrease of the constrictions diameter for the 0.5 fractile), the same cannot be said for the materials with a more widespread grain size distribution (WG and GG materials). For those materials whose uniformity coefficient is close to 4, the variation of the diameter of the constrictions for the 0.5 fractile has reached 25%. It can be higher than this in real materials whose uniformity coefficient can be a lot larger than 6. Thus, taking into account the density of the medium can turn out to be indispensable in quantifying the relevant phenomena.

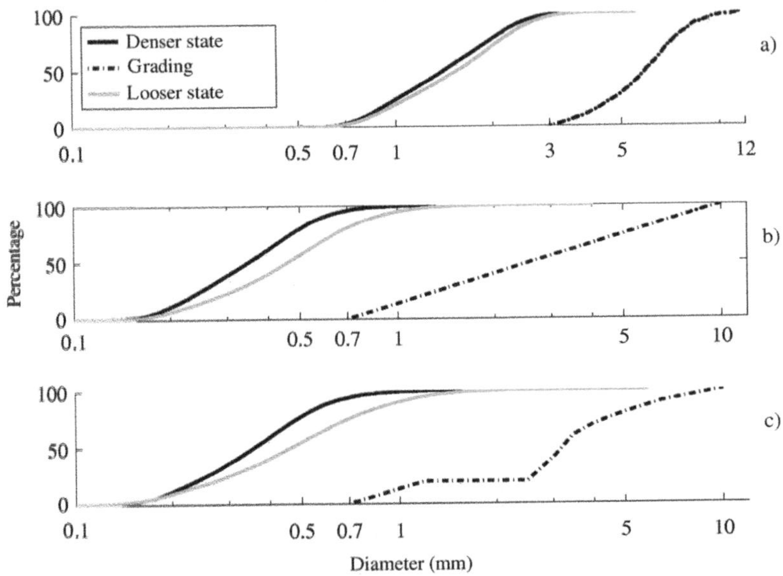

Figure 3.9. *Grain size distributions and constrictions for three materials in the most compact and the densest state: (a) for a UG material, (b) a WG material (c) and a GG material*

3.4. A probabilistic approach of constriction sizes

To simplify the analysis of the problem of fine materials filtration through granular materials, the studies that were carried out relied on the similitude between the average behavior of the filter and the average behavior of a sieve. The filtration characteristic is thus governed by the given opening of the mesh of the sieve, or the equivalent opening diameter for a granular material. This endeavor was soon followed by numerous authors, such as Silveira [SIL 65], Wittmann [WIT 79], Kenney *et al.* [KEN 85], Witt [WIT 86], Schuler [SCH 96], Giroud [GIR 03], Lone *et al.* [LON 05], and Indraratna *et al.* [IND 07]. In their approach, these authors relied upon several fundamental hypotheses:

– The coarse material that constitutes the filter is immobile; this ensures the condition of internal stability of the medium; only the particles at the exterior of the filter are likely to move inside the filter.

– The behavior of real materials is conveniently described if they are modeled by sets of spheres.

– The spheres of coarse material are positioned in space according to a geometrical configuration that is clearly identified and imposed beforehand.

– The network of constrictions is organized in the material in an equiprobable manner, in terms of sizes and according to a regular mesh.

As soon as we have enough data on the opening diameter, in a first instance, it might be useful to have a quick access to the totality of the constrictions curve to build certain devices for the mechanical engineer (filter criteria). Here, we have to be able to obtain this curve with very little information (i.e. often only with the grain-size distribution curve). As we have seen above, the grain-size distribution curve would not be enough, and knowing the density of the material should complement it. To draw the constriction curve, we need to rely on a probabilistic representation of the void space, which is supported by the solid space.

3.4.1. *Particle size distribution*

We must first consider the discretization of the coarse granular medium in N categories, each of them being characterized by a single diameter d_i. Each diameter has a certain occurrence within the medium, which is given by the distribution of the sizes issued out of the same categories. The grain-size distribution curve obtained through the sieving and weighting of the oversize on different sieves allows us to deduce this distribution, since it can be seen as a function of repartition associated with a probability density. We can thus speak of a distribution in volume by presupposing a constant absolute specific weight of the grains for the granular material. This distribution assigns a very important role to the coarser particles, because of their contributing volume.

It does not seem wise to try to build a network of constrictions on the basis of this distribution, given that the number of coarser particles is poorly represented in the medium. It seems to be better to use a distribution by numbers (Silveira [SIL 65], Kenney *et al.* [KEN 85], Witt [WIT 93], and Schulze [SCH 93]) because it is the number of particles within a class of particles that can explain the number of constrictions created from these particles, and with that their occurrence in the medium.

Thus, for the uniformity coefficients $U_c > 6$, the average arithmetic diameter of solid particles in a granular medium corresponds to diameters that are comprised between d_5 and d_{10}; it is comprised between d_{10} and d_{30} for $U_c < 3$ (Witt [WIT 93]). However, this approach will give a predominant role to the smaller particles in the medium, since they too interact with the coarser particles to form the constrictions. This aspect was pointed out through the study of the GG material that we have just described.

Humes [HUM 96] advances an intermediary approach as well, by using a distribution by lateral surface (or the external surface of the grain), since the possibilities for contact with the other particles thus forming constrictions are established through the external surface of the grains.

For the class of particles of a diameter i, the probability $p_{SA,i}$ in a lateral surface can be obtained from the following relation:

$$p_{SA,i} = \frac{\dfrac{p_i}{d_i}}{\displaystyle\sum_{i=1}^{n} \frac{p_i}{d_i}} \qquad [3.1]$$

where p_i represents the probability of particles with a diameter d_i in the grain-size distribution curve per mass (traditionally obtained by weighting the oversize on each sieve). The distribution of the particle size thus gives importance to the lateral surface offered by a category of particles to establish contact, in relation to the total lateral surface available in the material.

Although for different reasons, Giroud [GIR 03] proposed an approach that goes in the same direction: he suggests truncating the filter grain size distribution, retaining only the finest fraction, thus leading to a uniformity coefficient that is lower or equal to 3. This way, we may take out the excessive contribution of the coarser fraction in the characteristic of the filtration.

3.4.2. *Different geometrical configurations according to the density of the medium*

Obtaining the distribution of constriction sizes relies on a geometrical organization of particles that is postulated outside the mechanical considerations. Traditionally, in the case of dense materials, researchers (Silveira [SIL 65]) have considered three particles in mutual contact and the constriction is characterized by the disk that is tangential to those three particles (Figure 3.2(b)). The probability P_c associated with a constriction size d_c in the granular material is given by Kemeny [KEM 59], by considering the occurrence probability of the triplet (d_i, d_j, d_k) within a configuration of three particles. These diameters, whichever they are, belong to the grain-size distribution curve by lateral surface of the solid material. The probability P_c is thus written as:

$$P_c = \frac{3!}{r_i! r_j! r_k!} p_i^{r_i} p_j^{r_j} p_k^{r_k} \qquad [3.2]$$

where r_i, r_j, r_k correspond to the occurrence numbers of d_i, d_j, d_k in the group ($r_i + r_j + r_k = 3$). In this formula, as well as in the rest of this chapter, $p_{i;\,j;\,k}$ stands for $p_{SA,\,i;SA,\,j;SA,k}$.

In the case of loose materials, Silveira *et al.* [SIL 75] have proposed to describe the constrictions based on four mutually tangential particles. The size of the constriction is hereby described by the diameter of the disk that is equivalent to the maximal area A_v, being supported by four particles, each having the diameters (d_i, d_j, d_k, d_l) (Figure 3.10(a)). This value was corrected to account for the low probability of having a constriction that results from four particles whose centers belong to the same plane. Schuler [SCH 96] also considers four mutually tangential particles, but this time there will be two constriction sizes – d_1 and d_2 – that will be deduced later on (Figure 3.10(b)). The probability P_c, associated with a constriction size d_c in the granular material is thus written as (Kemeny [KEM 59]):

$$P_c = \frac{4!}{r_i! r_j! r_k! r_l!} p_i^{r_i} p_j^{r_j} p_k^{r_k} p_l^{r_l} \qquad [3.3]$$

where r_i, r_j, r_k, r_l correspond to the occurrence numbers of d_i, d_j, d_k, d_l, in the group ($r_i + r_j + r_k + r_l = 4$). Silveira's approach is chosen hereafter; the correction that is intrinsic to the method will allow us to consider the fact that, when dealing with real materials, the probability that a constriction arises from mutually tangential particles is very low.

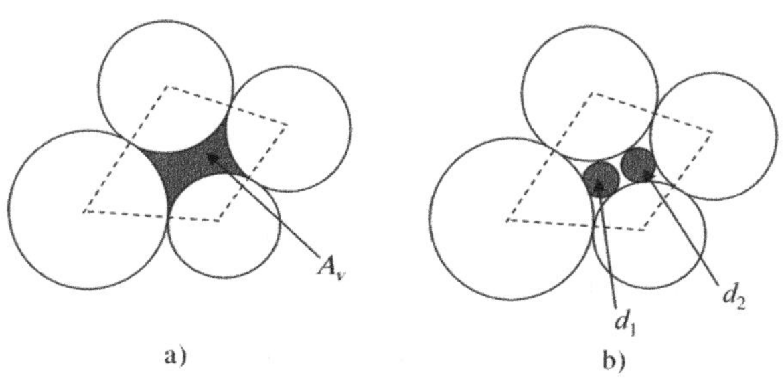

Figure 3.10. *The definition of constrictions for loose states: (a) Silveira's approach; (b) Schuler's approach*

3.4.3. *Distribution of constriction sizes: several models*

Given that the very complexity of the pattern of a polydispersed medium does not encourage the wish to approach the real organization of the medium based on these sole geometrical configurations, Locke *et al.* [LOC 01] have proposed to calculate d_c corresponding to a fraction P_c by using a balanced combination (which depends on the *RD*) of the three-particle (3P) and the four-particle (4P) configurations:

$$d_c(P_c, D_r) = d_{c3}(P_c) + P_c(1 - D_r)\left[d_{c4}(P_c) - d_{c3}(P_c)\right] \quad [3.4]$$

where d_{c3} and d_{c4} are the diameters corresponding to P_c in the configurations 3P and 4P, respectively. This method firstly consists of creating the two constriction curves for the configurations 3P and 4P on the basis of the grain-size distribution curve by the lateral surface area of the material (equations [3.2] and [3.3], respectively). For any given value P_c of the final constrictions curve, we can read on the two preestablished constrictions the corresponding diameters of these constrictions, namely d_{c3} and d_{c4} (Figure 3.11). For a given RD, the equation [3.4] gives the value d_c, which corresponds to the ordinate P_c.

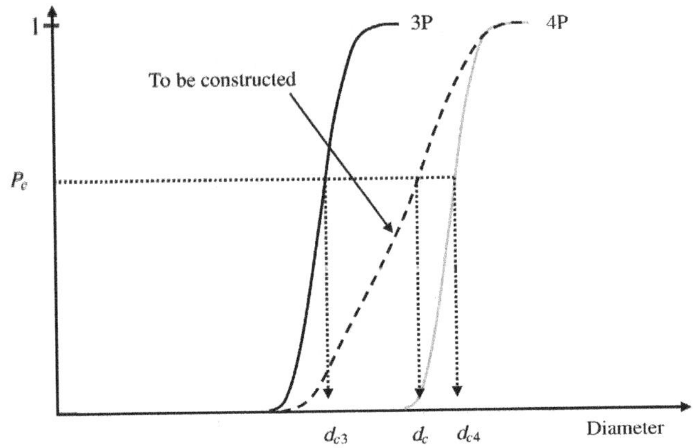

Figure 3.11. *Construction of an analytical constriction curve based on 3P and 4P configurations*

In this formulation, when *RD* = 100%, the medium is supposed to be represented only by the constrictions resulting from 3P configurations, whereas when *RD* = 0%, the two configurations are represented. As we have seen above, even in a compact case, a configuration with three mutually tangential particles is quite rare.

Reboul *et al.* [REB 10] have proposed a different formulation that rests on observations that can be easily detached from Figure 3.9(b)–(c):

– the smallest constriction diameter that can be conceived in this medium has the value d_{min}; this constriction is built upon the three smaller particles of the medium that are mutually in contact;

– the distribution of constriction sizes, for the looser situation, can be created with the aid of a formula that is slightly different from the formula proposed by Locke *et al* [LOC 01].

The construction schema of the analytical constriction curve then follows the above-explained method. If d_{cL} is the constriction diameter that is associated with the probability P_c in the case of a loosest state for the material, characterized by a void ratio e_{max}, then this diameter is given by:

$$d_{cL}(P_c) = 1.11\left(d_{c3}(P_c) + P_c\left[d_{c4}(P_c) - d_{c3}(P_c)\right]\right) \quad [3.5]$$

We are dealing with a rectification of equation [3.4] for $RD = 0\%$ by a multiplying coefficient of 1.1. This equation enables us to have a better predictive quality of the constriction curve in regard to the statistical data that were obtained on the materials UG, WG, and GG. With the aid of equation [3.5], we may recreate the constriction curve of the material when in its loosest state. For a given RD (or for the void ratio e) and a probability P_c, the constriction diameter d_c of the curve that needs to be built is equal to:

$$d_c(P_c, D_r) = d_{min} + \frac{e}{e_{max}}\left[d_{cL}(P_c) - d_{min}(P_c)\right] \quad [3.6]$$

where e is the void ratio that is associated with the *RD* (Figure 3.12).

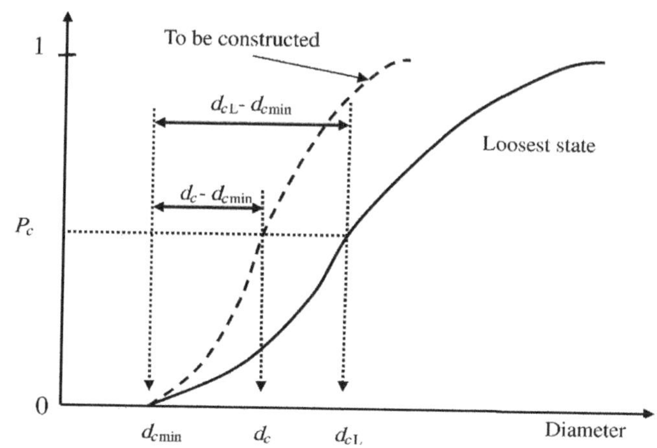

Figure 3.12. *The principle for the construction of a new model for the constriction curve starting from the curve obtained in the loosest state*

Contrary to the model proposed by Locke *et al.* [LOC 01], which does not require the identification of the parameters, this model requires us to identify e_{max}, which is nonetheless a classical physical characteristic. For the materials UG and WG, Reboul *et al.* [REB 10] have obtained an error rate that is lower than 5% between the constriction curves obtained by using this analytical model as well as the DEMDT method (Figure 3.13(a)–(b)). For the same materials, researchers have obtained an error rate of 15% when they used the formulation proposed by Locke *et al.* [LOC 01] (equation [3.4]).

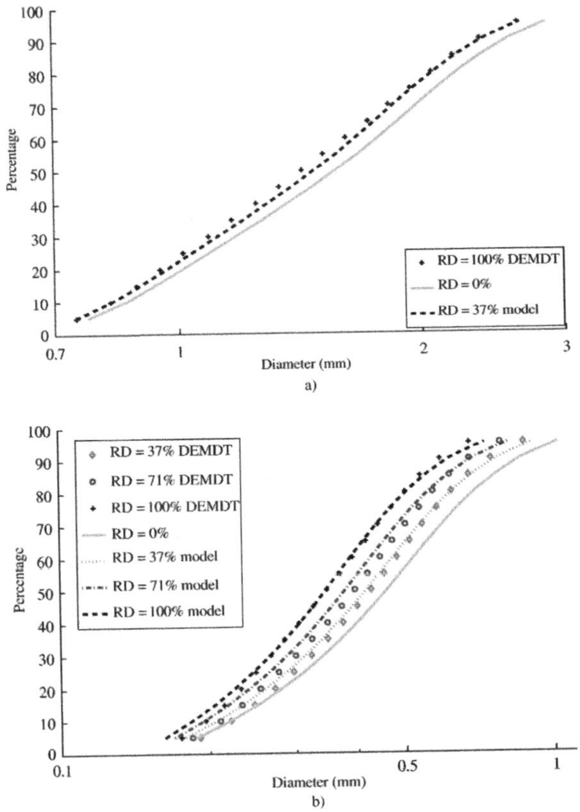

Figure 3.13. *Distribution model of the constriction sizes to 1 parameter for (a) UG material, and (b) WG material*

Whereas for the continuous grain size distribution, the process of densification seems to affect a wide range of constriction sizes; in the case of GG material the densification particularly affects the coarser particles that arise from the contact between a coarse particle and smaller particles. The result is a selective process that differs between the smaller constrictions (resulting from the contact between smaller particles) and the larger constrictions (implying at least one coarse particle). The simple formulation (equation [3.6]) cannot allow us to account for this complexity and thus the quality of the predictions suffers as a result (*vis-à-vis* the obtained results, using the DEMDT method). Reboul *et al.* [REB 10] have proposed a model with two parameters:

$$d_c(P_c, D_r) = d_{\min} + \frac{e_{\max} - e}{e_{\max}}(A.P_c + B)\left[d_{cL}(P_c) - d_{\min}(P_c)\right] \quad [3.7]$$

Reboul *et al.* [REB 10] have proposed a protocol for the identification of the constant A and B. With such a model, the average error rate obtained as a result of the processing of the GG material is 3% (Figure 3.14(a)–(b)). The effort required for processing this type of material is of crucial importance. On the one hand, when a filter receives fine particles, its grain size distribution will take on the shape of a GG material. On the other hand, when such materials are put into place on site, they are likely to develop an internal instability owing to the migration of the finest elements.

We have seen in this section that it is possible to accurately recreate the distribution accumulated from the size of the constrictions, for a material made up of spheres, on the sole basis of the knowledge of the grain-size distribution curve and the RD of the material. In the near future, the real challenge will be to see how far this formulation can apply to real, angular, and non-spherical materials.

3.5. Diameter of control constrictions

A filter is generally designed in such a way as to be traversed by a specific, controlled quantity of fine matter. Indeed, if as a consequence of the transport caused by an outflow, all of the fine particles remain blocked at the entrance of the filter, strong pore pressure can occur. On the contrary, if a consequent quantity of material goes through the filter, the result is a significant loss of matter for the upstream material, which can lead to its destabilization. To characterize the property of filtration, we are thus tempted to define a geometrical quantity that is associated with the control of this process, and which supports the possibility of creating a passage for a certain proportion of fine particles: this is called control diameter. However, moving from a non-filtering behavior to a completely filtering behavior is a relatively progressive movement, and the definition of a specific quantity, which is a characteristic of the filtering power of a material, can prove to be a delicate matter, difficult to establish in a generic and a non-specific case. Furthermore, the characteristic of filtration is difficult to dissociate from the geometrical properties related to the transported fine particles, since these two media interact strongly with one another.

For practical reasons, and analogously with the mesh opening of sieves or geotextiles, different authors have tried to bring together the control diameter and an opening diameter, which are equivalent to that of a sieve or a geotextile. This diameter is thus empirically obtained as a result of experimental trials. A probabilistic approach of the problem of filtration, carried out by Silveira [SIL 65] and Wittmann [WIT 79], provided a more theoretical definition of the diameter that characterizes the complete stopping of particles, a diameter that is not altogether equal to the control diameter.

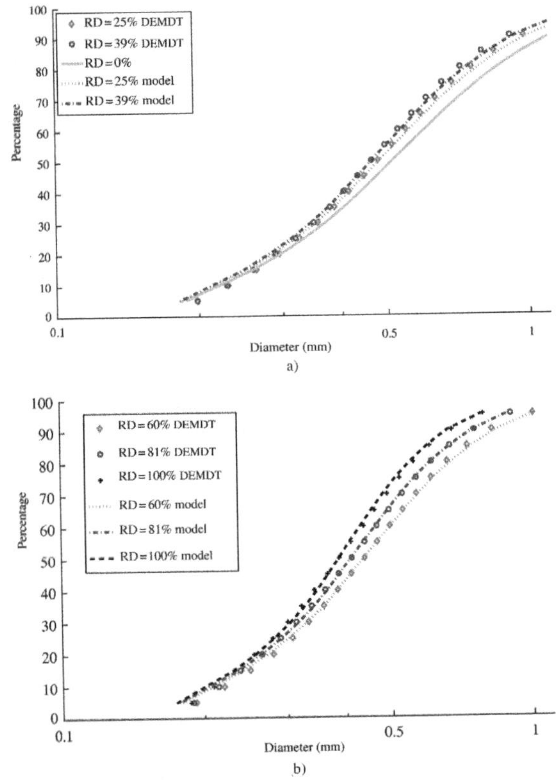

Figure 3.14. *Constriction curve model with two parameters for a loose: (a) middle dense GG material; (b) GG material that goes from middle-dense to dense*

3.5.1. *Probabilistic approach*

As we have explained above, the capacity of a filter to retain the fine particles that go through it, is linked to the smaller sizes of the constrictions that are likely to be encountered on their way. Since the variety of the sizes of the constrictions in the filter is given for one distribution, the length of the entire run through the filter increases, so does the probability that a particle bumps into a constriction of a size smaller than its own size. To understand this process, we may turn to a probabilistic approach of the individual transport of the particle with the diameter *d*.

A transported fine particle will have to face a series of differently sized constrictions when inside the filter. The probability of traversing a constriction of size d_c in the medium is very closely related with the constriction curve. Indeed, if the transported particle is of a larger size than the largest diameter of the constrictions, the probability of being blocked by a random constriction in the medium has a value 1. If the transported particle is smaller than the smallest diameter of the constrictions, then the probability of being blocked by a random constriction of the medium has the value 0. Thus, to ordinate P_c in the constrictions curve, we may associate the probability for a particle having the diameter *d* to be stopped by a constriction of the diameter d_c (Figure 3.15). Consequently, the probability *p* that a particle of diameter *d* encounters a constriction of diameter d_c has the value:

$$p(d_c > d) = 1 - p(d_c \leq d) = 1 - P_c \qquad [3.8]$$

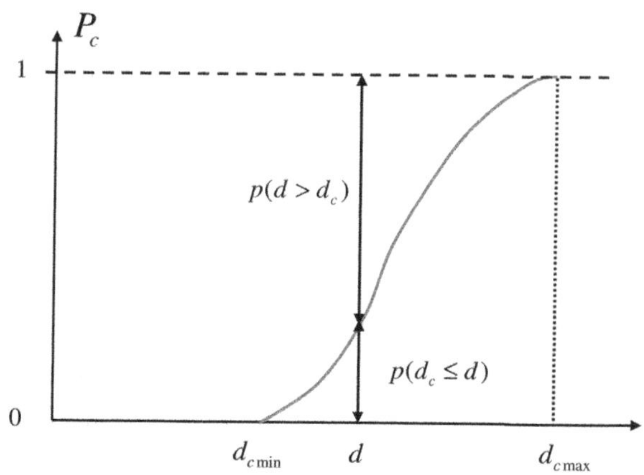

Figure 3.15. *Constriction curve and the probability that a constriction passes through a particle having a diameter d*

Once a transported particle has successfully overcome a constriction, it will continue on its track inside a pore and it is bound to come out of that pore through another constriction. Supposing that the network of pores is organized in a regular fashion (Figure 3.16), thus forming successive layers that need to be traversed, the probability that a particle having the diameter *d* passes a unity layer in the direction of the outflow has the value P_u. The analytical form for P_u may be found in Locke *et al.* [LOC 01]. It includes a term that is relative to the direct route (the black arrow in Figure 3.16), as well as a summation of the terms, which indicates that a lateral route going through a neighboring pore (or the neighboring pore of the neighboring pore) may be taken by the particle to advance through a unified layer in the sense of the outflow (the other arrows in Figure 3.16).

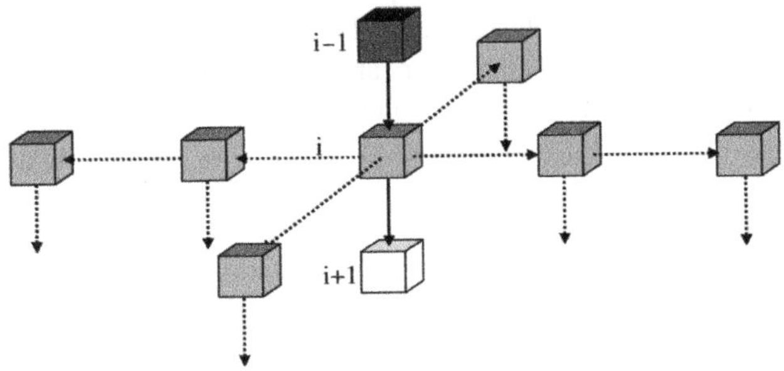

Figure 3.16. *The route of a fine particle through an orderly network of pores (made up of unit layers); a particle originating in layer i – 1 (dark gray pore) arrives in layer i (light gray pore), it can reach layer i + 1 (white pore) via a direct route (full arrow) or through a lateral route (dotted arrows)*

The probability that a particle might traverse *m* unit layers then has the value P_u^m and the probability $\bar{P}$ of being stopped after *m* unit layers has the value $1-P_u^m$. By isolating, *m*, the number of layers traversed by a particle with a diameter *d* before being stopped, has the value:

$$m = \frac{\ln(1-\bar{P})}{\ln(P_u)} \qquad [3.9]$$

To quantify *m* for a set of particles with a diameter *d*, we think that it is not important to force all the particles to satisfy the criterion, but rather that it suffices if the majority of the particles satisfy this criterion. Locke *et al.* [LOC 01] suggest that we take $\bar{P}$ (also called "confidence level") to equal 95%. Kenney *et al.* [KEN 85] have suggested 99%, Soria *et al.* [SOR 93] 98% as a result of a study showing that between 96% and 99% there is no significant difference between the filtering or non-filtering behavior of the medium.

The transcription of the result in dimensional terms requires the quantification of the thickness of a unit layer. Because the unit layer is linked to the average distance between two constrictions, i.e. to a poral characteristic, it surely follows that we attribute to it the value of the average diameter of a pore. However, as we have seen so far, it is difficult to define the entity "pore". Thus, several authors suggest that we take a value that is close to the arithmetic average of the solid diameters d_{50} issued from the grain-size distribution curve given by mass (Soria *et al.* [SOR 93]), by numbers (Locke *et al.* [LOC 01]), or by the lateral surface area [IND 07]. Certain recent calculations that imply sets of spheres tend to show that the nearest estimation of the diameter of the average pore, taken as the average size of the set of void spheres inscribed in the void space (by Delaunay tessellation), is d_{50}, issued from the grain size distribution given by numbers. This estimation is obtained from top to bottom, which is convenient given that the characteristic "inscribed sphere" gives an inferior limit to the local pore volume. This is a good quality estimation for broadly graded materials, but it can cause an overestimation of 100% for the quasi UG materials.

The approach presented hereby allows us to quantify the maximal distance that a particle can travel within a filtering medium, in the direction of the probabilities, and can allow us to define a control diameter.

3.5.2. *Control diameter: several definitions*

The probabilistic method indicates that there is a range of probabilities of the size of the constrictions under which a small decrease in the probability of having $d > d_c$ enables the particle with the diameter d to travel a very long distance before being stopped. This change in behavior takes place when this probability is comprised between 0.3 and 0.4 (Figure 3.17(a)). If we choose as the control diameter the size of the constriction that corresponds to the probability of having $d > d_c$ lower than a value that is equal to 0.3, the filter will have been entirely traversed (Figure 3.17(b)) and it will not have contributed to the process of filtration. In other words, a consequent loss in mass in upstream will take place before a filtration of fine particles intervenes. Thus, Indraratna *et al.* [IND 07] have chosen as a control diameter the size of the constriction corresponding to the fractile 0.35 on the constriction curve (ordinated on this curve). This control diameter thus depends not only on the grain size distribution, but also on the density of the filtering material currently under study. To obtain that, we need an analytical model for a constriction curve, such as the model we proposed in the previous chapter. This approach seems attractive as it is founded directly on the definition of a constriction curve, it is linked to the void network, and it is indirectly related to the grain-size distribution curve and allows us to consider the density of the medium.

The approaches that were developed to help define control diameter are generally empirical and are founded on a significant number of experimental trials. Kenney *et al.* [KEN 85] have associated control diameter with the largest diameter of the fine particles that can traverse a filter of a given thickness. This diameter is lower than 0.25 d_5, or 0.20 d_{15}, where d_x is the diameter corresponding to the percentage of soil going through the sieve x% of the grain-size distribution curve given in mass. Witt [WIT 86] has found that the diameter was close in value to 0.23 d_G, where d_G is the arithmetic average of the diameters of the solid grains in the filtering material whose value is d_5 for the grain size distributions $U_c > 6$ or comprised between d_5 and d_{10} if $3 < U_c < 6$. Foster and Fell [FOS 01] have found a superior limit for the control diameter, a limit that is equal to 0.20 d_{15} with an average of 0.16 d_{15}. Indraratna *et al.* [IND 07] have witnessed a very high agreement between their approach and that of Kenney *et al.* [KEN 85] if the control diameter issued from the latter is taken to be equal to 0.18 d_{15} for the very same compact materials.

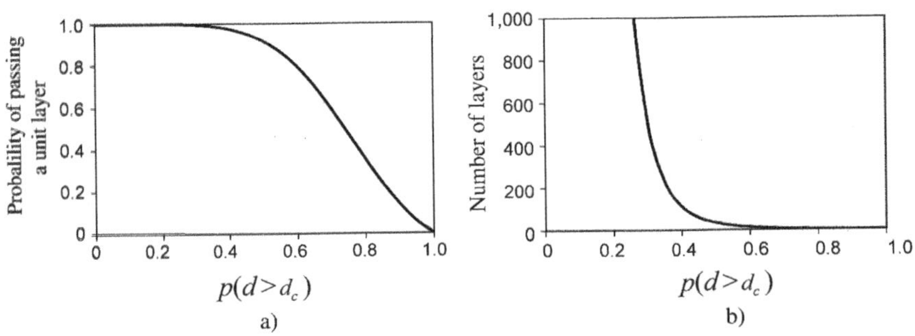

Figure 3.17. *The probability of the route taken by a particle with a diameter d before being stopped by a constriction of a diameter a_c: (a) traversing one unit layer; (b) total numbers of traversed layers*

Giroud [GIR 03] defines control diameter D_0 as a fraction of d_{15} that depends on the density of the filter and the uniformity coefficient of the grain-size distribution curve of the filter, truncated to solely keep the fraction of the fine particle that corresponds to a U_c of 3. This virtual truncation is appropriate for the development of retention equations. For the soils with a U_c' lower than 3, non-susceptible of being internally unstable, we have expressions that are similar to those of Pavchich [PAV 09]:

$$D_0 = (U_c')^{0.3} \text{ for the loose soils } RD < 35\% \quad [3.10]$$

$$D_0 = 1.5(U_c')^{0.3} \text{ for the middle-dense soils } 35\% < RD < 65\% \quad [3.11]$$

$$D_0 = 2(U_c')^{0.3} \text{ for the dense soils } RD > 65\% \quad [3.12]$$

where U_c' is the uniformity coefficient of the grain-size distribution curve taken as a whole, but approximated by a straight line. According to Pavchich [PAV 09], the control diameter is defined here as a fraction of d_{17}, a fraction that depends not only on the uniformity coefficient and the porosity of the material, but also on a segregation factor:

$$D_0 = 0.46\Gamma d_{17} (U_c)^{1/6} \frac{n^F}{1-n^F} \tag{3.13}$$

The parameter n^F is linked to the porosity of the filter, but its exact definition remains obscure. The parameter Γ accounts for a segregation phenomenon that takes place on site but that does not appear on small-scale samples created in the laboratory; $\Gamma = 1 + 0.05U_c$ for the materials having a $U_c \leq 25$.

Reboul [REB 08] has studied the possibilities of transport by using a purely geometrical method that is associated with a certain organization of the void space of the orderly network. When the diameter of a particle is slightly larger than the diameter corresponding to the distribution mode of the size of the constrictions, this particle is rapidly blocked by the medium. The control diameter could then be almost completely associated with the mode of distribution of the constriction size. However, there arises the problem of the physical definition of the control diameter for materials with discontinuous grain size distribution (of the type of GG material) for which there are two modes. It is worth pointing out that this type of grain size distribution did not fall within the framework of the approaches adopted by the majority of the authors we have quoted above. These materials, which may be quite unstable, were subject to specific developments (for instance, Sherard *et al.* [SHE 84], Lafleur *et al.* [LAF 89], and Fannin and Moffat [FAN 06]).

3.6. A continuous approach of the process of filtration

The classical methods for handling the problem of mass transport, within the field of the mechanics of continuous media, under fluid outflow, rest upon the definition of the three present entities: immobile particles that constitute a solid matrix of volumic mass ρ_{tr} and the interstitial fluid of volumic mass ρ_{ff} (Vardoulakis *et al.* [VAR 96] and Papamichos [PAP 04]). These theories rest upon the formulation of mass balance equations for each of the phases, complemented by a behavior equation or an evolution equation for the rate of transported material. This approach can be used in the study of filtration. For our purposes, we will suppose that the hydraulic flow is high enough to be able to suspend effectively all the particles that are likely to be transported. Therefore, there is no transport threshold effect in this work.

Let us note q_{er} as the erosion rate (volume flow of eroded fine particles), n as the porosity of the medium, and **v** as the Darcy vector of flow velocity at a given time t. The conservation of the transported mass is expressed thus:

$$\frac{\partial \rho_{tr}}{\partial t} + \nabla \cdot (\rho_{tr} \mathbf{v}) = q_{er} \qquad [3.14]$$

As for the conservation of the fluid mass, it is expressed thus:

$$\nabla \cdot (n\mathbf{v}) = 0 \qquad [3.15]$$

We associate the Darcy law with these equations:

$$n\mathbf{v} = -\mathbf{K} \cdot \nabla h \qquad [3.16]$$

where h is the hydraulic load and **K** the permeability matrix.

The erosion law can be formulated as follows:

$$q_{er}(t) = \beta \,|\, \mathbf{v} \,|\, \frac{M_{er}(t)}{V} \qquad [3.17]$$

where β is a parameter for the material, $M_{er}(t)$ the mass of erodible fine particles at a given time t, and V the total solid volume (Cividini and Gioda [CIV 04]).

Locke *et al.* [LOC 01] have proposed an original method for finding the value of $M_{er}(t)$ by relying on the constriction curve of the immobile solid material. They distinguish between three categories of erodible fine particles:

– The coarser particles whose infiltration distance in the coarse material corresponds to $m < 1$ (equation [3.18]). These particles will remain in the upstream of the filter and will not enter it. These particles will have a diameter that is superior to d_r and which can be found by considering $m = 1$ (Figure 3.18).

– Among the particles that are lower than d_r which will traverse the filter, some of them will have an infiltration distance lower than the thickness of the filter. These particles will be retained by the filter and they will not be displaceable later on. When we know the thickness of the filter or the spatial discretization element of filter, this thickness will correspond to a certain number of unit layers to be traversed and we can infer a diameter d_p (Figure 3.18).

– The particles lower than d_p that pass through the filter or through the discretization element without being stopped.

Thus, at each time increment, the diameters d_p and d_r are evaluated for the filter or for each discretization through the probabilistic approach. At time t, once the particles having the diameter comprised between d_p and d_r are stopped, they will be integrated in the grain-size distribution curve of the immobile solid material. Consequently, the constriction curve evolves toward narrower diameters. The permeabilities and the porosities are all updated with every instant. The explanatory algorithm of the successive increasing calculations inherent to this model is explained below; however, a more detailed account of this algorithm can be found in [REB 08].

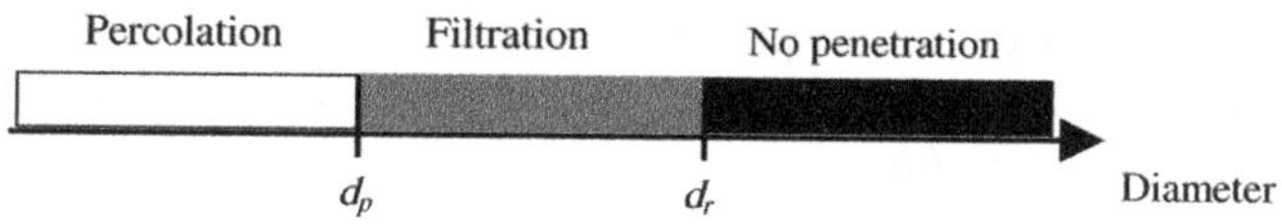

Figure 3.18. *The process of selective stopping of erodible particles according to their diameter*

– Initialize the variables: PSD^i, K^i, n^i, $t = 0$.

– Solve the hydraulic problem and calculate the load $h^i(\mathbf{x}, t)$ in every point of the domain, using the finite element method (equations [3.15] and [3.16]).

– Make explicit the constitutive law of erosion $q_{er}(t)$.

- Calculate the distribution sizes of the constrictions of the element.

- Determine the characteristic diameters of the element $i + 1$, d_p^{i+1}, and d_r^{i+1}.

- Calculate the mass of particles potentially erodible in the element i:

$$M_{er}^i = \sum_{j \le d_r^{i+1}} \rho_s (V_j^i - V_{retenus,j}^i) \quad [3.18]$$

- Determine the erosion rate q_{er}^i:

$$q_{er}^i = \beta \,|\, K^i \cdot \nabla h^i \,|\, \frac{M_{er}^i}{V^i} \quad [3.19]$$

– Solve the equation for the conservation of the mass of transported particles $\rho_{tr}(t + \Delta t)$ (equation [3.14]) using a spatial schema with finite differences and an explicit temporal schema.

– Deposit the transported particles from the element i into the element $i + 1$ and thus updating PSD^i, PSD^{i+1}, and $\rho_{tr}(t + \Delta t)$.

– Updating the variables K^i and n^i.

To illustrate this approach, we have studied a system composed of an element that corresponds to the base material, as well as two elements for the filter (Figure 3.19(a)). A hydraulic gradient between the superior and inferior frontiers of the system is imposed and entails a transport of fine material through the filter. The grain-size distribution curves given by the mass of the base material and the filter are given in Figure 3.19(b) with the initial constriction curve of the filter assumed to be in its loosest state. We will follow the evolution of grain-size distribution curves of the materials that constitute each element throughout time, as well as the evolution of d_p and d_r obtained through probabilistic analysis. We can observe in Figure 3.20 the decrease of fine particles in the base material. The majority of transported fine particles will be filtered through the element *filter* 1 and few of them will reach the element *filter* 2.

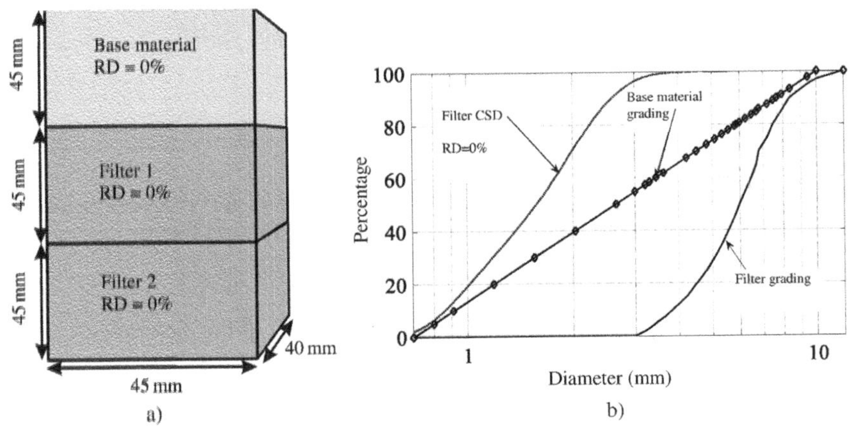

Figure 3.19. *Filtration of a base material through a coarse material: (a) the studied system; (b) the properties of the materials involved*

The result found here can be compared to a numerical study carried out by the DEM. To do that, an element of coarse material was created through the deposition (under normal gravity conditions) with the same grain size distribution as the one shown in Figure 3.19(b). This material is in its loosest state and the thickness of this sample corresponds exactly to the *filter* 1 element in Figure 3.19(a). A layer of fine material (or base material) of a thickness equal to the layer introduced through the continuous medium is directly positioned above the filter and submitted to gravity. The particles then move throughout the filter under the gravitational pull. Even if the effect of the fluid is not considered here, the stopping of a particle is conditioned by a geometrical criterion linked to the network of constrictions. This method therefore considers the major phenomena that need to be emphasized through the proposed comparison. When the system no longer evolves, the sizes of the particles that penetrate or do not penetrate the filter are analyzed to deduce the sizes of particles that correspond to d_p and d_r. The corresponding values are given in Table 3.1.

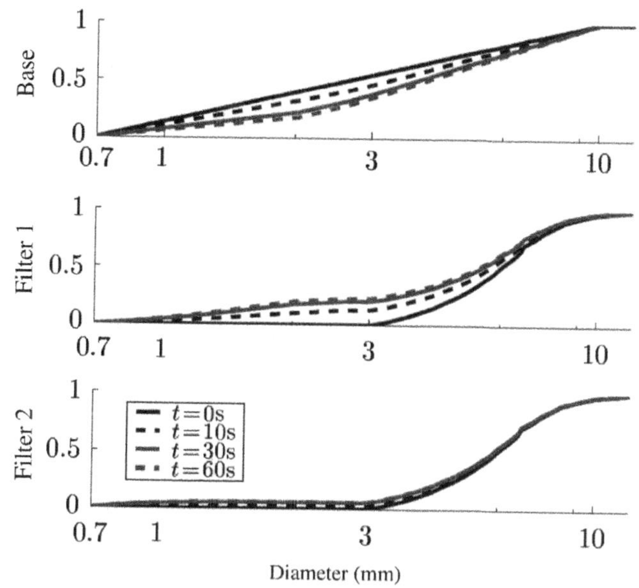

Figure 3.20. *Evolution of the grain size distribution of materials throughout time in the continuous approach*

	DEM	Continuous model	
	Filter 1	Filter 1	Filter 2
d_p	1 mm $\pm$ 0.2 mm	0.6 mm	0.8 mm
d_r	2.9 mm $\pm$ 0.3 mm	2.1 mm	2.8 mm

Table 3.1. *The values of the characteristic diameters d_p and d_r at the end of the process of filtration according to the continuous and the DEM approaches*

We can thus observe that, at the scale of the elements used, the characteristics of *filters* 1 and 2 of the DEM and the continuous approaches are not at all similar. The *filter* 1 of the model DEM enabled the entering of particles and the percolation of particles having a diameter larger than the diameter that could be obtained via the continuous approach. The analysis of the discrete sample revealed that there is a strong heterogeneity in the distribution of captured particles between the top and the bottom of the sample. The majority of the particles, especially the coarser particles, were trapped, in small numbers, in the upper part of the sample, thus producing a superficial clogging. The process of filtration does not manifest itself in a homogeneous way throughout the sample, which is the primary hypothesis, when compared to a continuous approach.

In a continuous medium, the impact of the entry of the material is completely "diluted" throughout the entire element. The homogenization affects all the sizes of the routes that will be shrinked randomly, and the medium will act as if it was more filtering by allowing particles of a smaller size to go in and out. Whereas at the scale of a real filter with a thickness that is higher than 1 m, this bias may not have an impact on the quality of the result given by the continuous approach, in the system studied here it produces a zooming effect. A more detailed analysis of the possibilities offered by this model is worth carrying out in more depth in the near future.

3.7. Conclusion

This chapter concerned itself with the understanding of the filtration processes within cohesionless granular materials, which are assumed to be internally stable.

The alluvium of fine elements comes from the pull out of the latter in the upstream of the filter. We based our analysis not only on the results shared in the specialized literature, but also on the analysis of numerical samples composed of spherical materials of different kinds: a quasi UG material, a WG material, or a GG material. Indeed, the DEM allows the access to multiple statistical data within the sample, in particular data related to the void space. However, it has certain limits for the user. The most constraining of these is linked to the time necessary for calculating, which increases with the number of generated particles and with the diameter of the particles within the system. To deal with this constraint, we can significantly reduce the number of particles if we do not wish to carry out any other comparison with real samples. In that case, we will manipulate samples equipped with periodical vertical, lateral conditions, thus vanquishing any boundary effect.

The constrictions, i.e. narrow passageways between the pores, play a major role in the process of filtration. They must constitute the central point of every study that is concerned with characterizing the filtering power of the granular material. The average size of the constrictions and the cumulated distribution of the sizes of these constrictions are characteristic of a certain grain size distribution and depend on the density of the medium. This influence can be major when the materials have a broad grain size distribution and must be considered particularly when we wish to define an average control diameter for a filter. The control diameter was seen as very close in nature to the mode of size distribution and constrictions. It will then be interesting to establish a relation between this mode and the characteristics of the grain-size distribution curve of the filter for a fixed density.

We can supply the reader with an analytical model of a constriction curve. We have seen that the incomplete grain size distributions need a more sophisticated model, because the densification process induces more selective effects for this type of material. In general, a specific study of these materials will have to be carried out because they are likely to have instabilities due to alterations of the hydraulic boundary conditions. Furthermore, the mechanisms at play are a bit more difficult to identify.

The use of such a model for processing real materials will require a specific study. As the angularity of grains affects the value of the maximal void ratio, the constriction curve for the looser configuration should depend on this angularity. A continuous approach of the filtration process may be integrated as an internal variable in the cumulated distribution of constriction sizes. Starting from this curve, which is characteristic of a coarse material, we can calculate two sizes (diameters) d_p and d_r, which explain why the fine particles were (or were not) stopped in the upstream of the filter or within the filter. The blocked particles modify the constriction curve and thus the values of the two diameters d_p and d_r. This way, the medium becomes more and more filtering with time. It is worth making an effort to better understand the scaling effects of the approach and the size of the meshing that is compatible with the phenomena we wish to emphasize. However, this approach seems promising and it will have to be further developed in the near future, to better discern its limits and its advantages.

3.8. Bibliography

[ALR 03] AL-RAOUSH R., THOMPSON K., WILLSON C.C., "Comparison of network generation techniques for unconsolidated porous media", *Soil Science Society of America Journal*, vol. 67, pp. 1687–1700, 2003.

[ALR 07] AL-RAOUSH R., ALSALEH M., "Simulation of random packing of polydisperse particles", *Powder Technology*, vol. 176, pp. 47–55, 2007.

[AUZ 96] AUZERAIS F.M., DUNSMUIR J., FERREOL B.B., MARTYS J., OLSON J., RAMAKRISHNAN T.S., SCHWARTZ L.M., "Transport in sandstone: a study based on three-dimensional tomography", *Geophysical Research Letters*, vol. 23, pp. 705–708, 1996.

[BAG 05] BAGI K., "An algorithm to generate random dense arrangements for discrete element simulations of granular assemblies", *Granular Matter*, vol. 7, pp. 31–43, 2005.

[BAL 04] BALHOFF M.T., THOMPSON K.E., "Modelling the steady flow of yield-stress fluids in packed beds", *American Institute of Chemical Engineers*, vol. 50, pp. 3034–3048, 2004.

[BRY 93] BRYANT S.L., KING P.R., MELLOR D.W., "Network model evaluation of permeability and spatial correlation in a real random sphere packing", *Transport in Porous Media*, vol. 11, pp. 53–70, 1993.

[CIV 04] CIVIDINI A., GIODA G., "Finite-element approach to the erosion and transport of fine particles in granular soils", *International Journal of Geomechanics*, vol. 4, pp. 191–198, 2004.

[COE 97] COELHO D., THOVERT J.-F., ADLER P.M., "Geometrical and transport properties of random packing of spheres and aspherical particles", *Physical Review E*, vol. 55, no. 2, pp. 1959–1979, 1997.

[FAN 06] FANNIN R.J., MOFFAT R., "Observations on internal stability of cohesionless soils", *Géotechnique*, vol. 56, no. 7, pp. 497–500, 2006.

[FIN 70] FINNEY J., "Random packings and the structure of simple liquids: the geometry of random close packings", *Proceedings of the Royal Society London A*, vol. 319, pp. 479–493, 1970.

[FOS 01] FOSTER M., FELL R., "Assessing embankment dams filters which do not satisfy design criteria", *Journal of Geotechnical and Environmental Engineering, ASCE*, vol. 127, no. 5, pp. 398–407, 2001.

[FRY 07] FRY J.-J., Context and framework of the report of the European Working Group on Internal Erosion, Assessment of the Risk of Internal Erosion of Water Retaining Structures: Dams, Dykes and Levees, Intermediate Report of the European Working Group of ICOLD, no. 114, Technische Universität München (TUM), Essen, pp. 1–12, 2007.

[GIR 03] GIROUD J.-P., "Filter criteria", *5th Anniversary of K. Terzaghi "Erdbaumechanik" (Soil Mechanics)*, in BRANDL H. (ed.), *Reports of the Institute for Soil Mechanics and Geotechnical Engineering*, TU Vienna, Austria, pp. 221–229, 2003.

[HUM 96] HUMES C., "A new approach to compute the void size distribution curves of protective filters", in LAFLEUR J., ROLLIN A.L. (eds), *GeoFilters96*, Bitech Publications, Montreal, Canada, pp. 57–67, 1996.

[IND 07] INDRARATNA B., RAUT A.K., KHABBAZ H., "Constriction-based retention criterion for granular filter design", *Journal of Geotechnical and Geoenvironemental Engineering, ASCE*, vol. 133, no. 3, pp. 266–276, 2007.

[KEM 59] KEMENY J., MIRKIL H., SNELL J., THOMPSON G., *Finite Mathematical Structures*, Prentice Hall, 1959.

[KEN 85] KENNEY T.C., CHAHAL R., CHIU E., OFOEGBU G.I., OMANGE G.N., UME C.A., "Controlling constriction sizes of granular filters", *Canadian Geotechnical Journal*, vol. 22, pp. 32–43, 1985.

[LAF 89] LAFLEUR J., MLYNAREK J., ROLLIN A.L., "Filtration of bradly graded cohesionless soils", *Journal of Geotechnical Engineering, ASCE*, vol. 115, no. 12, pp. 1747–1768, 1989.

[LOC 01] LOCKE M., INDRARATNA B., ADIKARI G., "Time dependent particle transport through granular filters", *Journal of Geotechnical and Geoenvironmental Engineering, ASCE*, vol. 127, no. 6, pp. 521–528, 2001.

[LON 05] LONE M.A., HUSSAIN B., ASAWA G.L., "Filter design criteria for graded cohesionless bases", *Journal of Geotechnical and Geoenvironmental Engineering, ASCE*, vol. 131, no. 2, pp. 251–259, 2005.

[MEH 08] MEHLHORN T., PROHASKA S., HOMBERG U., SLOWIK V., "Modelling and analysis of particle and pore structures in soils", *Proceedings of the Workshop on Internal Erosion*, Bauhaus-Universitt Weimar, Germany, pp. 51–58, 26–28 November 2008.

[MEL 89] MELLOR D.W., Random close packing of equal spheres: structure and implication for the use as a model porous medium, PhD Thesis, Open University, Milton Keynes, 1989.

[MOU 08] MOURZENKO V., THOVERT J.-F., VIZIKA O., ADLER P.M., "Geometrical and transport properties of random packings of polydisperse spheres", *Physical Review E*, vol. 77, pp. 1–14, 2008.

[NOL 92] NOLAN G.T., KAVANAGH P.E., "Computer simulation of random packing of hard spheres", *Powder and Technology*, vol. 72, pp. 149–155, 1992.

[ONO 90] ONODA G.Y., LINIGER E.G., "Random loose packings of uniform spheres and the dilatancy onset", *Physical Review Letters*, vol. 64, pp. 2727–2730, 1990.

[PAP 04] PAPAMICHOS E., "Hydromechanical coupling for erosion", *Revue Européenne de Génie Civil*, vol. 8, pp. 709–734, 2004.

[PAV 09] PAVCHICH M., "Transition filters", *European Workshop Group on Internal Erosion*, St Petersburg, Russia, pp. 1–8, 2009.

[REB 08] REBOUL N., Transport de particules dans les milieux granulaires, PhD Thesis, Ecole Centrale de Lyon, 2008.

[REB 10] REBOUL N., VINCENS E., CAMBOU B., "A computational procedure to assess the distribution of constriction sizes for an assembly of spheres", *Computers and Geotechnics*, vol. 37, nos. 1–2, pp. 195–206, 2010.

[RIC 99] RICHARD P., OGER L., LEMAITRE J., SAMSON L., MEDVEDEV N.N., "Application of the Voronoi tesselation to study transport and segregation of grains inside 2D and 3D packings of spheres", *Granular Matter*, vol. 1, pp. 203–211, 1999.

[ROU 05] ROUX J.N., CHEVOIR F., "Simulation numérique discrète et comportment mécanique des matériaux granulaires", *Bulletin des Laboratoires des Ponts et Chaussées*, vol. 254, pp. 109–138, 2005.

[SCH 96] SCHULER U., "Scattering of the composition of soils – an aspect for the stability of granular filters", in LAFLEUR J., ROLLIN A.L. (eds), *Proceedings of the GeoFilters96*, Bitech Publications, Montreal, Canada, pp. 21–33, 1996.

[SCH 93] SCHULZE B., "Neuere untersuchungen uber die injizierbarkeit von feinstbindemittel-suspensionen", *Grounting in Rock and Concrete*, A. A. Balkema, Rotterdam, pp. 107–116, 1993.

[SCH 94] SCHWARTZ L.M., AUZERAIS F.M., DUNSMUIR J., MARTYS J., BENTZ D.P., TORQUATO S., "Transport and diffusion in three dimensional composite media", *Physica A*, vol. 207, pp. 28–36, 1994.

[SCO 69] SCOTT G.D., KILGOUR D.M., "The density of random close packing of spheres", *Journal of Physics D: Applied Physics*, vol. 40, pp. 863–866, 1969.

[SHE 84] SHERARD J., DUNNIGAN L., TALBOT J., "Basic properties of sand and gravel filters", *Journal of Geotechnical Engineering*, ASCE, vol. 110, no. 6, pp. 684–700, 1984.

[SIL 65] SILVEIRA A., "An analysis of the problem of washing through in protective filters", *Proceedings of the 6th International Conference on Soil Mechanics and Foundation Engineering*, vol. 2, Montreal, Canada, pp. 551–555, 1965.

[SIL 75] SILVEIRA A., DE LORENA PEIXOTO JR. T., NOGUEIRA J., "On void size distribution of granular materials", *Proceedings of the 5th Panamerican Conference on Soil Mechanics and Foundation Engineering*, Buenos Aires, Argentina, vol. 3, pp. 161–176, 1975.

[SOH 68] SOHN H.Y., MORELAND C., "The effect of particle size distribution on packing density", *Canadian Journal of Chemical Engineering*, vol. 46, pp. 162–167, 1968.

[SOR 93] SORIA M.H.A., ARAMAKI R.T., VIVIANI E., "Experimental determination of void size curves", *Filters in Geotechnical and Hydraulic Engineering*, in BRAUNS J., HEIBAUM M., SCHULER U. (eds), Balkema, Rotterdam, pp. 43–48, 1993.

[THO 01] THOVERT J.-F., YOUSEFIAN F., SPANNE P., JACQUIN C.G., ADLER P.M., "Grain reconstruction of porous media: application to a low-porosity Fontainebleau sanstone", *Physical Review E*, vol. 63, pp. 1–17, 2001.

[VAR 96] VARDOULAKIS I., STAVROPOULOU M., PAPANASTASIOU P., "Hydromechanical aspects of the sand production system", *Transport in Porous Media*, vol. 22, pp. 225–244, 1996.

[WIT 86] WITT K.J., Filtrationsverhalten und Bemessung von Erdstoff-Filtern (in German), Veröffentl. D. Inst. F. Bodenmechanik und Felsmechanik, Universität Karlsruhe, Heft 104, 1986.

[WIT 93] WITT K.J., "Reliability study of granular filters", *Filters in Geotechnical and Hydraulic Engineering*, in BRAUNS J., HEIBAUM M., SCHULER U. (eds), Balkema, Rotterdam, pp. 35–41, 1993.

[WIT 79] WITTMANN L., "The process of soil filtration – its physics and the approach in engineering practice", *Proceedings of the 7th European Conference on Soil Mechanics and Foundation Engineering*, vol. 1, Brighton, UK, pp. 303–309, 10–13 September 1979.

[ZIE 68] ZIEMS J., Beitrag zur Kontakterosion nichtbindiger Erdstoffe, PhD Thesis, Technische Universität Dresden, 1968.

Chapter 4

Contact Erosion between Two Soils

4.1. Introduction

Contact erosion is a form of erosion that is internal to the hydraulic structure and takes place at the interface between two layers of different types of soil. This kind of erosion is sometimes called "contact suffusion", or "interface erosion", but in what follows we will keep using the term "contact erosion" or CE. There are two important factors that urge us to pay particular attention to the risks of CE in fluvial levees. On the one hand, the natural soils of the alluvial plains are generally very layered, composed of successive layers of sediments of different origins, and forming numerous interfaces between materials that have different grain size distributions. On the other hand, building a hydraulic structure, such as a fluvial levee and an embankment dam with shoulders, requires placing different layers of materials, each one having a specific function and thus different properties. A significant number of interfaces are thus naturally found within the body of the structure as well as inside its foundation. During the saturation of the structure, the subterranean flows and the infiltration ones will be carried out to traverse or to follow these interfaces that become areas with a particular propensity toward developing CE. Indeed, the layer of coarse soil is a lot more permeable than the one made of fine soil, and herein the hydraulic outflows are much more significant than in the other layers. Therefore, if the pores in the coarse soil are large enough and the outflow is sufficiently intense, the particles of fine soil may be detached by the flow, transported through the pores, and then redeposit themselves a bit farther away or eventually leave the structure.

Chapter written by Rémi BÉGUIN, Pierre PHILIPPE, Yves-Henri FAURE and Cyril GUIDOUX.

In order for CE to appear, we may thus conclude that there are two conditions that need to be met: first and foremost, the pores of the coarse soils must permit the passage of the fine particles through their constrictions[1], i.e. the geometric condition. Second, the hydraulic constraint induced by the flow must be enough for pulling out of the fine particles when in contact with the layer of coarse soil, i.e. the hydraulic condition. In terms of the safety of a hydraulic structure, observing the geometrical criterion enables us to protect the structure against all forms of CE. If this fails, we must then turn toward an adapted hydraulic criterion.

The strategy adopted in this chapter consists of starting from real hydraulic structures to localize the areas that could be sensitive to CE. The local physical phenomena that are at stake when CE takes place are then explained in more detail, before advancing a complete synthesis of the different works, both theoretical and experimental, carried out on the subject in the configuration of a coarse soil on top of a fine soil, as well as in the reversed configuration. The final section returns to the scale of the structure, with the aim of proposing possible failure scenarios.

4.2. Areas prone to CE in hydraulic structures

4.2.1. *General characteristics of constitutive materials*

The materials affected by CE are very different in nature. The layer of material called "coarse" is, because of its high permeability and larger pores, the main place for hydraulic outflow and for the potential transport of fine particles. This layer is typically made of gravel, sometimes mixed with sand or silt. In what follows, we will use, by referring to this coarse layer, the variable D to designate the characteristic size of particles and, when the highest precision is required, the x-centile D_x of the grain size distribution of the soil[2]. The layer of material called "fine" is the layer that is actually subject to erosion and can be made of sands, silts, clay, or a mixture of all of the above. Similar to the coarse soil, we will use for this layer the variables d and d_x to characterize their grain size distributions.

Here, on the one hand, we must distinguish between cohesion-less soils, i.e. the soils that do not have cohesion between the grains, such as sands, and cohesive (or coherent) soils where binding forces exist between the particles, such as clays and also certain silts. On the other hand, we must distinguish between a granular soil made of coarse particles that have high sedimentation velocities, and a fine soil

1 We call constrictions the areas having the thinnest sections within the poral space.

2 D_x defines a characteristic dimension for the particles as it appears in the material x% of the mass of particles having a size lower or equal to D_x.

made of fine particles that can be easily suspended within a liquid. In the case of CE, the layer of coarse soil is systematically a granular type of soil, whereas the layer of fine soil can have highly variable characteristics: granular and moderately fine in the case of a type of sand, fine and granular in the case of fine sand (silt-like sand), and, finally, very fine and cohesive in the case of silt or clay.

Alongside small and large rivers, the soil is made up of pebbles, sand, and silt, all transported by the water. Several authors have emphasized the presence of a gap in the grain size distribution of these alluvia, which is typically between 1 and 5 mm. This is especially the case of the soils present in the valley of the Rhine, the Rhône, or the Danube. A grain size subrepresentation can be equally observed in the moraine soils, between 0.1 and 20 mm. The presence of interfaces between layers of distinct grain size distributions is thus highly probable. Since the restriction of transport costs makes us use local materials, we find within hydraulic works materials that are similar to the soils of the foundation.

4.2.2. *Heterogeneities within the concerned hydraulic structures*

Hydraulic structures such as flood-protection dikes, which are rarely subjected to a considerable hydraulic head, are generally built with on-site materials whose properties have not been necessarily well controlled beforehand and which can then present significant heterogeneities and areas with high permeability. In this type of dike, which is schematized in Figure 4.1(a), CE is likely to develop at the interfaces of the more permeable areas, or at the contact with the foundation soils, which can be deposits of much finer silt-like soils.

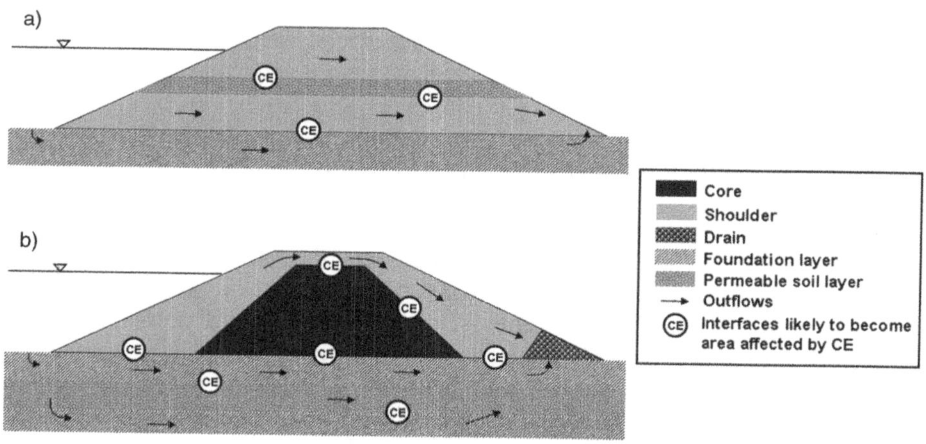

Figure 4.1. *Typical schema of (a) flood-protection dikes built with on-site materials and (b) zoned fluvial levees*

The presence of layers that are permeable enough to meet the two conditions, geometrical and hydraulic, thus leading to CE, remains quite unlikely at the time of building the structure. However, under the action of preferential outflows, heterogeneous layers may be sensitive to suffusion, i.e. the phenomenon of internal erosion that is described in Chapter 2, and which results in a progressive erosion and transport of the finer part of the soil. Such a process can last for several years, thus rendering the layer of soil more and more permeable. These areas, washed out of their finest particles, are likely to reach the conditions for the appearance of CE between a coarse layer and a fine layer.

In the case of sensitive structures, permanently subjected to significant hydraulic heads, a zoned structure is used, comprising a central core made of silt or compacted clay (with a low permeability whose function is to guarantee water tightness) (Figure 4.1(b)). Besides this core, upstream and downstream shoulders are added, which are meant to ensure the mechanical stability of the structure and are created with a more permeable coarse material. A drain layer is generally created at the base of the dike with a clean and calibrated gravel. In this situation, the highest risk of developing a CE is situated at the interface between the core made of fine soil and the foundation soil, which is rather more gravel-like. As previously discussed, this gravel-like soil is not always, right from the beginning, in the conditions favorable for the apparition of CE, but it can gradually become so under the effect of suffusion within the levee. The potential existence of stratified layers in the foundation soil equally corresponds to the interfaces that are potentially sensitive to CE. Yet, another risk regards the drain elements, which, in case they do not observe the rules of geometrical filters with neighboring layers, can become a spot for CE development. Finally, if the upstream hydraulic head surpasses the maximum permissible head of the structure, after a flood for example, an overflow is likely to take place throughout the shoulders or the superficial rocky embankments, thus generating CE between these protection layers, which are very coarse, and the core.

4.2.3. *Possible configurations of CE within a hydraulic structure*

As we have inventoried in Figure 4.2, the areas previously identified as potentially subjected to CE can be described according to different elementary configurations, depending on the orientation of the interface, the flow direction, and the positioning of the layers. These sensitive interfaces are most frequently horizontal, since they are the result of the successive deposit of sediment layers or of layers made up of the materials that build up a structure. However, certain interfaces, such as the walls of a core, may be inclined or even vertical. At the same time, the natural hydraulic flow is made preferably on the horizontal plane to traverse the structure. However, areas with a higher permeability may cause the outflow to bend down and to traverse soil layers perpendicularly. In terms of classification, the most frequent configuration of CE is the one of a horizontal interface, subjected to an outflow tangential with two possible positionings for the two layers, the fine material being thus situated beneath or above the coarse material. The differences in behavior are thus due to the influence that the gravity has on the fine soil subject to erosion: gravity pull is stabilizing in the configuration of coarse soil on top of fine soil, and destabilizing in the reversed situation. In the same manner, when the interface is not horizontal, gravity will play an inhibiting or activating role with regard to the erosion, depending on the inclination angle of the layers. We must also bear in mind that this influence of gravity is much more limited, almost negligible, in the case of cohesive soils where the internal bonding forces are predominant compared to weight. On the contrary, as erosion develops in the fine soil, intermittent and localized re-organizations of the grains are likely to take place in the coarse layer under the effect of gravity.

In this chapter, we will only retain two configurations that are typical of CE: "fine soil under coarse soil" and "fine soil on coarse soil". In both of the cases, the two materials are disposed in horizontal layers and they are subjected to a tangential outflow. Erosion and transport only regard, here, the particles that constitute the fine material, whereas the coarse soil remains mainly static[3]. Furthermore, we will not examine the configuration where the outflow is normal, at the interface between the two layers. This situation is related to a process of filtration that has been detailed in Chapters 2 and 3. In certain specific configurations, the coarse soil may be entrained before the fine soil. It is particularly the case of a process, called "piping" [SEL 88], that was studied in the Netherlands. This phenomenon appears in a certain type of dike, which is very common in the Netherlands; these dikes are made of clay, and built on a foundation made of sand. The sand is progressively entrained through backward erosion, at the contact of the layer of clay that remains stable. By hollowing, a pipe is created regressively, which may quickly cause the failure of the dike. Several phenomena are involved, resembling rather the phenomenon of piping flow erosion (Chapter 5) or river bed erosion (Chapter 7). This mechanism is not actually related to CE, so it is not examined in this chapter.

Finally, there is one last particular case of CE that will not be developed here: it regards the situation of a smooth or slightly rough surface (wall, sheet-pile, masonry coating) or of a traversing object (pipes, electric cables), which can be found in the body of a dike. This type of transition within the internal structure of the work induces preferential water channeling at the interface between the transition and the soil that constitutes the dike. These localized outflows can then generate an erosion of the entire material or a suffusion of its finest fraction.

---

3 Except intermittent and localized collapses in the coarse layer, as a result of the development of basins in the underlying fine soil.

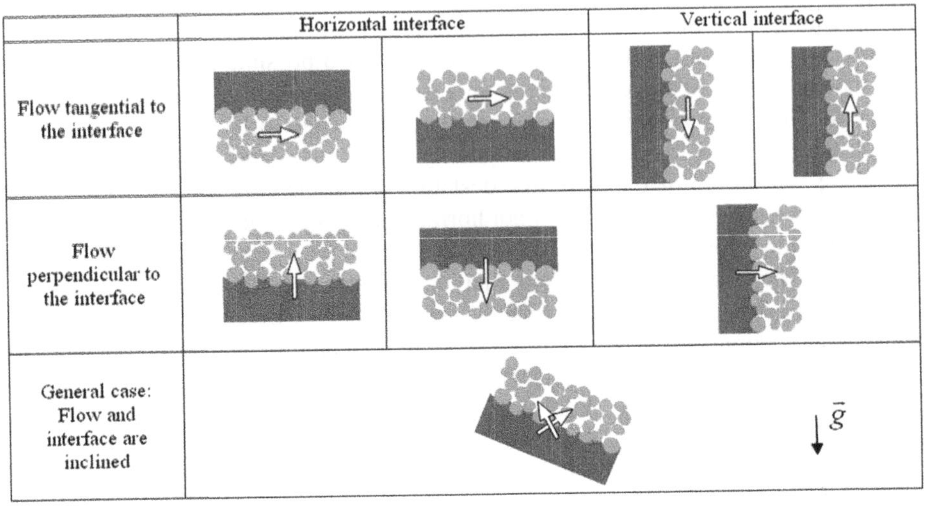

Figure 4.2. *Elementary configurations of contact erosion depending on the orientation of the interface and of the outflow (white arrows) in relation to gravity (fine soil in dark gray, coarse soil in light gray)*

4.3. Description of CE mechanisms on a local scale

4.3.1. *CE in the general context of surface erosion*

The "classic" situation of surface erosion of a soil by a homogeneous fluid flow is specifically examined in Chapter 7. It has become the object of a large amount of theoretical research, experimental research carried out in the laboratory, as well as field measurements. In what follows, we will systematically refer to this basic configuration of surface erosion to emphasize the expected similarities and differences in the specific framework of CE.

The presence of a granular layer in the area where hydraulic outflow takes place constitutes the main specificity of CE; several major consequences ensue from this fact, and it is worth considering them. First, hydraulic outflow is made throughout the porous matrix of the granular layer. This poral geometry induces locally strong heterogeneities in the outflow, particularly in the transition area, where the grains are in contact with the layer of fine soil. A further incidence: a significant fraction of the interface of the fine soil is in contact with grains. Thus, it escapes the action of the water, but it is finally subjected to the action of the grains, which locally apply the efforts transmitted by the granular skeleton. Finally, the transport of eroded particles is conditioned by the geometry of the porous medium, which, in turn, increases the probability of particle capturing and of depositing between the grains. It is worth noting that, to distinguish the size scales between the two layers in contact, we call the elements that constitute the coarse soil as well as the fine soil – "grains" and "particles", respectively. We may then conclude that CE is not an elementary mechanism of erosion, but, more specifically, it is a combination between various phenomena. CE includes complexities that characterize surface erosion, hydraulic outflow in a porous medium, and the transport and sedimentary deposit of fine particles in a granular filter.

4.3.2. *Local measures in the contact area between two soil layers*

The authors of this chapter carried out an optical matching technique (associated with laser-induced fluorescence) to better study the hydrodynamics in the contact area between two layers of material that have different grain size distribution. On the one hand, the poral space of the model porous media, made of glass beads, is filled with a type of oil that has a refractive index that is perfectly adjusted to the one of the glass beads. On the other hand, the lighting made by a planar laser beam, one fine fluorescent particle inseminated in the oil, induces a shifted re-emission of the wavelengths that can be isolated by means of optical filter interposition. Figure 4.3(a) presents a typical image where the luminous dots correspond to the tracers and the shadowy areas correspond to the beads. Starting from two of the successive images obtained with a high-speed camera, a particle image velocimetry (PIV) processing technique reveals the 2D velocity field of the liquid in the poral space (Figure 4.3(b)). A second image processing allows us to go back to the position of grains in the contact zone between the two layers (Figure 4.4(a)) by creating a binary mask[4] of the repartition of each of the liquid and solid phases. We may deduce the vertical porosity profiles and the profiles of pore velocity component of the liquid, when found at the transition point between the two layers (Figure 4.4(b)) as well as the velocity component distributions and the shear stress distribution (Figure 4.5). These experiments were carried out, on the one hand, with layers of millimetric beads having three different size ratios (D/d = 1.9, 2.4, and 3.2) and on the other hand with a layer of fine sand (d_{50} = 0.3 mm) covered by a layer of beads having the diameter D = 9.7 mm. In this latter case, the image processing can only be implemented in the bead layer. These results illustrate the different points that will be developed in the rest of this section, at the same time keeping in mind that, because of the uniform grain size distribution of the beads, their packing presents some effects of ordering in superposed layers, which are much stronger than in a real coarse soil. However, without questioning the generality of the issue, the consequence is that the local heterogeneities of porosity and velocity, as we can clearly see in Figure 4.4(b), are directly correlated with the same porous medium.

4 Matrix of the same size as the image whose coefficient has the value 0 or 1, depending on whether the pixel in the image corresponds to the part of a pore or to the part of a bead.

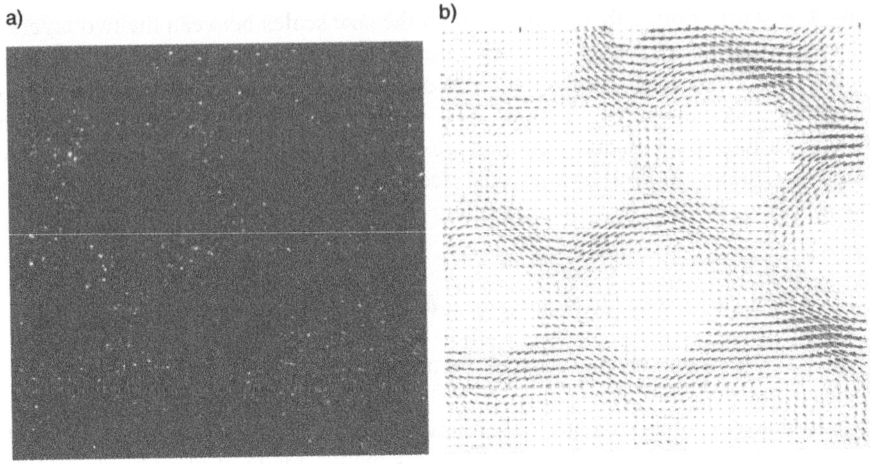

Figure 4.3. *(a) Typical image of tracers in the pore space and (b) 2D velocity field of the liquid flow obtained by means of PIV processing and time averaging*

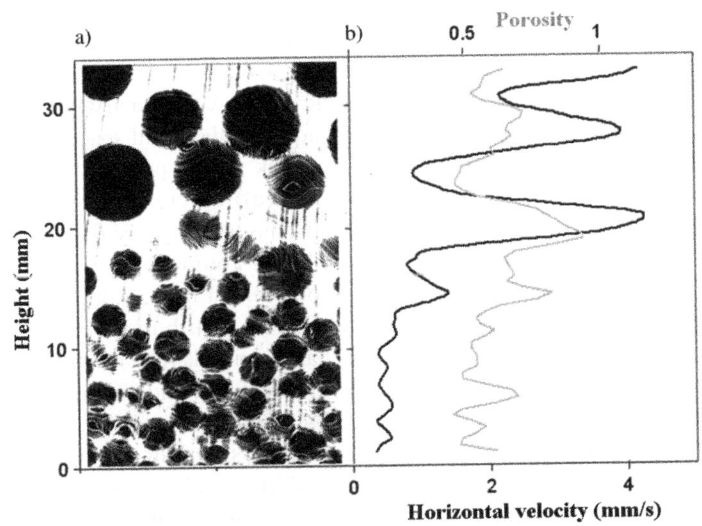

Figure 4.4. *(a) Positions of the balls at the interface and (b) corresponding profiles of the horizontal velocity of the liquid and that of the porosity (spatial averages per width)*

4.3.3. *Flow regimes and bottom shear stress*

Studying surface erosion of a soil requires a precise evaluation of the efforts induced by the fluid outflow at the surface of the soil layer. By limiting ourselves to the aspects related to the fluid, it is, first, convenient to define the hydraulic regime by introducing the associated Reynolds number, $Re = UL/\nu$, where U and L are, respectively, the characteristic velocity and length of the flow, and ν is the kinematic viscosity of the fluid. A flow with a small Reynolds number corresponds to the viscous regime where the diffusive transport by means of viscosity forces is widely spread. Such flows are observed at a low velocity, in small size systems and/or for liquids with high viscosity. In practice, this regime can never be found in surface flows where the velocities are relatively high $Re \gg 1$. These surface hydraulic flows are generally turbulent, characterized by strong temporal fluctuations in velocity. When the velocities are moderate, we can, however, notice a laminar regime where the outflow remains stable and may be viewed as a simple sliding between layers of fluid without mixing at larger scale. The transition between the two regimes depends to a large degree on the outflow conditions (geometry, entry conditions, rugosity of the solid surface, etc.) and may appear on a wide range of Reynolds numbers.

In the context of surface erosion where the hydraulic situation is a turbulent free surface flow, as described in more detail in Chapter 7, the hydrodynamic near the interface between the soil and the water can be mainly divided into two layers whose characteristics are discussed below [TRI 88]:

– *A viscous sublayer* that is immediately close to the interface where the tangential shear stress exerted by the outflow on the wall is constant and of purely viscous origin. It is noted $\tau_0 = \rho_w U^{*2}$, where we define a friction velocity U^* whose typical values are of the order of centimeters per second. The existence of this sublayer is conditioned by a low rugosity of the interface. Should this not be the case, the rugosity is important enough to generate turbulence in the interface.

– *An upper boundary layer* where viscosity forces very quickly become completely negligible. The flow is turbulent while at the same time remaining influenced by the interface that limits the maximal size of eddies that can exist within the flow. Several measures have shown that the turbulent shear stress still remained constant in this layer, equal to the bottom shear stress τ_0 that was exerted at the wall. This area is called the logarithmic turbulent boundary layer because the mixing length theory by Prandtl allows us to predict a logarithmic profile of velocity.

In the case of CE, the hydrodynamic is very strongly modified since the outflow is made through the granular matrix whose characteristic size of pores d_{pore} is a fraction of the average size D of grains. The associated Reynolds number, $Re_D = UD/\nu$, has typical values that are quite low, of the order from 10 to 100, for a characteristic velocity U estimated from the average flow that traverses the granular layer. It corresponds to the so-called Forchheimer regime, and significantly differs from the Darcy regime (laminar and stationary flow) characterized by a proportionality between the head loss and the average flow rate [BEA 88, HLU 06].

Beyond a Reynolds number of 10, a linear relation between these two quantities is no longer measured, and Forchheimer was the first one who proposed to add to the load loss a quadratic contribution in velocity, which accounts empirically for the inertial terms that correspond to non-regular head losses imposed by the presence of constrictions and cavities in the porous medium [BEA 88, HLU 06]. However, this visco-inertial outflow remains laminar and stationary in a first instance. By increasing the Reynolds number, the outflow becomes non-stationary and progressively transits toward turbulence, characterized by the emergence and development of instabilities. The turbulent regime is finally achieved for Reynolds numbers of the order of several hundreds [HLU 06]. The Forchheimer regime encompasses the nonlinear laminar regime (stationary and non-stationary) as well as the transition regime. Several empirical laws have been proposed to deal with this nonlinear dependency between head loss and average velocity [BEA 88]. We will only focus here on Fand relation, which is an improvement of the older Ergun relation, the latter is valid only for the Reynolds numbers that go as high as 120 and is well adapted to the case of quasi-spherical grains with an almost uniform size distribution [FAN 87]:

$$i \approx 182\frac{(1-n)^2 \nu}{n^3 g D^2}U + 1.92\frac{(1-n)}{n^3 g D}U^2 \qquad [4.1]$$

where g is the gravity and D is the average size of grains that constitute the porous matrix with a global porosity n. The hydraulic gradient i is a dimensionless number built as the ratio between the gradient of hydrodynamic pressure generated by the outflow and the specific weight of the water. The average shear stress in the porous medium may thus be deduced from the hydraulic gradient, as it will be discussed in section 4.4.

The spatial velocity fluctuations play a major role since the existence of high maximal values suffices to activate erosion locally and intermittently. The velocity distributions and the stress distributions discussed in section 4.3.2 agree with several studies, numerical and experimental, and show that the distribution of longitudinal velocities (i.e. in the outflow direction) is characterized by a non-centered shape or exponential with a remarkable contribution to high velocities (Figure 4.5). This aspect, compatible with the lognormal distributions, can be observed just as well in the Darcy regime as in the Forchheimer regime, at least in stationary condition. This distribution is imposed by the spatial variability of the sections offered locally to the outflow and depends on the geometrical structure of the layer. The non-centered and large tail shape of the distributions seems to be interpretable by the presence in the coarser pores of a viscous sub-layer of variable thickness and of a central flow of inertial nature that generates higher velocities [JOH 00]. To conclude, there is a specificity characteristic of a hydraulic outflow in a porous medium both in the Darcy and Forchheimer regimes that is revealed both by the strongly non-centred nature and the large tail of the distributions of longitudinal velocities and shear stresses.

The very strong grain size distribution contrast between the two layers engenders a discontinuity in permeabilities, particularly when one of the layers is constituted of a very fine soil and, because of that, is not very permeable. This difference in the spatial scales poses difficulties for the description of the present media: we will prefer an approach of the type of continuous medium for the layer of fine soil, while the granular layer will not be described as a homogeneous porous medium as soon as we will try to locally characterize the hydraulic outflow in the transition area between the two layers. We must then provide a description in terms of discrete medium for the granular layer to account for very strong local heterogeneities in the porosity. The latter directly impose significant spatial variations of velocity because of the constriction passages that separate the poral cavities from the very tortuous corresponding streamlines. The porosity and velocity profiles presented in Figure 4.4(b) clearly illustrate these spatial heterogeneities that are linked to the discrete character of the contact zone between the two layers.

The hydrodynamic in the contact zone between the fine layer and the granular layer is mainly characterized by its spatial heterogeneity, and it can no longer be described by a single average approach. The structure of the outflow in boundary layers can only be extended to the scale of each of the pores locally delineated by the fine soil and the first layer of coarse grains. More specifically, the presence of a viscous sublayer underneath the surface of the fine layer depends on the average size of the particles that constitute it and of the rugosity that may develop at the interface during the process of erosion. It is therefore difficult to link the average shear stress locally induced to the surface of the fine soil to the macroscopic measurable quantities such as pressure gradient or total flow rate. Furthermore, the hollows of the porous, such as erosion medium, induce very strong spatial fluctuations in the velocity, with a strong dissymmetry in the distribution of the longitudinal component of the velocity and a non-negligible probability for high velocities to emerge. This poses the question of the ability of an average measurable quantity to account for behaviors that are mostly generated by extreme values.

4.3.4. *Erosion threshold of the particles of fine soil*

The effort made by the fluid shearing on the fine particles modifies the local mechanical balance. We must thus quantify and determine the threshold beyond which this balance is broken. This calculation needs to exploit the characteristics of the outflow detailed in the previous chapter to determine the hydraulic stress exerted on the scale of the elementary entity of fine soil that is susceptible to destabilization.

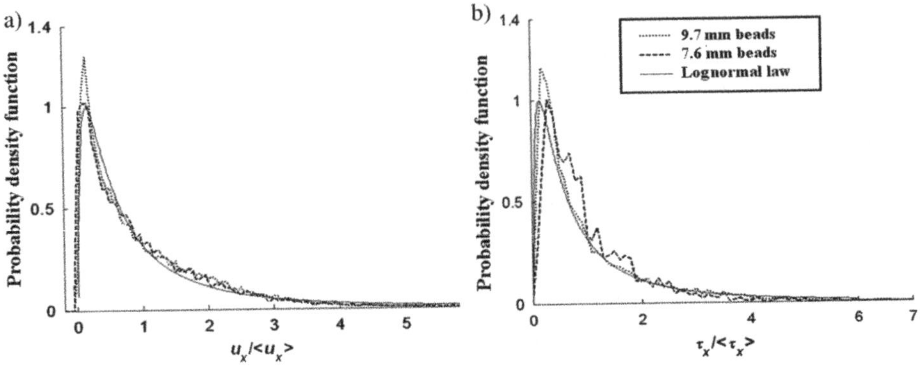

Figure 4.5. *a) Distribution of the longitudinal velocities of the liquid, and (b) distribution of the longitudinal shear stresses at the interface (experimental data of section 4.3.2)*

The approach developed by Shields [SHI 36] consists of defining a dimensionless number, called Shields number, which compares the tangential stress applied to the saturated weight of the fine soil at the interface:

$$\Theta = \tau_0 / (\Delta \rho g d) \qquad [4.2]$$

For the classic scenario of turbulent free surface flows, the Shields number takes on the expression $\Theta = U^{*2} / [(s-1)gd]$, where s is the relative density of particles in relation to the fluid density. The diagram that was experimentally obtained by Shields for cohesionless soils with a uniform grain size distribution links the critical value Θ_c measured at the erosion threshold to the particle Reynolds number $Re^* = U^* d/\nu$. The threshold Θ_c depends on the grain size distribution of the soil and of the outflow regime (see Chapters 7 and 8).

The same approach is not directly applicable in the case of a cohesive soil for which the expression of cohesion terms still remains partially unknown. It appears directly that for the dominant cohesive terms of gravity, the Shields number is no longer suitable: we notice an increase in Θ_c as soon as cohesion becomes significant, i.e. in practice, when the characteristic size of particles decreases. This could be physically interpreted as a modification of the scale of the characteristic length on which erosion takes place, which goes progressively from the size of the particles, which corresponds to the process of individual erosion, to the one of an aggregate, containing an increasing number of particles and setting into motion very different mechanisms when the particles are being pulled out by means of soil fragmentation. Several approaches consist of adding to the simplifying hypotheses, in an empirical fashion by means of analytical calculations, a contribution of the cohesion forces to the stabilizing terms of gravity and allowing them to obtain a more general expression of the Shields criterion (section 4.4) [CLA 06, TER 08].

However, these improvements remain limited to the framework of an erosion mechanism on an elementary scale of the particles and not by fragmentation as observed for a very cohesive soil. In the field of geomechanics, the resistance to the shear of these coherent soils is usually measured with experimental devices such as the "vane shear stress" and the "fall cone", i.e. a penetrometer with a conical end. Several research works have sought to link this value to the critical shearing stress τ_c, deduced from the hydraulic parameters of an erosion test. However, the shear stresses obtained in soil mechanics is of the kilo-Pascal order, whereas those identified during the erosion test are of the Pascal order. A possible explanation could be that these tests measure the overall value of soil resistance, whereas the process of erosion is limited to the pulling out of the particles at the surface, particles that are slightly linked to the matrix of the soil [LEO 04]. Consequently, in the actual state of the knowledge on the subject, the characterization of the erosion threshold of cohesive soils must go through experimental tests that allow us to measure directly the hydraulic parameters of the outflow that initiates erosion.

Besides cohesion, there are other parameters that intervene. For example, the density of the fine soil, although quite slight in the case of fluvial erosion because of the fine particles deposited via sedimentation, is higher consequently in the case of CE in a hydraulic structure, where the materials are generally compacted when they are put into place. We may expect to notice in a real situation that the erosion thresholds are slightly higher and that the kinetic is less intense [ROB 98]. The shape of particles as well as the grain size distribution plays an equally important role. In particular, the choice of a characteristic size for the particles is not obvious if we seek that it is representative of the properties of the soil. Numerous propositions exist in relevant literature, based on physical or pure empirical arguments (section 4.4). Grain size distribution can also be considered in a statistical approach, by way of introducing a distribution of the erosion threshold, which follows more or less directly from size distribution. One last parameter that influences the erosion threshold as well is the local slope of the soil which, for a fine granular soil, favors or limits the pullout of the particles, depending on the direction of the slope.

In the specific framework of CE, erosion emerges within the pores created by the coarse grains coming into contact with the fine soil. The surface of the soil exposed to the outflow, and thus to erosion, is only a fraction of the total surface. This fraction will depend on the density of the grains in contact and their degree of burial, connected to the way in which they are placed and on the weight of all the overlying layers. This limitation of the erodible surface, associated with an average overload on the local contact interfaces, produces a stabilizing effect with regard to the erosion. The erosion of fine soil is initiated as soon as the maximal velocities close to the interface are high enough to impose a local shear that exceeds the pullout threshold. However, as we have seen in section 4.3.2, the distributions of velocities and tangential shear stress in a porous medium are not symmetrical, and they are very widespread toward highest values. We may thus suppose that, for the same average outflow velocity, the maximum values of the obtained velocities in the porous granular layer will be higher than in the case of the outflow of a pure fluid. The fine soil could thus be eroded eventually, if it is in contact with the granular layer, whereas it would keep its integrity when in contact with a homogeneous outflow in identical macroscopic hydrodynamic conditions. However, this is not the case since the flow is in fact slowed down near the surface within a transition zone.

4.3.5. *Kinetics of erosion and transport of the fine particles*

Once the erosion has started because of an outflow that significantly exceeds the pullout threshold, the particles of fine soil are regularly entrained and the soil is progressively hollowed. The transport of these particles will depend on their size and will be made either by means of bedload transport or through transport by suspension.

Indeed, the particles, which are coarse enough and whose sedimentation velocity is elevated enough, displace themselves, once pulled out, by rolling on the surface of fine soil with velocities that are well below the velocities of the fluid. Beyond the erosion threshold, a large number of eroded particles are in motion and a soil thickness that can find itself in motion, on several sheared layers between a superficial entrainment velocity that is induced through the action of the hydraulic outflow and a null velocity at a larger depth. This mode of transport is called bedload transport and several empirical laws have been advanced with the aim of linking the mass flow q to the hydraulic stresses that are exerted within the classic framework of surface erosion. Laws of the same type, yet specific to CE, will be discussed in section 4.4. We should equally note that the conjunction between erosion and the bedload transport may engender a sorting in the grain size distributions of the particles of fine soil. Indeed, the coarser particles remain located at the interface, while the finest particles are either entrained in suspension or gradually attracted by the gravity toward the inferior layers of the soil. A superficial paving is thus created with the coarser particles, which are more difficult to erode. This paving protects the soil layer underneath and that moderates the kinetics of erosion. In the specific case of CE, the presence of the coarse layer of soil only adds, in practice, obstacles on the trajectory of the transported particles. This increases the possibilities of particle deposit on the surface of coarse grains by means of bonding, or by means of geometrical capturing in the constrictions or by hydrodynamic capturing in the stagnant areas of the outflow. The experiments discussed in section 4.3.2 offer the possibility to visualize the contact area throughout the layer of beads placed above a layer of sand. The bedload presents a strong spatial heterogeneity (Figure 4.6), with bedload much less mobile depending on the time. The deposit areas are also located where the outflow is slower without going as far as developing structures as definite as ripples. Furthermore, we have noticed that certain particles that must have been previously ejected far enough from the interface when they were first pulled out, were transported to the interior of the granular layer, following preferential paths of the outflow, without re-descending back to the bedload layer.

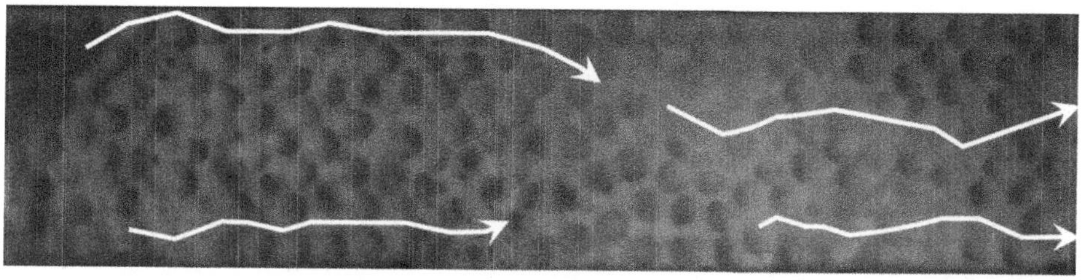

Figure 4.6. *Surface of a sand layer subjected to erosion observed from above through a layer of index-matched beads, localization of preferential paths (white arrows) for bedload transport of sand*

In the case of a cohesive soil, the particles are *a priori* eroded by soil fragments. If the cohesion forces are extremely strong, such a fragment may come to keep itself in a floc and behave almost as a particle of an equal size. However, these fragments are generally very weak and will almost instantly dislocate by creating a cloud of fine particles. These fine particles have negligible sedimentation velocities when compared to the outflow velocity, so that the transport is made exclusively through suspension with a concentration diluted enough to reduce any risk of clogging in the layer of granular soil. We expect only a light deposit of fine particles on the surface of the pores, through bonding. Clogging can, however, take place somewhere else in the interior of the structure if finer grain size distributions, associated with the slowing down of the outflow, are encountered. If this is not the case, the particles are transported up to the exit of it and they will deposit outside the structure.

To specifically quantify the process of erosion, one of the commonest approaches consists of supposing a linear relation between the erosion rate ε (kg/s/m^2) and the excess of the hydraulic shear stress applied to the soil in relation to the stress threshold τ_c, mentioned previously: $\varepsilon = k_{er}(\tau - \tau_c)$, where k_{er} (s/m) is the erosion coefficient that characterizes the kinetic of the process and thus the sensitivity of the soil to erosion [PAR 65]. This linear relation therefore does not have a physical justification, but it accounts for the erosion measurements taken in several configurations. Other authors have proposed higher exponents for this relation, of the type $\varepsilon = k_{er}(\tau - \tau_c)^b$, which enables us to adjust more faithfully certain measurements for slight shear stresses [KNA 07]. However, the linear relation correctly represents the majority of the experimental results and has the advantage of simplicity, which, all in all, makes this the most widely used relation.

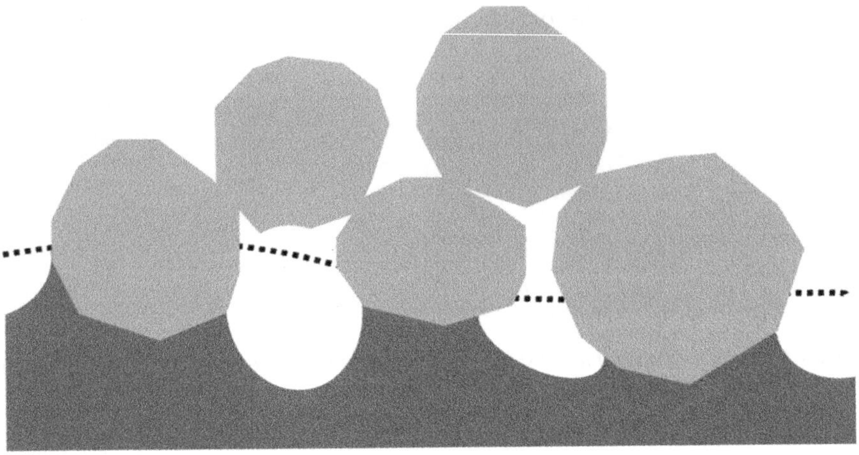

Figure 4.7. *Schema that illustrates the progressive hollowing in fine soil (represented in dark gray) of the erosion basins at the contact of porous cavities between the coarse grains (light gray). The dotted line corresponds to the initial layer of the surface of fine soil. In this example, the stability of the central grain is hardly ensured anymore*

A final point that needs discussion regards the development of these processes of transport. Indeed, for the classic surface erosion, the soil surface can remain relatively flat and hydrodynamic conditions at the interface evolve very little over time. For CE, we may well understand that the only areas where the fine soil neighbors large size pores are subject to erosion, while those underneath the grains remain protected. Therefore, as is illustrated in Figure 4.7, the fine soil is gradually hollowed at the level of the pores, which results in the modification of the local geometry of the contact zone between the two layers and of significantly changing the hydrodynamic of that area. This hollowing effect is particularly strong in the case of a cohesive soil where erosion basins may form. Supposing there is a constant local outflow, the velocity should decrease, resulting in a moderating effect *vis-à-vis* the erosion process. However, as we have seen in Figure 4.4(b), the local velocity is strongly correlated to the porosity of the medium and the increase in porosity induced by the hollowing of a basin, rather risks accelerating the local kinetic of erosion.

The subsequent evolution will depend on the relative position of soil layers:

If the fine soil is above the granular soil, a gap forms gradually between the two soils and carries almost all the water flow. The balance of the superior layer of fine soil depends on its level of cohesion and we must expect to notice the massive destabilizations of entire pans of soil whose frequency and size will vary considerably depending on the nature of the soil, its degree of consolidation, and the effective stresses that apply. For a fine yet granular soil, the situation is not really one of CE, but could rather resemble the sedimentation in a porous medium, which exceeds the scope of this chapter.

If the fine soil is below the granular soil, the hollowing of erosion basins entrains the unbalance of grains, which causes rearrangements and localized settlements. As a result of one of these events, the contact surface in the area concerned goes back to a position closer to its initial position, before erosion, since the basins are filled with grains. However, even if locally, erosion presents a fluctuating temporal evolution of this kind, overall behavior integrates these temporal variations on the entire contact surface and can reach quite a stationary regime and eventually sinkholes. Finally, it is worth noting that these destabilizations, localized in the contact area between the soil layers, can have repercussions on the scale of the structure by inducing in every superior part a progressive settlement of the entire facility (see section 4.7).

4.4. CE of a fine soil under a coarse soil

The configuration of a fine soil under a coarse soil was historically the most used for the study of CE, most certainly because it is more easy to carry out. The first research works carried out on CE use a very pragmatic approach, consisting of identifying a hydraulic structure in the interfaces of the soils susceptible to CE, and then taking samples to recreate the interface in the laboratory, in an intact or reworked manner, in a testing device such as the one represented in Figure 4.8. An outflow is generated in the sample, in the interior of the coarse soil layer, generally through the intermediary of a constant hydraulic head that is progressively increased up to the initiation of erosion. This method allows a quick response to the sensitivity of the interface tested for CE. In the case of an affirmative response, the minimal hydraulic stress, or "critical stress", initiating the process of erosion can be estimated with fine precision. From an operational point of view, it is sufficient for the engineers in charge of the safety of the work to guarantee that the hydraulic outflow, of the considered interface area, does not exceed these critical conditions that are susceptible of initiating CE.

At the research level, numerous authors have used experimental devices, similar to the one in Figure 4.8, to study the sensitivity to CE of an interface made of a fine soil underneath a coarse soil, depending on the characteristics of the two materials present. These works have allowed us to emphasize the conditions for the initiation of CE of a geometrical origin and/or of a hydraulic origin, depending on the grain size distributions of the soils, as it is represented in Figure 4.9. The geometrical origin is directly linked to the filtering criteria (see Chapter 3) that determine, between two soil layers, the maximal pore opening of the coarse layer, thus preventing the transport of any particles from the fine layer through the filter, regardless of the hydraulic outflow.

On the contrary, if the pores are large enough in the coarse soil, it is the hydraulic force applied on the layer of fine soil that will (or will not) generate soil erosion and the transport of particles. The intermediary situations require simultaneously a double condition, geometrical and hydraulic, to be able to prevent CE. Several empirical relations were put forward [BRA 85, BEZ 87] for linking these conditions to the constituting parameters of the soils in contact, such as grain size distribution and porosity. Other authors [WOR 92, DEN 94] have looked to model CE on the basis of the physical process that is at stake; they have also looked to quantify, in terms of volume or mass of eroded soil per time unit, the development of CE. These models, besides bringing information on the erosion initiation threshold, inform us about the kinetics of the phenomenon. However, the majority of the research have considered the erosion of fine, sand-like soils, with particles having diameters comprised between 80 μm and 1 mm, whereas little research was carried out on silt-like soils or clay soils, which, as we have seen in section 4.3, display very different behaviors with regard to the erosion.

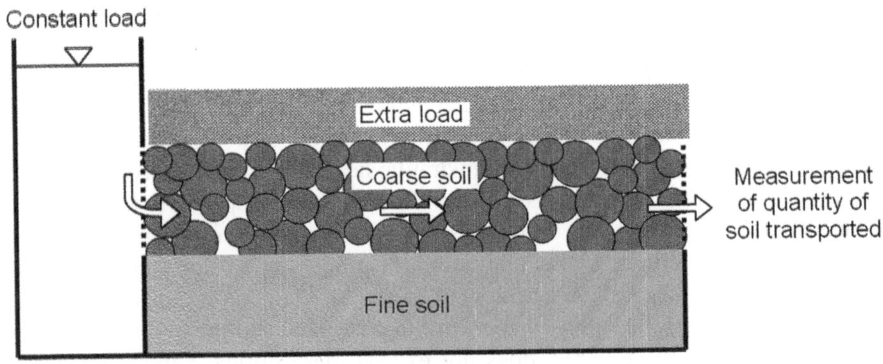

Figure 4.8. *Typical experimental device that enables the study of CE*

	Grain size ratio D_{15}/d_{85} →				
Braun (1985) with soil porosity $n = 0.4$	Geometric criterion	7.5	Geometric AND hydraulic criteria	25	Hydraulic criterion
Wörman (1992) for a soil with $D_{15} = 0.88D_H$				14.6	
Den Adel (1994) for a soil with $d_{85} = d_{50}/0.9$		8.1		11.7	
Sherard (1984)		4 or 9			

Figure 4.9. *Summary of the boundaries defined by different authors regarding the geometrical influence of the grain size distribution of the two layers on the process of CE*

4.4.1. *Determination of the threshold of CE initiation*

The CE trials that we have cited in relevant literature have all followed a similar experimental global protocol. The layer of fine, more or less compacted, soil is placed at the bottom of the device, before placing the coarse layer on top of it. An overload possibly is then applied by layers of clay or ballast, or by a membrane containing a fluid under pressure, the latter having the advantage of enabling us to impose a normal uniform stress on the surface of the coarse soil. The application of this overload enables us to bring together the true stresses that exist in hydraulic structures. The sample is saturated with water, and a constant hydraulic head is applied upstream. The layer of coarse soil, being by far more permeable, drains the majority of the outflow. It is thus necessary to determine if the layer of fine soil suffers an erosion, and eventually to quantify that erosion. If the device is equipped with a small window on a lateral wall, we may visually observe the displacement of the fine particles, but this observation is limited to the wall, which is an area influenced by the boundary effect (Figure 4.10). Unfortunately, it is thus impossible to observe the interface in its entirety throughout the experiment.

We then proceed to evaluate the quantity of soil particles that were transported by the outflow at the exit of the device. This is an important point: what we measure is the quantity of soil that is transported through the exit, and this quantity integrates all the erosion, deposit, and clogging phenomena that might take place in the sample. To carry out this measurement, we may use, in the case of a sand-like soil, a decantation tank or a sieve. If the fine soil that is tested is not granular, the low sedimentation velocity of the particles does not allow us to use decantation efficiently. In this case, we have used a turbidity meter that allows us to estimate through a previous calibration the concentration of the particles of the effluent. This method also has the advantage that it enables measurements to be taken while the particles are being transported. The applied hydraulic head is maintained for a different duration according to the authors, ranging from several minutes to 24 hours. We may thus characterize the behavior of the sample in relation to short-term CE and the extrapolation of these results in the long term must be made with caution. At each step, the identification of the appearance of erosion is carried out in a different way according to each author. This could be limited to the visual observation of a presence of particles at the exit [BRA 85], or to the measuring of an arbitrary transport value [BEZ 87], or yet to the existence of a non-null turbidity after 30 minutes of stationary outflow [GUI 10]. We may note that the notion of an erosion threshold is relatively ambiguous and certain authors even consider that there is no such thing as an erosion threshold, and that very small, but non-null, transport exists as soon as an outflow is generated [WOR 92].

Figure 4.10. *Interface between a gravel and a sand-like silt, before and after CE. The outflow is directed from right to left*

If no erosion is identified at the maximum of the hydraulic head capacities of the device, we may conclude that the geometrical conditions that generate CE are not met. The boundary limits proposed by the authors are only valid up to the maximum tested head. These conditions are summed up in Figure 4.9, and they are coherent with the geometrical filter criteria, established for filtration problems (detailed in Chapter 3). Starting from the grain size distribution contrast expressed by the term D_{15}/d_{85} comprising between 4 and 9, CE is geometrically possible and will depend at the same time on the geometrical conditions as well as on hydraulic conditions. The first research works carried out on CE are those undertaken in Russia by Istomina [IST 57]. Their results give a geometrical criterion that is different because it is expressed according to d_{10} of the fine soil, under the form: $D_{10}/d_{10} < 10$. A transition zone was thus identified for the geometrical conditions that allow for erosion to form, but with $D_{15}/d_{85} < 11.7$, 14.6, or 25 depending on the author (Figure 4.9): the size of the constrictions of coarse soil is thus just enough to allow particles to go through the fine soil, but these constrictions influence the process by capturing some particles, for example through friction or shock on the coarse grains. In this transition zone, the threshold is influenced at the same time by the geometrical conditions as well as hydraulic conditions, which makes its estimation complex. The laws proposed in the relevant literature do not generally apply to this field.

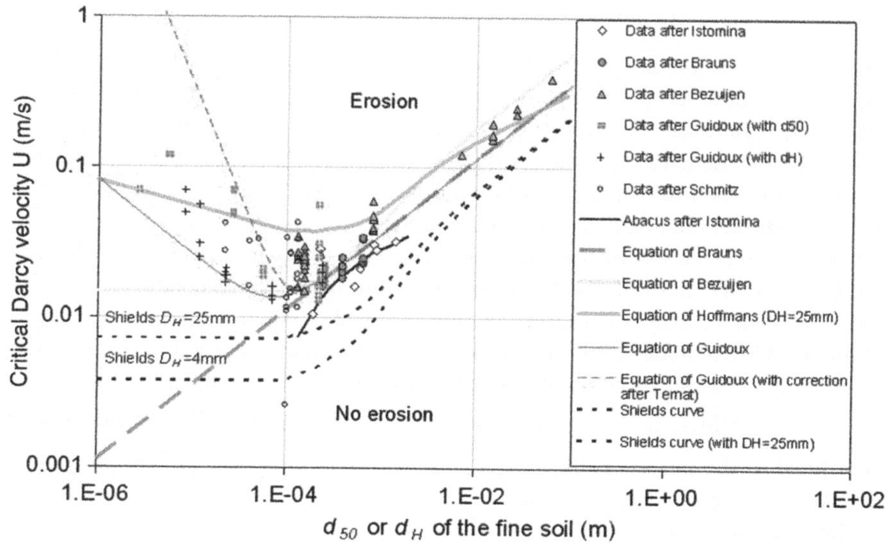

Figure 4.11. *Summary of the current research works concerning the initiation threshold of CE in the case of cohesive and non-cohesive fine soils underneath a coarse soil*

Starting from a large enough grain size distribution contrast, the conditions for the initiation of erosion are thus purely hydraulic. In this field, several authors [IST 57, BRA 85] note that, in the conditions of the initiation of CE, the so-called "critical" gradient depends at the same time on the fine soil and on the coarse soil whereas the average outflow velocity in the coarse soil is, at first approximation, independent of the grain size distribution of the coarse soil. The hydraulic threshold for the initiation of CE is expressed as a "critical" velocity and thus depends only on the grain size distribution characteristics of the fine soil. To establish this critical velocity, Istomina proposed an abacus based on the experimental results (Figure 4.11). After 30 years, Brauns noted that this critical velocity corresponds to a densimetric Froude number[5] that is practically constant, comprised between 0.65 and 0.7 [BRA 85]:

$$Fr^* = \frac{U_{\text{crit}} / n_D}{\sqrt{(s-1)gd_{50}}} = 0.65 \text{ to } 0.7 \qquad [4.3]$$

This relation [4.3] allows us to provide the following expression of critical velocity:

$$U_{\text{crit}} = Fr^* n_D \sqrt{(s-1)gd_{50}} \qquad [4.4]$$

We can thus note that the erosion threshold expressed in velocity depends on the critical Froude number, on the porosity of the coarse soil, and on the relative density of particles, parameters that do not vary a lot during the process of erosion, and on the diameter of the particles of fine soil d_{50}. In the Netherlands, CE was examined by several authors each having a different approach. Bezuijen proposes to interpret the erosion thresholds obtained by using the Shields criterion given by the relation [4.2] [BEZ 87, DEG 83]. He considers that the processes at play at the moment where CE is initiated can be assimilated to those observed for a free surface flow and he assumes that the critical shear stress for the fine soil is equal to the stress proposed by Shields. To calculate the shear stress within the coarse-grained layer according to the average measurable quantities, he uses, always by analogy to the case of free surface outflow, the friction velocity $U^* = \sqrt{\tau_o / \rho_w}$ that he considers to be proportional to the pore velocity U_p ($U_p = U/n_D$). The hypothesis of proportionality introduces an empirical coefficient e, $U^* = eU_p$. Bezuijen makes here the implicit hypothesis of a turbulent regime in the pores and uses the average measurable quantities of the outflow without really considering the characteristics of an outflow in a porous medium described in section 4.3. This reasoning enables us to link the critical Froude number discovered by Brauns to the Shields parameter: $\text{Fr}^* = \Theta_c / e$. On the basis of experimental results, Bezuijen obtains the following relations for the coefficient e:

– if d_{50} is comprised between 0.1 and 0.2 mm, $e = 0.75(\text{Re}_D)^{-2}$, with $\text{Re}_D = \text{UD}_{15}/\nu$.

– if d_{50} is comprised between 0.7 and 1 mm, $e = 0.22$.

---

5 The Froude adimensional number mainly appears in free surface flows and it characterizes, in a fluid, the relative importance of the forces related to the velocity in relation to the gravitational forces.

Bezuijen thus notices a slight dependence *vis-à-vis* the grain size distribution of the coarse soil introduced by the Reynolds number Re_D, in the case where $0.1 < d_{50} < 0.2$ mm, but not in the case where $0.7 < d_{50} < 1$ mm. By completing its database with the experimental results obtained by Brauns [BRA 85], he proposes to linearly interpolate the values of e for the range $0.2 < d_{50} < 0.7$ mm. These research works apply to the slopes of certain shoreline dikes, which are very common in the Netherlands, built of sand and protected by a granular shell. This is why, on the one hand, a modified relation was advanced to account for potential inclination of the interface as well as the possibility to have a gradient perpendicular to the interface, and, on the other hand, why tests were carried out with outflows that have a stationary component as well as a cyclic component for stimulating the breaking of waves, without, however, noticing any significant difference in the critical erosion velocity [BEZ 87]. As a conclusion of this experimental campaign, a synthesis abacus was proposed, which enables the prediction of the initiation of CE according to all the investigated parameters.

This research work was more recently resumed by Hoffmans in a different context, the one of scour in river [HOF 08]. To prevent erosion at the bottom of the water flow, it is indeed common to put in place a layer of coarser granular material on the fine soil that was put in place. The significant velocities and the turbulences of the free surface flow are then toned down by this granular filter. This hydraulic stress, though decreased, may generate a CE at the interface between the fine soil and the granular filter. To characterize the initiation of CE in this context, Hoffmans has resumed the results previously obtained by Bezuijen [BEZ 87] and sought to improve the relation [4.4] by pushing a bit further the analogy with the free surface flows and by expressing the empirical parameter e according to the Chézy equation. The latter expresses the friction velocity according to the average velocity of the outflow, through the bias of a friction parameter C_f: $U = C_f / (g^{1/2}U^*)$. Hoffmans uses an empirical relation for the friction coefficient: $C_f = \alpha_{15} g^{1/2} (D_{15} / d_{50})^{1/6}$ with α_{15} being a dimensionless coefficient, estimated at 1.28 after the previous experimental results. The dependence of C_f to the grain size distributions of the coarse soil, introduced by the term $(D_{15} / d_{50})^{1/6}$, is different from the one introduced by the empirical coefficient e, proposed by Bezuijen. Furthermore, he proposes to consider the turbulence of the free surface flow with modifications that remain, however, appropriate in the context of scour in river.

The use of the Shields criterion (equation [4.2]) is an interesting approach and seems to give a coherent tendency in relation to experimental results. By conserving this criterion, we propose to calculate the shear stress in the granular medium without making analogies with free surface flows that were previously proposed. The shear stress may be directly calculated starting from the gradient in the porous medium by making a balance of the pressure forces between upstream and downstream. We can deduce from this the average shear stress in the sense of the outflow, applied to the fluid/solid interface [BEA 88, WOR 92]:

$$\tau_x = n_D \rho_w g i / A_S \qquad [4.5]$$

where A_S (m^{-1}) is the specific surface of the granular medium. It can be estimated after the grain size distribution curve of the soil: $A_S = \alpha_{\text{shape}}(1 - n_D) / D_H$ [KOZ 53], where α_{shape} is a grain shape factor, having a value ranging from 6 to 8 depending on the type of granulate; D_H is the effective diameter of the grain size distribution defined by $D_H = (\sum_j F_j / d_j)^{-1}$, with F_j being the fractions of the class of particles with an average diameter d_j. The specific surface A_s can also be deduced from the intrinsic permeability k that was experimentally measured and also from the Kozeny–Carmann formula: $A_S = \sqrt{c_0 n_D^3 / k}$, where c_0 is the Kozeny parameter, generally considered equal to 0.2.

Thus, we can assess the average shearing stress in the porous medium, exerted on the solid–liquid interface, generated by a gradient i. By considering that the critical shear stress for the fine soil is that of Shields, we can thus estimate a critical gradient after formula [4.5]. We deduce from this a critical velocity value by using the outflow law proposed by Fand in formula [4.1]. Thus, this calculation introduces a dependence in relation to the coarse soil grain size distribution. We may observe in Figure 4.11 that the critical velocity calculated in this way (Shields curves for the two values having an effective diameter D_H) is inferior to the experimental data, but it has a similar tendency. By averaging a multiplicity factor of about 2, we obtain a correct prediction of experimental values. This coefficient undoubtedly considers the difference between the average shear stress in the porous medium and the stress at the level of the interface between the two types of soil. This interface is indeed a grain size transition zone where the porosity can be different and has a bearing on the outflow. On the other hand, we have emphasized in section 4.3 that the shear stress was distributed according to a lognormal law, and it is highly likely that the erosion is uniquely produced for the extreme values of this distribution.

The different research works discussed in this section are related to predicting an initiation threshold of CE for non-cohesive soils on the basis of experimental results. Different laws are proposed, mainly coming from the Shields criterion applied to a fine soil. There is consensus on the choice of the average outflow velocity U as the macroscopic parameter for identifying the most adapted threshold. These laws differ in the manner in which they link this average velocity to the critical shear stress deduced from the Shields criterion. They all propose a similar dependence of the critical velocity U_{crit} on the size of particles (see relation [4.4]), but different dependences of U_{crit} on the grain size distribution of the coarse soil, which is a less influential parameter. The effective vertical stress that exists at the interface is not mentioned by any author as having an influence on CE, and our own research has shown that there are a few effects of this parameter in the range 10–200 kPa [GUI 10]. The area of fine soil subjected to erosion is situated between the grains of the coarse layer, and it is therefore not directly subjected to this vertical stress.

Figure 4.11 summarizes the set of experimental results and the proposed laws. Despite the differences existing between authors regarding the experimental protocols and the methods of detecting erosion initiation, we can observe quite a significant homogeneity of results. The set of these hydraulic thresholds is comprised between 1 and 10 cm/s for the fine soils with $d_{50} < 1$ mm, which represents the majority of the cases encountered. This spread of the erosion threshold over one single decade is finally quite insignificant for an engineer in relation to the uncertainty that can exist on the evaluation of on-site permeability, for example. Similarly, the proposed threshold laws give quite close critical velocities, and a dependence coherent with d_{50}, for the particles larger than 200 μm.

4.4.2. *Influence of cohesion*

Cohesive soils have not been considered in the relevant literature. We can only quote the work of Istomina [IST 57], which suggests a critical gradient comprised between 0.6 and 0.8 for these soils, with a saturation degree higher than 95%, and in the case where the minimum diameter of the pores of the coarse soil is lower than 3 mm, and the works of Pravedny, cited by Istomina [IST 57], which proposes an abacus. Cohesive soils are characterized by the existence of a binding force that appears between the particles typically smaller than 40 μm. A small percentage of these fine particles in a granular soil can create bonds between the coarser particles and give a global cohesive character to the soil. To characterize these cohesive soils with a widespread grain size distribution, the d_{50} does not seem fully adapted. Guidoux [GUI 10] advises using the effective diameter d_H defined by Kozeny [KOZ 53] for estimating the permeability of a soil. This diameter, with the hypothesis of spherical particles, allows us to preserve the specific surface of the initial grain size distribution. The binding forces are mainly linked to surface properties; this choice seems to be more suitable for the fine soils with a widespread grain size distribution. These soils are eroded rather by means of floc detachment, and being able to reach up to several millimeters in diameter (see section 4.3), the Shields criterion given by equation [4.2] (balance for a particle between gravity and the hydrodynamic force) is no longer adequate for these soils.

However, for the soils whose cohesion remains weak, the binding forces between the particles marked $F_{binding}$ can be added in the expression of the Shields number (equation [4.2]) to the stabilizing forces of gravitational origin $F_{\text{gravitational}}$. A corrective term that takes into account these bonds is then added to the Shields criterion as follows [CLA 06, TER 08, GUI 10]:

$$\Theta_{\text{with bond}} = \Theta_{\text{without bond}} \left(1 + \frac{F_{\text{binding}}}{F_{\text{gravitational}}} \right) \qquad [4.6]$$

The critical velocity calculated, for example, according to the Brauns hypothesis thus becomes:

$$U_{\text{crit}} = Fr_{\text{crit}} n_D \sqrt{(s-1) g d_{50}} \sqrt{1 + \frac{F_{\text{binding}}}{F_{\text{gravitational}}}} \qquad [4.7]$$

The binding forces between the particles can have diverse origins. The theoretical determination of the term F_{binding} seems particularly difficult at the moment. Guidoux assumes that this term is proportional to the particle diameter d; the term $F_{\text{gravitational}}$ is proportional to d^3; $F/F_{\text{gravitational}}$ takes, in this case, the form $\beta_{\text{binding}}/d^2$, where β_{binding} is an empirical adjustment coefficient that takes into account the numerous influences previously mentioned. Other hypotheses can be made for the binding force term, as a force proportional to $1/d$ [TER 08]. This corrective term causes a significant increase in the theoretical velocity for erosion initiation in the finer particles, which seems to better correspond to the experimental results (Figure 4.11).

However, the binding forces are not solely dependent on the diameter of the particles d, but they also depend on the mineralogy of the presented clays, on the chemical activity of the medium, and on other parameters. On the other hand, the erosion of these soils is made mainly through aggregates, and the element that must be considered is no longer the isolated particle but rather this entire floc of particles of a much larger size. The corrective term that we have previously mentioned thus allows us to respect the observed tendency of the experimental results for the not-so-cohesive soils, but it cannot be expanded to very cohesive soils. When cohesion becomes predominant, the first feasible approach consists of using the geotechnical characteristics of the soil, such as non-drained cohesion and the internal friction angle, determined by standardized tests of soil mechanics. However, as we have seen in section 4.3, it is very difficult to relate these parameters to the erodibility of specific soils. Nowadays, it seems therefore necessary to determine the sensitivity to erosion of a cohesive soil directly by means of a CE test.

4.4.3. *CE kinetics*

Contrary to previous studies, our experimental device has not only enabled us to study at the same time granular and cohesive soils, but also allowed us to follow the evolution of the quantity of transported soil on an ongoing basis. The results obtained show that erosion evolves over time. In Figure 4.12, we can observe peaks of turbidity, more or less visible, after each change in the outflow velocity. These peaks are explained by a washing out of the particles that are most exposed to the outflow: the most unstable and the most easily transportable. This initial transport flow is a classic phenomenon in surface erosion. After a certain value of the outflow velocity, the turbidity increases significantly and very sharply before decreasing only slightly with time. The hydraulic conditions are then met for the development of erosion. If the outflow velocity is increased again, turbidity becomes high again. There is, then, a certain relation between the hydraulic head and the quantity of transported soil that may allow us to characterize the development velocity of erosion if the hydraulic conditions are known. This kinetics is generally characterized by an erosion rate ε, expressed as the mass of eroded soil per surface unit and per time unit (kg/s/m^2), or as eroded volume per surface unit (m/s) that corresponds to an erosion depth. This data can allow us, for example, to estimate the urgency of a repairing or reinforcement measure in a hydraulic structure when CE is suspected.

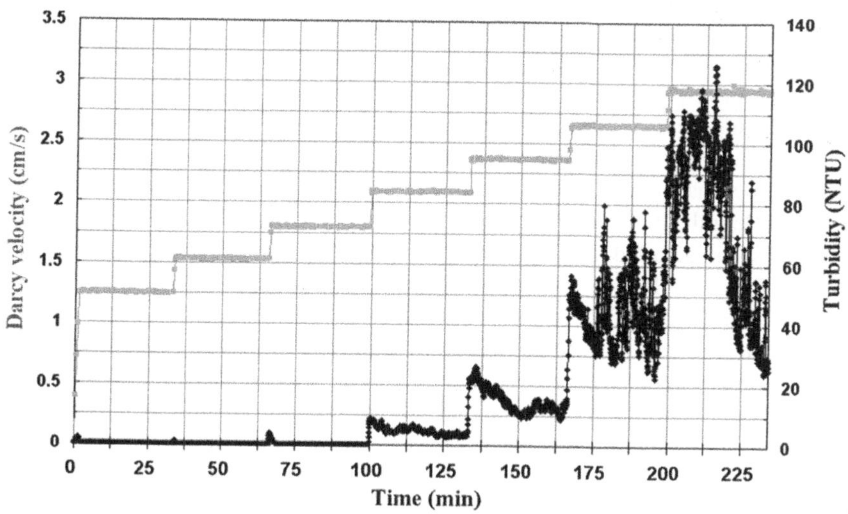

Figure 4.12. *Typical experimental result of turbidity and outflow velocity throughout time during a CE testing*

This turbidity measurement carried out at the exit of the device allows us to characterize the transport of particles for a given length of the eroded interface, along the flow direction. But it does not provide us directly with the erosion rate as defined previously. Indeed, an eroded particle must eventually be transported throughout the poral network of the coarse soil, and it can sediment, be trapped in a constriction, and then be reset into motion. Measuring the transport flux at the exit thus includes all these phenomena. These effects are even more conspicuous as the ratio between the diameter of the coarse soil particles and those of fine soil is small, and even more so when the sedimentation velocity is higher than the outflow velocity.

The finest particles, smaller than 80 μm, are generally transported in suspension in the outflow. If the outflow is quick enough, and the geometrical effects are limited, we can assume that the measured quantity of transported particles at the exit provides a good estimation of the erosion rate in the expression $\varepsilon = q / A_e$, with q being the quantity of transported soil measured at the exit of the device and A_e the interface surface subjected to CE. This surface, which is the fraction of the fine soil surface actually in contact with the fluid, can be estimated due to the porosity: $A_e = n_D A_{\text{interface}}$. This estimation is based on the assumption that the sedimentation and the capturing of the particles are limited and that the erosion over the entire surface is homogeneous.

The coarser particles, larger than 200 μm, have high sedimentation velocities and they are generally transported by means of bedload at the level of the interface (section 4.3). The presence of this mobile sheet having a few particles in thickness, called bedload sheet, inhibits the erosion. The quantity of transported soil in an interface point thus increases progressively in the direction of the outflow, until it reaches a maximal quantity for a certain eroded length, called "saturation length". Above this length, the transported quantity does not increase anymore [CLA 06]. This process is detailed in Chapter 7. In the case of CE, several authors have estimated that this length would be in the order of ten coarse grain diameters [DEN 94]. The length of the experimental device must be higher than the saturation length to determine the maximum transport capacity. For these coarse particles, the kinetics is thus characterized by a transport law expressed in terms of transport capacity q (kg/s/m), per width unit and not through an erosion law, expressed in erosion rates per surface unit. As a consequence of this saturation phenomenon, erosion will generally develop on the upstream of the interface that is exposed to the outflow, an area where the transport capacity is not yet reached. Downstream, this capacity is reached: the relation between the eroded particles and the particles deposited must be null because the transport flux cannot increase any further.

4.4.4. *Modeling*

In the literature, some empirical laws that are applicable to granular soils transported by bedload were proposed. First of all, Wörman has studied CE in the context of the modeling of an overflow in an embankment dam equipped with a granular shell as superficial protection of the dam [WOR 92]. In this situation, a CE is likely to develop at the interface between this granular shell and the core. Wörman proposes a semi-empirical transport law based on the Shields criterion, by estimating the shear stress according to the hydraulic gradient, and the shape parameter $\alpha_{\text{shape}} = 8$ as presented in section 4.4.1:

$$a(q^* / n_D)^b = (\text{Re}_d)^c \left(\frac{e_{\text{void}}}{8} \frac{i}{(s-1)} \frac{D_H}{d_{85}} \right) \tag{4.8}$$

where $q^* = q/(Ud_{85})$ is the adimensional transport flux; $\text{Re}_d = Ud_{85}/\nu$ the Reynolds number; e_{void} the void ratio; and finally a, b, and c the three empirical parameters that are adjusted on Wörman's experimental results. In contrast to the previous works, this transport law does not have a threshold, and a non-null transport, even if it remains very weak, is predicted for a small outflow velocity. However, the potential evolutions of the process of erosion over time are not taken into account.

This transport modeling proposed by Wörman is based on the empirical adjustment of a power law that relates the adimensional transport to the Reynolds number and the Shields number. Scheuermann proposes a different model [SCH 06], adapted from the two-phase transport model elaborated by Vardoulakis in the context of sand and hydrocarbon extraction [VAR 96]. It is a phenomenological model that considers both particles at rest and fluidized particles. In this model, the term characterizing the detachment of particles ε is modeled according to the following constitutive law:

$$\varepsilon = \rho_S \lambda (1 - n_d) n_D \beta c_{\text{par}} U \tag{4.9}$$

where c_{par} is the concentration of transported particles and β an unknown coefficient. Erosion is thus governed by the quantity of transported fluidized particles $c_{\text{par}}U$, and is proportional to the parameter λ, which characterizes the sensitivity to erosion of the fine soil. After Scheuermann, this parameter must be brought nearer to a spatial frequency of the starting points of erosion [SCH 06]. Hydraulic outflow is modeled by means of a Navier–Stokes flow between the grains of the coarse layer, through Darcy–Brinkman equations in the superficial layer of fine soil, and through a Darcy law within the fine soil. This modeling allows us to obtain density profiles of the fluidized particles, of porosity, and of the quantity of transported soil according to time. It thus allows us to analyze the temporal evolution of the process of erosion.

For the modeling of bedload transport, Den Adel proposes a statistic approach [DEN 94]. The transfer from a motionless state of the particles to a mobile state is defined by a probability of detachment, and the stopping of particles is defined by a blocking probability. The probability that a particle is destabilized for a given outflow velocity and the probability of having a certain outflow velocity are considered to be distributed according to lognormal laws. The combination of these distributions allows us to obtain the conditional probability of detachment used in this model. The kinetics of detachment of these particles is obtained by defining a characteristic timescale of the medium: $t_{\text{char}} = (D_{15}n_D)/U$. The saturation effect due to bedload transport is considered by means of a corrective term that is proportional to the quantity of transported particles. This model is thus adjusted according to the experimental results of Van Der Meulen cited in Den Adel [DEN 94]. When erosion depends only on hydraulic conditions, this model, along with the experimental results, gives an accurate prediction. When $D_{15}/d_{50} > 200$, we, notice the formation of ripples, which modify the process and make the model non-usable. This modeling, however interesting, is based on a statistical approach to model the detachment of the particles that seems particularly adapted to the case of CE (section 4.3), and thus brings a quantification in space and time of the transport of particles.

The quantitative erosion models discussed previously do not really consider an erosion threshold even if, for the Scheuermann model for example [SCH 06], a transport must, however, exist to generate erosion. These models are transport laws for granular sediment whose transport preferential mode is bedload, and thus they do not apply to the case of fine soils, clays, or silts, generally transported through suspension and characterized by an erosion law.

In the latter case, we have proposed using a classic threshold law to characterize surface erosion, of the form [PAR 65]:

$$\begin{aligned} &\varepsilon = k_{\text{er}}(\tau - \tau_c)^b \quad if \quad \tau > \tau_c \\ &\varepsilon = 0 \quad if \quad \tau \le \tau_c \end{aligned} \qquad [4.10]$$

where τ_c is the threshold stress for the initiation of erosion, corresponding to the hydraulic conditions detailed in section 4.4.1, and k_{er} and b the two parameters characterizing the sensitivity of the soil to erosion. We must, in this case, estimate the shear stress in the porous medium, for example with the expression proposed by Wörman [WOR 92], in equation [4.5]. In the case of surface erosion, numerous authors have chosen to use a linear law (b = 1), which seems to be very coherent with the experimental results and enables us to avoid the adjustment of the two parameters to measurements that are generally quite scattered. This linear law is also successfully used in the case of hole erosion tests, as will be detailed in Chapter 5. The soil's sensitivity to erosion is thus characterized by its erosion threshold τ_c and by the parameter k_{er}, called erosion coefficient. Figure 4.13 compares a linear threshold law to the experimental results of CE for a sandy silt.

However, regarding erosion thresholds, we have already noticed that the proposed expression of the shear stress did not allow us to directly apply the Shields criterion (equation [4.2]), but a proportional coefficient of almost 2 was necessary to obtain a good correlation with the measurements (section 4.4.1). As a consequence, using the k_{er} and τ_c values measured in a classic configuration of surface erosion to model a situation of CE currently seems difficult to envisage.

These different erosion models predict a quantity of transported or eroded soil, depending on the hydraulic load. The erosion rate or the transport rate is thus taken to be constant, i.e. not evolving over time. However, due to the turbidity meter, the tests that we have carried out have shown that this erosion rate had a tendency to decrease over time. We can attribute this tendency to a progressive paving process at the upper surface of the fine soil, as mentioned in section 4.3. The smallest particles, or the particles situated in a part of the interface that is more exposed to the outflow, are primarily eroded. The particles that are too coarse to be entrained, or to simply traverse the pore network of the coarse soil layer, remain in place and they become slowly predominant at the interface. The resistance to erosion at the interface increases and the erosion rate decreases. This process may tend toward a final erosion rate that is lower than the initial rate, or even tend toward the stopping of erosion. Wörman [WOR 96] was interested in the influence that the uniformity of the sand has on the process of CE. He has consequently carried out experiments by considering sands with widespread grain size distributions. The results obtained show a decrease in the quantity of sand transported over time: after 15 hours of experimentation, the transported quantity is decreased by two orders. Wörman attributed this evolution to a progressive clogging of the pores of the coarse soil layer by the coarsest particles of the fine soil layer. Consequently, Wörman proposes a mathematical expression that represents the evolution of this clogged surface, depending on the progression of erosion. This clogging may appear for a very small percentage of particles that are large enough to be captured. Even if they are in negligible quantities at the beginning, they can play an important part in the development of erosion by means of a process of grain size sorting similar to the paving process examined previously. We are therefore dealing with a geometrical influence on erosion even for high diameter ratios D_{15}/d_{85}. The criteria that specify the geometric influence (Figure 4.9) are usually expressed depending on the d_{85} of the fine soil, but this diameter does not necessarily reflect this very small percentage of particles, potentially captured, which will be represented by a d_{95}. Consequently, the hydraulic conditions, considered to be favorable to CE, can become stable conditions, after carrying out that paving of the soil. The initiation thresholds mentioned previously are thus valid only for the duration of the experiment, but they can increase if longer erosion durations are considered.

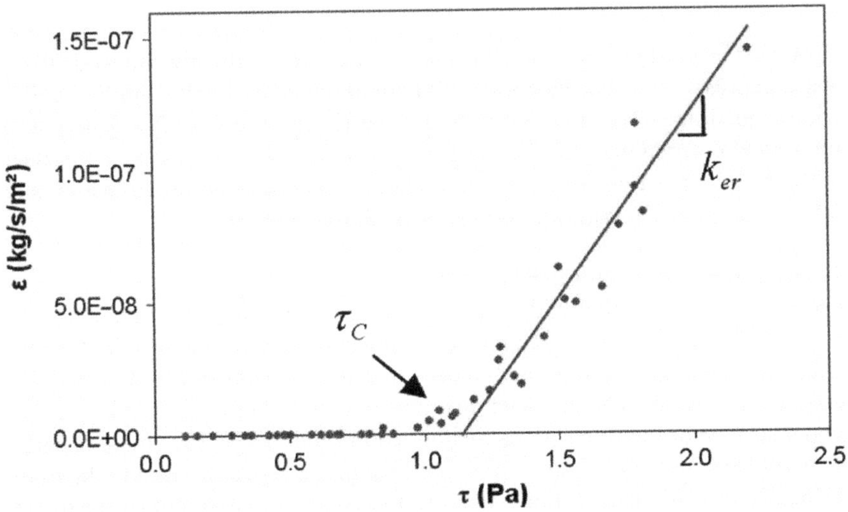

Figure 4.13. *Application of a threshold erosion law to the CE experimental results for a sandy silt [BEG 11]*

4.5. CE of a fine soil on a coarse soil

4.5.1. *Conditions for the occurrence of CE*

The fluvial levees built within large alluvial valleys, such as the ones of Rhône or Rhine, usually have a core made up of fine materials, put into place over a layer of natural foundation, often made up of gravel-like materials (Figure 4.1). The interface between these two layers corresponds to the classic and quite frequent configuration of CE of a fine soil on top of a granular soil, subjected to a tangential outflow. Numerous studies have examined the geometrical criteria of the filter, thus guaranteeing the stability of the interface, regardless of the hydraulic stress in the case of an outflow normal to the interface [SHE 84]. These filter criteria are commonly used in pratice and guarantee the absence of CE. In the previously mentioned context, it is possible that these criteria were not respected at the time for the construction, or that they cannot be respected with the materials that could be found on site. Because of that, the hydraulic conditions will control the initiation of erosion. To our knowledge, only the work of Schmitz [SCH 07] as well as our own [BEG 11] have specifically examined the influence of a tangential outflow in this configuration.

The experimental device used by Schmitz is close in nature to the one in Figure 4.9, with the layer of fine soil underneath the coarse soil. Schmitz has tested silty soils, which were more or less cohesive [SCH 07]. In the same manner as in the reversed configuration, he emphasizes a critical velocity of the initiation of erosion, marked by a sharp increase in the mass of transported soil. Starting from this level of hydraulic load, cavities appear in the fine soil layer and erosion develops up to the point where it causes sinkholes at the top of the sample. The values obtained for these critical velocities are generally comprised between 1 and 5 cm/s, that is they are of the same order as the values found in the reversed configuration (Figure 4.11). Schmitz's results show that an increase in the overload applied to the sample delays the beginning of erosion and increases the value of the critical gradient. The increase in the vertical effective stress in the fine soil strengthens the soil's resistance to shear: $\tau_{\lim} = c + \sigma' \tan\phi$, where ϕ is the angle of internal friction. This gain on shear resistance seems to improve the soil's resistance to erosion, when no influence of the vertical stress was emphasized in the reversed configuration (section 4.4). The tests with three different types of gravel do not show any obvious tendency of the influence that the grain size distribution of the coarse soil has on the critical velocity and no conclusion could be drawn.

To model this erosion threshold, Schmitz adopts two successive assumptions. First, the one of a solid behavior of cohesive soil. The hydraulic shear stress is estimated by assuming a turbulent regime within the pores with a viscous sublayer and a turbulent boundary layer. The velocity profile at the center of the pore is thus described by means of a law derived from the mixing length theory by Prandtl. The hypothesis made is that the velocity at the center of the pore is an accurate approximation of the average pore velocity within the porous medium. The critical velocity $U_{\text{crit sol}}$ is deduced from this relation for a bottom shear stress τ that equals the resistance to shear obtained by a measurement carried out on the soil with a vane test. Schmitz then goes on to advance the hypothesis of a fluid behavior for cohesive soil at the level of the interface. This assumption is justified by observing a rather muddy consistency of the soil after the experiments. By making this hypothesis, Schmitz no longer considers the pulling out of particles but rather the generation of unstable waves at the interface between the two fluids of different densities, beyond the difference in velocity between the two fluids marked as $U_{\text{crit liqu}}$. At the moment of erosion, the soil being in a state of transition between these two hypotheses, solid and liquid, Schmitz finally uses a combination of these two critical velocities.

This combination of the two conditions can be discussed from a physical point of view. It is, indeed, difficult to imagine the soil going from one state to another.

Actually, since the agreement between this model and the experimental results is not conclusive, we must keep our reservations concerning its use in terms of prediction without a more convincing validation. However, the approach is interesting from two points: the use of the value obtained from vane tests to estimate the critical shear stress and the description of a fine soil as a viscous fluid even though a yield stress fluid could be more suitable.

4.5.2. *Comparison with the configuration of a fine soil under a coarse soil*

In the case of cohesive soils, the gravitational forces become negligible compared to the binding forces. As a result, CE is roughly similar in the two configurations, the direction of gravity being the only parameter that distinguishes them. The threshold law of surface erosion that was detailed in section 4.4.4 seems suitable for modeling this erosion phenomenon. Actually, no significant difference between the values of the parameter τ_c has been observed during the CE tests in the two configurations for cohesive soils [BEG 11]. The results obtained by Schmitz are of the same order of magnitude. However, there is a discrepancy that appears between the two configurations when erosion lasts a long time: the cavities that form in the configuration of the fine soil under the coarse soil are filled because of the successive collapses of coarse grains, thus causing locally somehow a regular cycle of erosion. On the contrary, in the reversed configuration, these cavities are not necessary filled by the gravel. Depending on the mechanical characteristics of the fine soil, these cavities can either collapse or grow bigger up to the point where they generate preferential outflows (Figure 4.14). At this point, the process of erosion moves away from the configuration of fine soil under coarse soil since a gap has been created between the two soils.

a) Fine soil below coarse soil

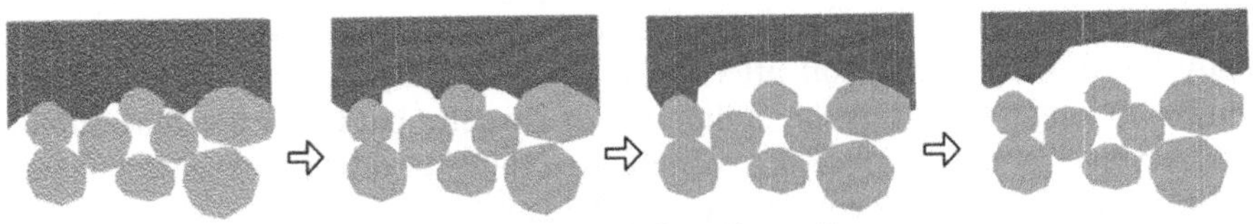

b) Coarse soil below fine soil

Figure 4.14. *Local geometrical evolution in the two configurations of CE*

In the case of granular soils, since the grains are subjected to gravity without *a priori* any forces to stabilize them, detachment and collapse of particles can take place, even without the action of an outflow. However, erosion cannot develop as long as the particles, once fallen in the pores of the coarse soil, are not transported by the outflow. We then come to the process of suffusion (see Chapter 2). If the flow velocity exceeds the sedimentation velocity of particles, their transport is very likely [GOL 09]. If not, the particles depose and accumulate in the pores within the coarse soil, until they fill them completely thus stopping the erosion. The erosion threshold therefore corresponds to a condition of the initiation of transport particles within the coarse soil, rather than to a condition of the detachment of particles.

Furthermore, in the course of the experiments carried out on soils with widespread grain size distributions, we were able to observe a segregation phenomenon quickly followed by clogging, similar to the one observed by Wörman in the reversed configuration. As a result of the erosion of fine soil and the subsequent generation and collapse of cavities, the soil is reworked several times, and washed out by the outflow. The finest particles are entrained at first, and if there are particles that are coarse enough to plug the pores of the gravel layer, the latter remain in place and progressively generate a layer with a coarser grain size distribution than the initial one of the fine soil. Figure 4.15 presents an interface between a granular soil and a gravel, subjected to CE during several hours: a part of the initial fine soil was sorted and is now composed of particles that are relatively coarse and a decrease in the erosion rate over time was observed, down to a null turbidity. The stopping of the erosion can be attributed to the formation of this "filtering" layer. As in the case of the configuration of a fine soil under a coarse soil, a large enough diameter d_{95} can stop erosion. However, the quantity of soil that needs to be eroded in order for this filtering layer to appear can be significant, and that can cause beforehand unacceptable disorders within the hydraulic structure. The phenomenon is thus very complex to model, coupling the occurrence of arches in the fine soil, their collapse, the specificity of the outflow in the cavities, and, finally, the transport of particles. However, several research works are currently under way regarding the process and they aim to model this process and attempt to predict its kinetics.

4.6. Possible scenarios that may lead to failure of a hydraulic structure

In the configuration of CE where the fine soil lies under the coarse soil, the layer of fine soil is eroded up to the point where cavities appear in-between the coarse grains (Figure 4.14). The local porosity increases progressively, which may have two opposite effects. First, the velocity reduces as a result of the enlargement of the cross section; but the velocity can also increase due to an intensification of the flow in this area that is more porous compared to the rest of the coarse layer. In either of these cases, the development of this cavity, more or less rapid, will lead to the local collapse of the grains of the coarse layer. These grains will fill in the cavity, and they will locally reduce the porosity. As a consequence, the process of CE is almost cyclic in this case and, cycle-after-cycle, it causes a progressive subsidence of the grains in the coarse layer. These settlements propagate within the structure above the zone of CE and can have different consequences regarding the geotechnical properties of the soils and of the structure. They may eventually create a sinkhole, at the top of the dike or global settlements of the work at its surface. This kind of disorder is relatively easy to detect, in order to take appropriate action (Figure 4.16). These settlements can also generate decompressed areas, with low density and poor mechanical resistance, or even cavities. This can generate a loss in the stability of the structure or also the initiation of a pipe flow erosion, consequences which could cause the brief failure of the entire structure. CE can also have as a repercussion the clogging of granular filters or of drains located in the upstream of the erosion zone. The particles eroded and transported will then accumulate in the areas where the hydraulic and geometric conditions no longer allow their transport. This clogging can induce an increase in the pore pressure and the local mechanical stability may be no longer ensured. In this configuration, the CE is thus a process that is *a priori* slow, progressive, which can be identified visually by the apparition of fine particles in the downstream of the structure, or settlements and sinkholes on its surface. However, in certain conditions, CE can also cause a sudden failure of the work before any disorder is visible outside.

Figure 4.15. *Interface subjected to CE for a long time: a part of the soil was sorted and only the largest particles remained in place*

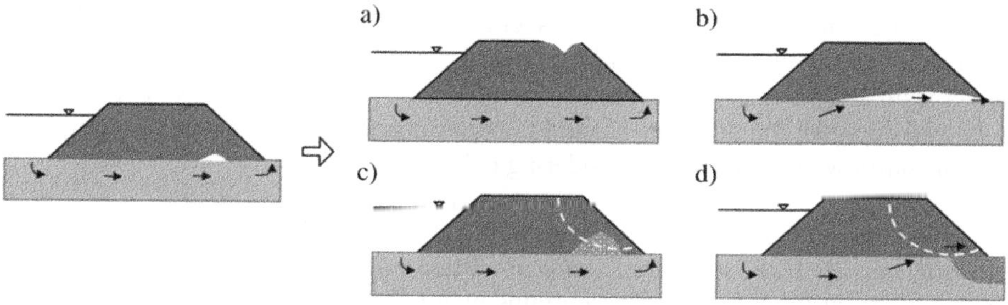

Figure 4.16. *Consequences of CE on a hydraulic structure. The black arrows represent a subterranean outflow throughout a more permeable layer (light gray) under a structure that is less permeable (dark gray): (a) observation of a sinkhole at the upper surface, (b) initiation of a pipe flow erosion, (c) generation of a decompressed area (low density, poor resistance) and eventually loss of stability, (d) clogging of the permeable layer and pore pressure increase, risking a loss of stability*

In the configuration of a fine soil on a coarse soil, the consequences of CE will be similar to the reversed configuration in the case of a non-cohesive soil. During the development of CE, a significant loss in the materials induces settlements, which propagate through the structure in relation to the soil's mechanical properties. Consequently, sinkholes can form at the surface, or decompressed areas, causing preferential outflows or stability issues (Figure 4.16). On the contrary, for a cohesive soil, a very specific scenario can arise in this configuration with a fine cohesive soil under a coarse layer: the cavities caused by erosion are not filled in with gravel, and if the properties of the fine soil allow for it, the roof of this cavity will not collapse. The outflow is then channeled in this area free of grains, which is more permeable than the rest of the coarse layer. The phenomenon becomes unstable: the cavity expands under the action of erosion and drains an increasingly larger part of the outflow, and this, in turn, accelerates the process of erosion. A pipe flow erosion (see Chapter 5) begins, thus potentially causing a rapid failure of the dike.

4.7. Conclusion and perspectives

The interfaces between layers of soil are numerous in the hydraulic structures and in their foundations. These are privileged spots for the initiation of erosion, due to the abrupt transition, at the interface, between two grain size distributions and two values of permeability. What is at stake in this CE phenomenon is thus very important from a practical point of view, and even more so given that external indexes of such an erosion are commonly observed on many existing structures. The knowledge of the favorable conditions for CE initiation, in terms of the grain size distribution of layers (geometrical conditions) and the characteristics of the outflow (hydraulic conditions), is therefore most useful for civil engineering.

Despite the wide variety of situations encountered, the majority of the research works carried out up to now regard the specific case of the erosion of a layer of sand underneath a layer of coarse material subjected to a tangential outflow. There is thus a knowledge gap for different situations regarding the orientation of the interface and the outflow, as well as for considering cohesive soils.

This CE process can be considered as a surface erosion, similar to the erosion existing in river beds, but significantly modified and perturbed by the presence of coarse grains at the interface. These grains mainly influence the outflow, which can no longer turn into the classic velocity profile of a free surface flow but becomes an outflow in a porous medium, mainly characterized by a large variability in direction and magnitude. The coarse soil also influences the transport of the eroded particles, by eventually blocking some of them. The majority of the authors, who have studied CE, thus use some classic surface erosion laws, such as the Shields criterion regarding the erosion threshold, adapting them, more or less empirically, to this specific case of CE. Quantitatively, the set of results in terms of CE threshold velocity is remarkably coherent despite the important discrepancy between the experimental protocols and the choices of modeling: for fine particles smaller than 1 mm, the critical velocities for initiation of erosion are all comprised between 1 and 10 cm/s.

4.8. Bibliography

[BEA 88] BEAR J., *Dynamics of Fluids in Porous Media*, Dover Publications, New York, 1988.

[BEG 11] BEGUIN R., Etude multi-échelle d'érosion de contact au sein des ouvrages hydrauliques en terre, Doctoral Thesis, University of Grenoble, p. 320, 2011.

[BEZ 87] BEZUIJEN A., KLEIN-BRETELLER M., BAKKER K.J., "Design criteria for placed block revetments and granular filters", *Proceedings of the 2nd International Conference on Coastal and Port Engineering in Developing Countries*, Beijing, China, 1987.

[BRA 85] BRAUNS J., "Erosionsverhalten geschichteten Bodens bei horizontaler Durchstromung", *Wasserwirtschaft*, vol. 75, pp. 448–453, 1985.

[CLA 06] CLAUDIN P., ANDREOTTI B., "A scaling law for eolian dunes on Mars, Venus, Earth, and for subaqueous ripples", *Earth and Planetary Science Letters*, vol. 252, pp. 30–44, 2006.

[DEG 83] DE GRAAUW A., VAN DER MEULEN T., VAN DER DOES DE BYE M., *Design Criteria for Granular Filters*, publication no. 287, Delft Hydraulics Laboratory, Emmeloord, 1983.

[DEN 94] DEN ADEL H., KOENDER M.A, BAKKER K.J., "The analysis of relaxed criteria for erosion-control filters", *Canadian Geotechnical Journal*, vol. 31, no. 6, pp. 829–840, 1994.

[FAN 87] FAND R.M., KIM B.Y.K., "Resistance to the flow of fluids through simple and complex porous-media whose matrices are composed of randomly packed spheres", *Journal of Fluids Engineering – Transactions of the ASME*, vol. 109, no. 3, pp. 268–274, 1987.

[GOL 09] GOLTZ M., Determination of critical filter velocity in suffisive soil, European Working Group on Internal Erosion, Saint Petersburg, Russia, 2009.

[GUI 10] GUIDOUX C., FAURE Y.-H., BEGUIN R., HO C.-C, "Contact erosion at the interface between granular filter and various base-soils with tangential flow", *Journal of Geotechnical and Geoenvironmental Engineering*, vol. 136, no. 5, pp. 655–775, 2010.

[HLU 06] HLUSHKOU D., TALLAREK U., "Transition from creeping via viscous-inertial to turbulent flow in fixed beds", *Journal of Chromatography A*, vol. 1126, pp. 70–85, 2006.

[HOF 08] HOFFMANS G.J.C.M., DEN ADEL H., VERHEIJ H.J., "Horizontal granular filters", *4th International Conference on Scour and Erosion ICSE-4*, paper C-16, Tokyo, Japan, pp. 480–485, 2008.

[IST 57] ISTOMINA V.S., *Soil Stability to Seepage*, Moscow, 1957.

[JOH 00] JOHNS M.L., SEDERMAN A.J., BRAMLEY A.S., GLADDEN L.F., ALEXANDER P., "Local transitions in flow phenomena through packed beds identified by MRI", *AIChE Journal*, vol. 46, no. 11, pp. 2151–2161, 2000.

[KNA 07] KNAPEN A., POESEN J., GOVERS G., GYSSELS G., NACHTERGAELE J., "Resistance of soils to concentrated flow erosion: a review", *Earth Science Reviews*, vol. 80, pp. 75–109, 2007.

[KOZ 53] KOZENY J., *Hydraulik: Ihre Grundlagen und praktische anwendung*, Springer-Verlag, Vienna, 1953.

[LEO 04] LEONARD J., RICHARD G., "Estimation of runoff critical shear stress for soil erosion from soil shear stress", *Catena*, vol. 57, pp. 233–249, 2004.

[PAR 65] PARTHENIADES E., "Erosion and deposition of cohesive soils", *Journal of Hydraulics Division, ASCE*, vol. HY1, pp. 105–139, 1965.

[ROB 98] ROBERTS J., JEPSEN R., GOTTHARD D., LICK W., "Effects of particle size and bulk density on erosion of quartz particles", *Journal of Hydraulic Engineering*, vol. 124, no. 12, pp. 1261–1267, 1998.

[SCH 06] SCHEUERMANN A., VARDOULAKIS I., PAPANASTASIOU P., STAVROPOULOU M., "A sand erosion problem in axial flow conditions on the example of contact erosion due to horizontal groundwater flow", in EHLERS W. (ed.) *IUTAM Symposium on Theoretical and Numerical Methods in Continuum Mechanics of Porous Materials*, pp. 169–175, 2006.

[SCH 07] SCHMITZ S., Zur hydraulischen Kontakterosion bei bindigen Basiserdstoffen, PhD Thesis, Bundeswehr University, Munich, Germany, 2007.

[SEL 88] SELLMEIJER J.B., On the mechanism of piping under impervious structures, PhD Thesis, Delft, Netherlands, 1988.

[SHE 84] SHERARD J.L., DUNNIGAN L.P., TALBOT J.R., "Basic properties of sand and gravel filters", *Journal of Geotechnology Engineering – ASCE*, vol. 110, no. 6, pp. 684–700, 1984.

[SHI 36] SHIELDS A., Anwendung der Ähnlichkeitsmechanik und Turbulenzforschung auf die Geschiebebewegung, PhD Thesis, Versuchsanstalt für Wasserbau und Schiffbau, Berlin, Germany, 1936.

[TER 08] TERNAT F., BOYER P., ANSELMET F., AMIELH M., "Erosion threshold of saturated natural sediments: modeling and experiments", *Water Ressources Research*, vol. 44, no. 11, W11434, 2008.

[TRI 88] TRITTON D.J., *Physical Fluid Dynamics*, Oxford University Press, 1988.

[VAR 96] VARDOULAKIS I., STAVROPOULOU M., PAPANASTASIOU P., "Hydro-mechanical aspects of the sand production problem", *Transport in Porous Media*, vol. 22, pp. 225–244, 1996.

[WOR 92] WÖRMAN A., OLAFSDOTTIR R., "Erosion in a granular medium interface", *Journal of Hydraulic Research*, vol. 30, no. 5, pp. 639–655, 1992.

[WOR 96] WÖRMAN A., "Constitutive equation for filtration of well graded base soil with flow parallel to base/filter interface", in LAFLEUR J., ROLLIN A.L. (eds), *Proceedings of GeoFilters 1996*, Montreal, Canada, 1996.

Chapter 5

Concentrated Leak Erosion

5.1. Introduction

The internal erosion of a soil is a complex phenomenon and also a major source of problems related to the safety of earthfill hydraulic structures, i.e. particularly dikes and dams. This phenomenon that starts out locally and develops gradually is indeed difficult to predict and handle. It can be one of the leading causes for the failure of these structures.

Several mechanisms of internal erosion were identified in hydraulic facilities and, among these, we may count the phenomenon of erosion, localized along a pipe ([FEL 07, FRY 97], see Chapter 1). In this particular type of erosion, water infiltrates along a preferential path and it circulates while pulling out and entraining the soil particles toward the exterior of the structure, thus causing elongation and enlargement of the pipe.

The situations that lead to the emergence of this preferential path between upstream and downstream can vary greatly: the presence of a crack of geological origin, or of a crack caused by settlement, or by a hydraulic fracture in a core made of clay. Among the causes, we can also count animal burrows, tree roots, open contact between two very different types of soil or between a soil and a rigid structure, and the final stage of retrogressive erosion.

Among the different kinds of internal erosion, the phenomenon of concentrated leak erosion represents the final stage mechanism, the erosion leading to failure. This

---

Chapter written by Nadia BENAHMED, Christophe CHEVALIER and Stéphane BONELLI.

mechanism must not be confused with backward erosion (Figure 5.1), introduced by Terzaghi in 1927. This phenomenon is called "pipe flow with erosion".

Figure 5.1. *Two types of internal erosion in a hydraulic structure. Backward erosion must not be confused with concentrated leak erosion*

The pipe flow with erosion can be examined in the laboratory by means of hole erosion tests (HETs). The development of the HET as well as its use were the object of significant investigations [LEF 85, ROH 86, RED 00, WAN 02, WAN 04b]. This device proved to be simple, robust, and well adapted for the characterization of the phenomenon[1]. After it was first used in Australia [WAN 04a] and in the US [FAR 07], the use of this device was finally introduced in France [COU 09].

This chapter is dedicated to the experimental characterization of pipe flow with erosion in the laboratory by means of the HET, and the use of the HET for determining two geotechnical parameters: the critical shear stress and the erosion coefficient.

5.2. General points

5.2.1. *Experimental reproduction of pipe flow with erosion*

The HET is a test that allows us to recreate in the laboratory the phenomenon of pipe flow with erosion as well as enabling its modeling. The principle consists of creating a millimetric hole that traverses a decimetric sample of soil, which is either intact or remolded, to recreate a preexistent fault, and to subject it to a sufficient hydraulic load to cause its erosion. The concern of this test is to be able to determine measurable quantities that are characteristic of the phenomenon of pipe flow with erosion. Examples of these measurable quantities are the erosion threshold which, if exceeded, causes erosion (critical shear stress τ_c) and the erosion kinetic (erosion coefficient C_e or erosion index I_e). This test also allows us to identify the key parameters (geotechnical, physical, chemical, etc.) that influence the development of erosion.

1 The HET must not be confused with the pin hole test designed by Sherard *et al.* [SHE 76] and standardized by the ASTM. The pin hole test is only concerned with dispersive soils, whose characterization is performed with distilled water. It does not provide any quantitative result.

Figure 5.2 shows a schema of the typical equipment of HET developed by Wan and Fell [WAN 02]; the soil sample is at the center of the device that serves the circulation of water. Because of an instrumented follow-up, the measured parameters (water flow and hydraulic gradient) as well as the estimated parameters (hole diameter), we can determine the erosion parameters, i.e. the critical shear stress and the erosion coefficient.

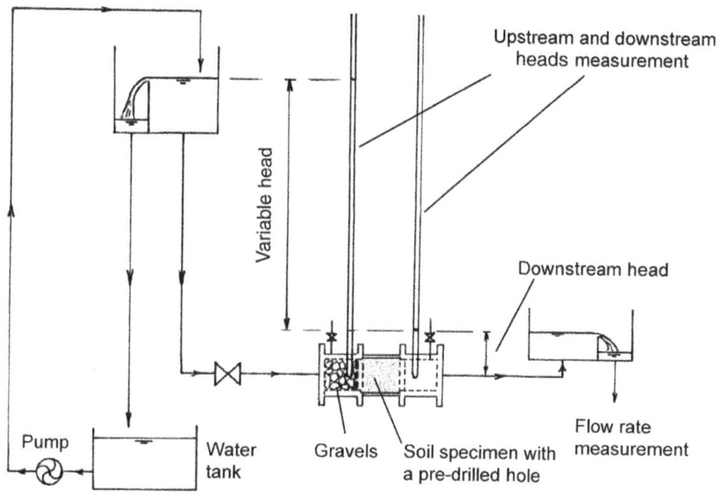

Figure 5.2. *Equipment schema of the hole erosion test [WAN 02]*

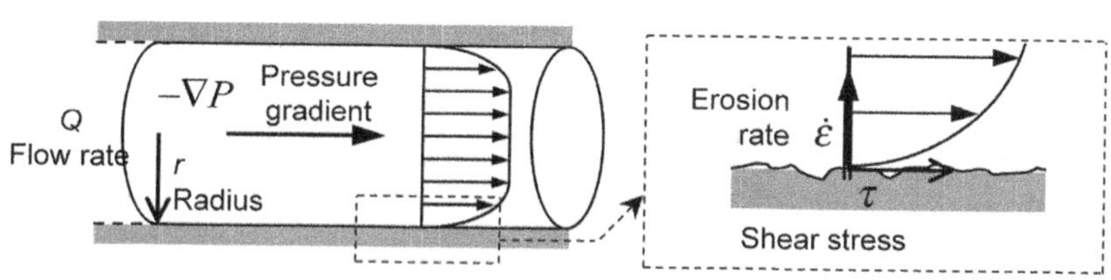

Figure 5.3. *Interface schema: configuration of the axisymmetric outflow with erosion on the wall and the transport of the eroded particles*

5.2.2. *Erosion parameters: shear stress and erosion rate*

The goal of interpreting erosion tests such as the HET is to determine the sensitivity of materials to erosion by means of "erosion laws" or "erosion curves" that explain the relation between the two physical quantities represented in Figure 5.3 as the case of pipe flow with erosion:

– the shear stress τ that the water exerts on the interface as it flows (SI unit: Pa);

– the erosion rate $\dot{\varepsilon}$ that represents the mass of eroded dry soil per surface unit and per time unit (SI unit: kg m^{-2} s^{-1}).

5.2.3. *Erosion law: shear stress threshold and erosion coefficient*

During erosion tests, and in particular during the tests that will be detailed in the later part of this chapter, an empirical law of erosion has generally been observed; this law has the following expression:

$$\begin{cases} \dot{\varepsilon} = 0 & \text{if } \tau \leq \tau_c \\ \dot{\varepsilon} = C_e(\tau - \tau_c) & \text{if } \tau > \tau_c \end{cases} \qquad [5.1]$$

where C_e (SI unit: s m^{-1}) is the erosion coefficient and τ_c (SI unit: Pa) the critical shear stress or shear stress threshold.

By definition, the erosion rate $\dot{\varepsilon}$ is equal to the mass of eroded soil per unit of the lateral surface of the pipe and per time unit. This rate is expressed as follows:

$$\dot{\varepsilon} = \rho_d \frac{dR}{dt} \qquad [5.2]$$

where ρ_d (kg m^{-3}) is the volume mass of dry soil, which is the mass of dry soil divided by the total volume that the soil inhabits, and dR/dt (m s^{-1}) the enlargement velocity of the hole.

The erosion law is schematized in Figure 5.4. It is worth noting that although such a law is only founded on experience, it can also be comprehended from a physical point of view.

The critical shearing stress, τ_c represents the erosion threshold. If the shear stress exerted by the fluid is higher than the critical shear stress, the erosion is significant; otherwise, it is low or inexistent. The critical constraint can be linked to the local forces that resist erosion [DEY 99, RED 97, BON 07, TER 08]. However, we are dealing with a phenomenological model that cannot account for the displacement of the "first grain". This model does not include the heterogeneity of the soil's resistance to erosion, nor does it integrate the heterogeneity of the local strains exerted by the outflow. For example, in Figure 5.4 we may notice that an outflow that leads to a constraint τ equal to 40 Pa will not cause any erosion in the model, but in reality, it will however correspond to an erosion in the order of 1 cm h^{-1}. This may have no consequences on a short-term situation, but it can become problematic in the long run.

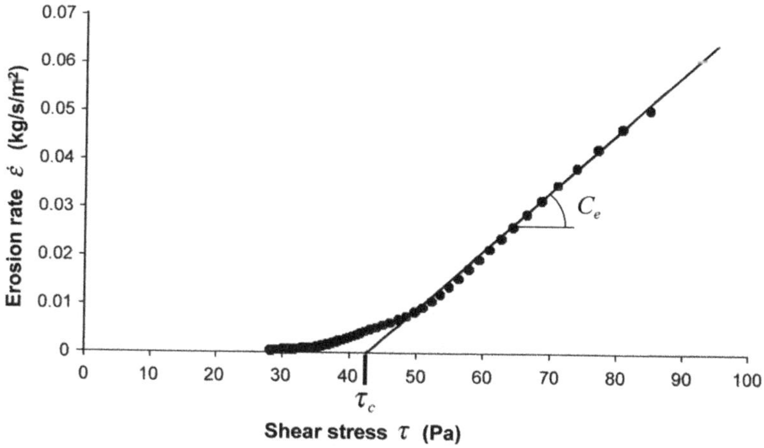

Figure 5.4. *The relation between erosion rate and shear stress obtained in a HET with an imposed flow (sand-like clay: 50% proclay and 50% Hostun sand), and the modeling done according to the threshold law with two parameters (equation [5.1])*

The erosion coefficient expresses the erosion kinetic. This model assumes a linear relation between the erosion kinetic and the shear stress exerted by the fluid. We have no knowledge of any research works examining the relation between the erosion coefficient and the physical measurable quantities at the scale of the grain. Contrary to what the C_e unit (s m^{-1}) seems to suggest, there is no inverse velocity to be considered here.

In an application context, Wan and Fell [WAN 04a] have researched the materials in the core of the dams. Their study examines the resistance to erosion of 14 types of constitutive materials. An erosion rate index I_e was introduced to characterize the resistance to erosion as follows:

$$I_e = -\log_{10}(C_e) \text{ with } C_e \text{ in s m}^{-1} \quad [5.3]$$

A simple method was then proposed with the aim of qualifying the kinetics of pipe flow with erosion in earthfill dams, starting from a classification of soils as it is rendered in Table 5.1. The development of this index and the ensuing classification enables the engineer to make a quick classification of his or her results. In particular, the index I_e offers a wide range of values, generally comprised between 0 and 6, where a low value signifies a quick erosion of the considered soils.

Group no.	Erosion index	Description
1	< 2	Extremely fast
2	2–3	Very fast
3	3–4	Moderately fast
4	4–5	Moderately slow
5	5–6	Very slow
6	> 6	Extremely slow

Table 5.1. *Qualitative classification of soils in relation to their erodibility, according to the erosion rate index I_e [WAN 02]*

As a result of the research carried out by Wan and Fell [WAN 02, WAN 04b], significant improvements of the experimental device, the test protocol, and the interpretation of the tests were proposed [REI 06, BON 07, BON 08, PHA 08, CHE 10, BEN 12, HAG 12]. The rest of this chapter is concerned with the detailed presentation of these improvements, as well as their use in the context of a study on some parameters that regard the influence that the nature of the soil has on the sensitivity to erosion.

5.3. The device and the protocol of the HET

5.3.1. *Equipment*

The testing device, which is a cylindrical cell made of Perspex, comprises three parts (Figure 5.5):

– The main part (center) is meant to receive the soil sample that will be tested (which has the length L) and can vary in length from 130 to 150 mm, with a diameter of 80 mm.

– The upstream part can be connected either to a water supply or to an under pressure water tank. A pressure regulator allows us to control the water pressure within the circuit.

– The downstream part is connected to a turbidimeter with the aim of characterizing the effluent.

Two pressure sensors are placed at the upstream and the downstream of the device to measure the inbound and outbound pressures. A flow meter is also installed on the device, in the upstream or the downstream of the sample, depending on the version of the device. A honeycomb is then placed at the upstream of the sample, with the aim of homogenizing the turbulence.

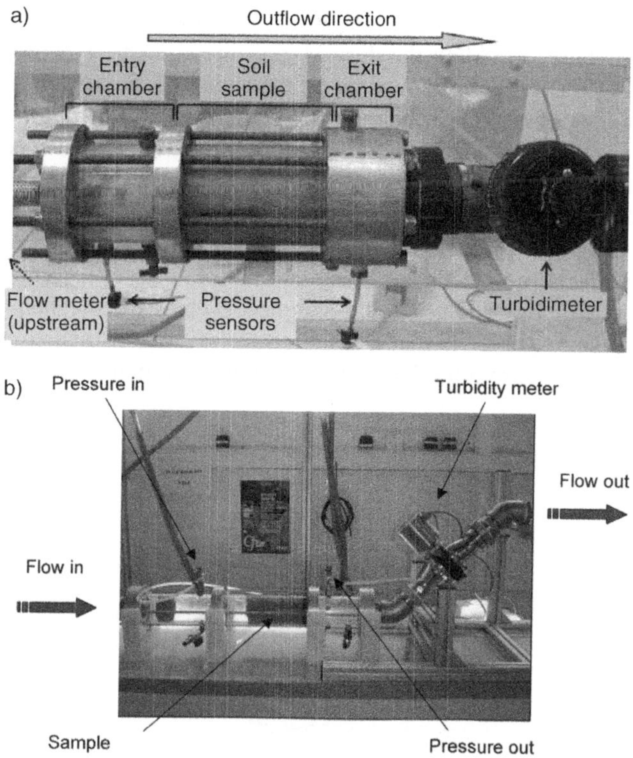

Figure 5.5. *HET devices: (a) after [CHE 10] and [HAG 12]; (b) after [BEN 12]*

5.3.2. *Preparation of the specimen*

A methodology for preparing the samples, either intact or remolded, as well as a methodology of the test procedure were elaborated and validated by exploratory tests.

In regard to the intact material, the soil specimen to be tested is directly sampled from the core of intact soil, by means of swabbing in the main cell of the testing device. A waxing of the side surface of the sample, is sometimes necessary for avoiding the excessive outflows between the mold walls and the sample.

For the fabrication of remolded samples, the values according to which the specimen will be created need to be previously determined: these are the values of the dry volume mass ρ_d and the moisture content W. These values are generally set either according to the characteristics of the on-site soil or in relation to the values obtained at the Normal Proctor Optimum, starting from the Proctor test.

As a general rule, the remolded soil is mixed with the water whose quantity is determined previously to obtain the desired moisture content. The mixture obtained is then preserved for 48–72 h (depending on the nature of the soil) in an airtight bag, to ensure a good equalization of the moisture content. The preparation of the sample is then made directly in the central part of the test cell, in several successive layers that are manually compacted by means of a drop hammer, each higher part of the layer being scarified before being added to the material. The height and the number of layers are fixed in advance, according to the desired dry density and the length of the specimen. The height of the layers is controlled at every compacting stage to ensure a good homogeneity of the sample. Then, with the help of a drill, a longitudinal, traversing hole is made at the center of the test tube so as to simulate a preexisting fault (Figure 5.6). This tube is 3–6 mm in diameter, depending on the operating procedure.

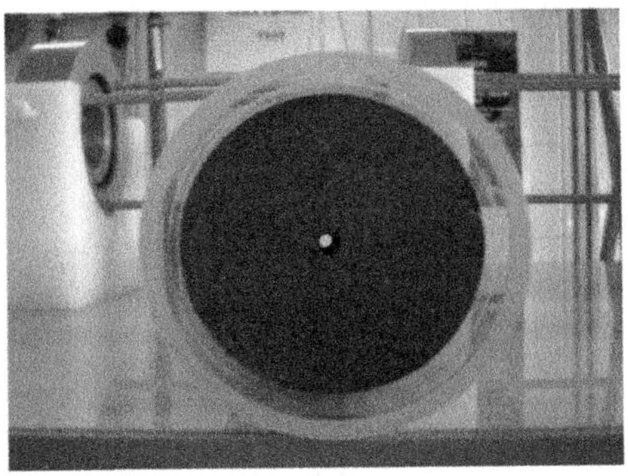

Figure 5.6. *Front view of initial hole in a sample of remolded soil*

5.3.3. *Test procedures*

Once the prepared sample is placed inside the testing device, we may continue by filling the experimental device with water, in the upstream and downstream parts simultaneously. The test starts out by setting the water into motion and making it circulate along the hole drilled in the test tube, from the upstream toward the downstream. This hole will then increase at the contact with water circulation.

There are thus two procedures that enable us to operate the testing device: with an imposed flow, or imposed pressure.

5.3.3.1. *Operating the device by prescribing the flow rate*

Having introduced water in the entire system, we can now set a flow rate Q at a given value. If this value is significant enough to cause erosion, the hole is enlarged as a result of this erosion; this in turn causes a decrease of the flow rate, hence the erosion shear stress. Therefore, when the flow is constant, this erosion must stop at the end of a certain period of time.

The effluent is characterized by its turbidity T in nephelometric turbidity units (NTU). The value of the flow rate is maintained long enough so that the turbidity decreases to a low value (< 5 NTU), and so that the pressure drop ΔP becomes constant. The outflow, the pressure drop, and the turbidity are measured throughout the entire test. The recording of data is done every second by a Profibus data collecting system.

An example of recorded measurements is given in Figure 5.7. Since the hole is eroded, its radius grows larger. The flow being maintained at a constant level, the difference in pressure decreases during the test. The erosion law we may deduce from this is the law expressed in Figure 5.4.

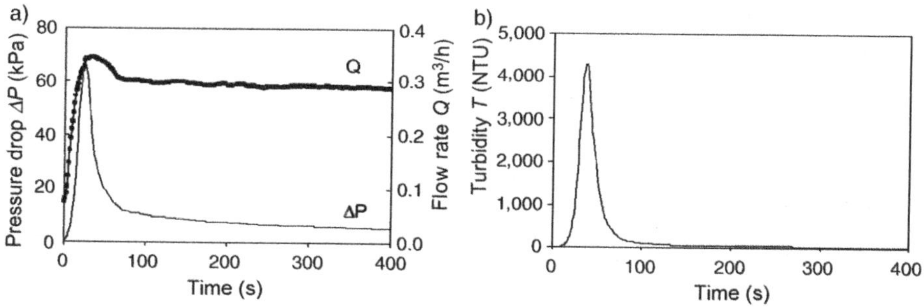

Figure 5.7. *Evolution of the pressure difference, of the outflow, and of the turbidity during a test with an imposed flow rate (sand-like clay: 50% proclay and 50% Hostun sand)*

The appearance of the curve that corresponds to the difference in pressure presents a gradual drop that reflects the increase of the initial diameter of the hole in the test tube during the process of erosion. We may also note the coherence between the turbidity curve and the pressure drop curve.

5.3.3.2. *Operating the device by prescribing the pressure drop*

After introducing water in the system, erosion is obtained by keeping a constant pressure difference between the extremities of the sample. Throughout the test, the data measured by the flow meter (flow Q), the pressure sensors (pressure difference ΔP), and the turbidimeter (turbidity T) are stored on a computer, at a frequency of 1 Hz.

An example of such measurements is given in Figure 5.8. Since the hole is eroded, its radius grows larger. Since the pressure difference is maintained at a constant level, the outflow increases during the test, as it can be observed in Figure 5.8(b).

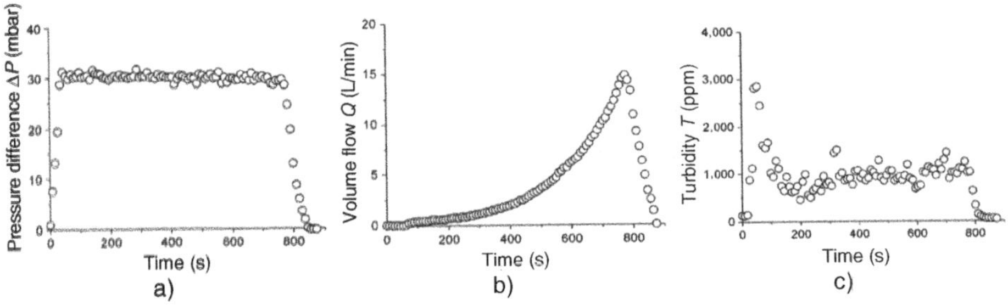

Figure 5.8. *Evolution of the measured parameters throughout a typical test with an imposed pressure difference: (a) pressure difference, (b) flow, and (c) turbidity*

5.3.3.3. *End of the test procedure*

At the end of the test (whether it is a test with an imposed pressure drop or flow rate), the sample of soil is taken out of the device. Melted paraffin wax is poured in the eroded hole (Figure 5.9). Once cooled and extracted, the volume of the obtained spark plug allows us to calculate the final average radius of the hole. If, instead of the mold, a cutout is made, it will clearly emphasize the enlargement of the initial hole and also account for its regularity.

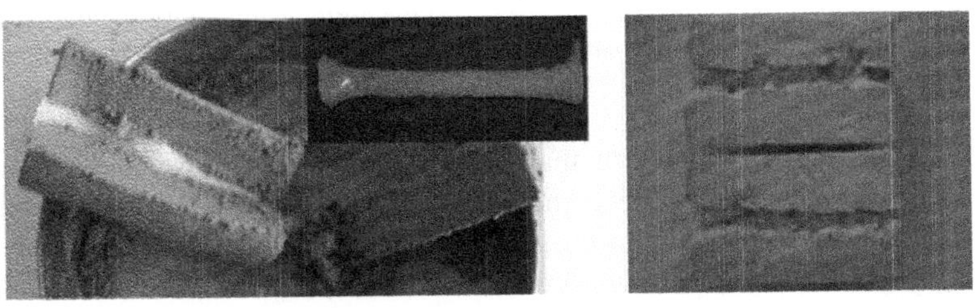

Figure 5.9. *Paraffin wax mold and cutouts of eroded samples*

5.4. Methods of interpretation

After detailing the experimental devices and the measurements carried out in these tests, the procedures used for interpreting the results will be detailed. Several methods of interpretation may be approached: we may identify the erosion parameters with a turbidity measurement, by analyzing the relationship between flow rate and pressure difference. Comparing or combining these methods may constitute topics for a research that is yet to be carried out. This, however, shows that the HET has reached a level of maturity that enables it to be deployed, at least in an exploratory way.

5.4.1. *Determining the shear stress*

To characterize the measurable quantities of shear stress τ and erosion rate over time, the pierced and then eroded hole made in the sample of the test cell can be modeled by a straight cylindrical pipe with current radius $R(t)$ and length L. This pipe is subjected to a pressure difference ΔP (fixed or variable, depending on the operating procedure) that causes a water outflow. This outflow causes a tangential shear stress τ on the walls of the hole (Figure 5.2).

The outflow within the pipe is assumed to be quasi-stationary. Thus, the outflow of a given time t can be assumed to be the same as the outflow set in a pipe with radius $R(t)$. The final outcome of the forces carried out on the volume of the fluid in the direction of the outflow enables us to express the shear stress τ according to other parameters:

$$2\pi R L \tau = \pi R^2 L |\nabla P| \quad [5.4]$$

where $|\nabla P| = \Delta P/L$ is the pressure gradient. The quantities $2\pi RL$ and $\pi R^2 L$ represent the saturated surface where the shear stress force τ is exerted and the volume of the hole. From this results:

$$\tau = \frac{R\Delta P}{2L} \quad [5.5]$$

Parameters such as ΔP, L, or ρ_d are either noted/recorded throughout the test or known beforehand. The parameter that remains to be determined to calculate the shear stress and the erosion rate is thus the hole radius $R(t)$ throughout the test. This is the object of several methods of interpretation [WAN 02, WAN 04a, WAN 04b, BON 06, BON 07, BON 08, PHA 08, CHE 10, BEN 12, HAG 12].

5.4.2. *Interpretation by means of turbidity*

When the experimental device of hole erosion is equipped with a turbidimeter, the turbidity signal may be used to estimate the evolution of the hole radius over time during the test, and thus to determine the erosion parameters τ and $\dot{\varepsilon}$.

5.4.2.1. *Obtaining the radius from the turbidity signal*

During the test, the data measured by the flow meter, the pressure sensors, and the turbidimeter are stored on a computer. These measurements, along with the data of the initial and final diameters of the hole, allow us to calculate the evolution of the quantity of eroded soil and to deduce from it the average radius of the hole over time [PHA 08, HAG 12].

Indeed, the turbidity signal is linked to the load of the soil particles in the eroding fluid when found in the sample. It accounts for the transparency of the water that flows in the downstream of the test tube in time. Thus, turbidity will depend on the concentration of soil particles in the fluid, i.e. the quantity of eroded soil. However, the turbidity signal is not simply a linear function of solid concentration measured from any effluent.

Turbidity T will be, among other things, a function that increases with the number of particles at the downstream of the test tube, and thus an increasing function of the mass concentration of soil in the effluents, for a given type of particle. As a consequence, we can make the hypothesis according to which for a given texture the turbidity is proportional to the soil concentration in the fluid C_{soil} (SI unit: kg m^{-3}):

$$T = \alpha C_{\text{soil}} \quad [5.6]$$

with α a factor of proportionality that must be determined for every texture of the soil (in 10^{-6} m^3 kg^{-1} in the case of turbidity expressed in ppm). This hypothesis assumes, among other things, that the soil is dispersed in the same way throughout the fluid, independently of the pressure applied. This implies that the shear stress force causes the erosion rate to vary, but not the characteristics of the eroded material (usually the floc size). This is a hypothesis that has been verified so long as the shear stress forces did not become too large or as long as the material did not have the tendency to form aggregates; the latter is the case of the materials presented in this chapter.

The turbidity in the downstream of the test tube can thus be linked to the mass of eroded soil. Indeed, during a set time dt, the quantity of eroded soil is defined by:

$$dm = Q\,C_{\text{soil}}\,dt = \frac{1}{\alpha} Q T\,dt \text{ given that } \frac{dm}{dt} = \frac{1}{\alpha} Q T \quad [5.7]$$

where Q is the outflow flow rate and C_{soil} the concentration of eroded soil in the water.

The evolution of the radius is thus given by writing the relation between the mass variation of the sample and the volume variation of the hole $V_{tr} = \pi L R^2$:

$$\frac{dm}{dt} = \rho_d \frac{dV_{tr}}{dt} = 2\pi\rho_d LR\frac{dR}{dt} \qquad [5.8]$$

Knowing the initial diameter as well as the coefficient α, (α being in turn determined by the knowledge of the final diameters) allows us to go back and readjust equations [5.7] and [5.8] to the temporal evolution of the radius $R(t)$.

5.4.2.2. *Test results and the exploitation of these tests*

Once the temporal evolution of the radius has been determined, the shear stress and the erosion rate can be calculated with equations [5.2] and [5.5]. They are represented in Figure 5.10(a) for a typical test with an imposed pressure difference. Since the pressure difference at the extremities of the sample is fixed, and since, because of the erosion, the radius of the hole increases with time, the shear stress as well as the erosion rate will also increase.

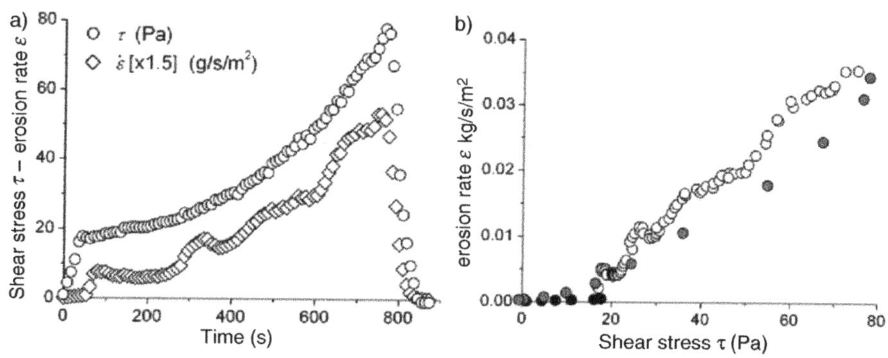

Figure 5.10. *Erosion curves for a typical test with imposed pressure difference: (a) erosion parameters obtained over time, (b) erosion rate according to the shear stress: at the beginning of erosion (black dots), in a quasi-permanent regimen (white dots) and at the end of the test (gray dots)*

Figure 5.10(b) represents in more detail the erosion rate depending on the shear stress for a typical test with an imposed pressure difference. The test can be executed in three steps:

– An increase in pressure at the very beginning of the test (black dots): during this period of time, erosion takes place very late as the shear stress is not yet high enough to cause it. The surface of the hole being shaped by the piercing, these points cannot be considered relevant.

– The actual test (white dots): shear stress and erosion rate increase significantly. This is the erosion law we have been researching.

– The abrupt end of the test (gray dots): the hypothesis of the quasi-stationary nature of the phenomenon can no longer be respected and the turbidity signal can relax for a certain period of time. These values are not significant either.

Consequently, throughout the exploitation of the results of researching an erosion law, we will only consider the second stage. Having processed the data for a particular test, the results can be analyzed for a given texture of soil. Thus, Figure 5.11 represents the erosion data $\dot{\varepsilon}$ and τ calculated for a given soil texture (50% Armorican kaolinite (KA) clay and 50% Fontainebleau sand, characteristics that are given in the following part) with 11 tests having four different values of pressure difference (we are still in the framework of an operating procedure that relies on an imposed pressure difference).

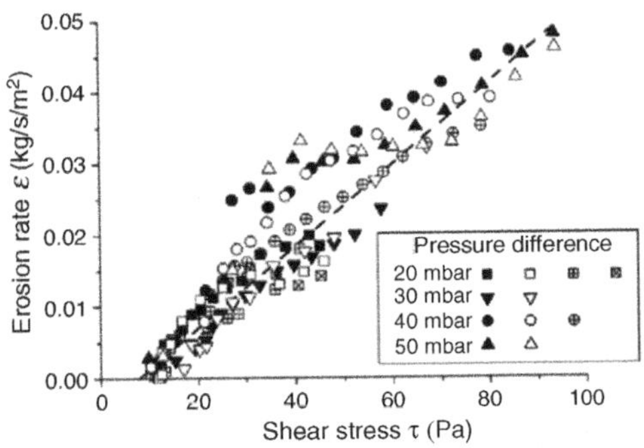

Figure 5.11. *Erosion curve for a sand-kaolinite clay mixture with a mass fraction of clay of 50%: tests carried out for various pressure differences and repeatability tests; dotted line: linear adjustment of data*

We may successively note that:

– The test is repeatable, since the tests that were carried out at the same pressure give comparable results.

– The test does not depend on the applied pressure difference ΔP since the data are placed on the same curve, regardless of the value of ΔP.

– There is a characteristic erosion law of the tested material that can even be expressed as equation [5.1].

Consequently, the tests that are exploited by means of the turbidity signal allow us to characterize the sensitivity to erosion of a soil, by using the data of a critical shear stress τ_c and an erosion coefficient C_e.

5.4.3. *Interpretation using the flow/pressure relation*

The experimental device of hole erosion is equipped with pressure sensors and a flow meter. There are, therefore, in relevant literature, numerous relations that allow us to describe the relation between the flow and the pressure difference for a circular, rectilinear pipe [BRO 02, POP 00, SCH 87]. We may thus deduce the pipe radius.

Let us consider a rectilinear pipe of length L and current radius $R(t)$, with an initial value R_0. The flow is supposed to be turbulent. We may mark t as the time, $Q(t)$ as the flow rate (initial value Q_0), $\Delta P(t)$ as the pressure difference between the entrance and the exit of the pipe (initial value ΔP_0). The average outflow velocity (initial value V_0) is $V = Q/[\pi R^2(t)]$. The shear stress exerted by the flow on the soil is $\tau(t) = R(t)\Delta P(t)/(2L)$ (initial value τ_0). The analysis of the two-phase pipe flow with erosion equations leads to results that allow us to justify several simplifying hypotheses [LAC 08].

The non-dimensional number C_eV, called the "erosion number", characterizes erosion kinetic in relation to the outflow velocity. We have observed in practice, on the tests that were carried out, that $C_eV < 10^{-3}$.

When this number is small ($C_eV \ll 1$), i.e. a situation called "low erosion kinetic", the following hypotheses become valid [BON 08]:

– The outflow is intense enough compared to the erosion flow so that we may consider this outflow as diluted. Indeed, the eroded soil particles do not have enough time to concentrate in water, and we may consider that the presence of these particles does not modify the outflow or the density of the fluid.

– The outflow is quick enough compared to the erosion velocity so that we may neglect the unsteady terms of the outflow. We may consider a stationary outflow. However, the erosion law gives the system a transient character.

– The erosion is slow enough compared to the outflow so that we may neglect the momentum due to the displacement of the soil–water interface.

We may define a characteristic time of the system as follows:

$$t_{er} = \frac{2L\rho_d}{C_e \Delta P_0} \qquad [5.9]$$

This characteristic time allows us to make the time non-dimensional, and to regroup the various curves under the same graph. During our research, we come across exact solutions of the system of equations for turbulent outflow with erosion.

There are two particular cases that allow us to explain analytical expressions: erosion with a constant pressure drop and erosion with a constant flow rate. The use of non-dimensional measurable quantities allows us to simplify the following expressions:

$$\tilde{t} = \frac{t}{t_{er}}, \quad \tilde{\tau}_c = \frac{\tau_c}{\tau_0}, \quad \tilde{R}(\tilde{t}) = \frac{R(t)}{R_0} \qquad [5.10]$$

$$\tilde{Q}(\tilde{t}) = \frac{Q(t)}{Q_0}, \quad \Delta\tilde{P}(\tilde{t}) = \frac{\Delta P(t)}{\Delta P_0}, \quad \tilde{V}(\tilde{t}) = \frac{V(t)}{V_0} \qquad [5.11]$$

5.4.3.1. *Erosion with constant pressure difference*

We consider a HET with a constant pressure drop ($\Delta P(t) = \Delta P_0$). We may deduce the following relations from the equations for pipe flow with erosion (Chapter 6):

$$\tilde{R}(t) = \left[\tilde{Q}(t)\right]^{2/5} = \left[\tilde{V}(\tilde{t})\right]^{2} \qquad [5.12]$$

$$\tilde{R}(\tilde{t}) = \tilde{\tau}_c + \exp\left[\tilde{t} + \ln(1-\tilde{\tau}_c)\right] \qquad [5.13]$$

Equation [5.12] allows us to estimate the radius knowing the flow. Adjusting equations [5.12] and [5.13] allows us to identify the erosion parameters (τ_c, C_e).

The analysis of 17 HETs with an imposed pressure drop, carried out on several types of soil, demonstrates the validity of this model [BON 07, BON 08, WAN 02]. Table 5.2 comprises the ranges of the geotechnical parameters and the ranges of the test parameters. The critical shear stress ranges from 6 to 128 Pa. The erosion coefficient ranges from 3.51×10^{-5} to 1.47×10^{-3} s m^{-1}. The initial Reynolds number (which is very low at the beginning of the test) ranges from 1,904 to 8,144: knowing that this number increases during the test, the outflows are turbulent. The erosion number ranges from 10^{-4} to 1.1×10^{-2}: the erosion kinetics are small for the majority of the tests, but this hypothesis reaches its limit in several tests, when the erosion index I_e is lower than 3.

Figure 5.12 shows the evolution of the non-dimensional flow depending on the non-dimensional time. Without non-dimensionality, it would have been impossible to reassemble these curves in the same figure. Figure 5.13 regroups the 17 results of the test in the master curve defined by equation [5.13]. This result is a validation element of the simplified modeling.

		Minimum	Maximum
Porosity	n	0.20	0.55
Initial moisture content	W (%)	8	38.7
Critical shear stress	τ_c (Pa)	6.42	128.22
Erosion coefficient	C_e (s m^{-1})	3.51×10^{-3}	1.47×10^{-2}
Initial pipe radius	R_0 (mm)	3	
Length of sample	L (mm)	117	
Initial shear stress	τ_0 (Pa)	6.91	129.00
Initial outflow velocity	V_0 (s m^{-1})	0.63	2.71
Initial Reynolds number		1,904	8,144
Initial erosion number	C_eV_0	10^{-4}	1.1×10^{-2}
Characteristic erosion time	t_{er} (s)	73	2.183

Table 5.2. *The range of geotechnical parameters of 17 soils and the range of test parameters of 17 HETs with an imposed pressure difference*

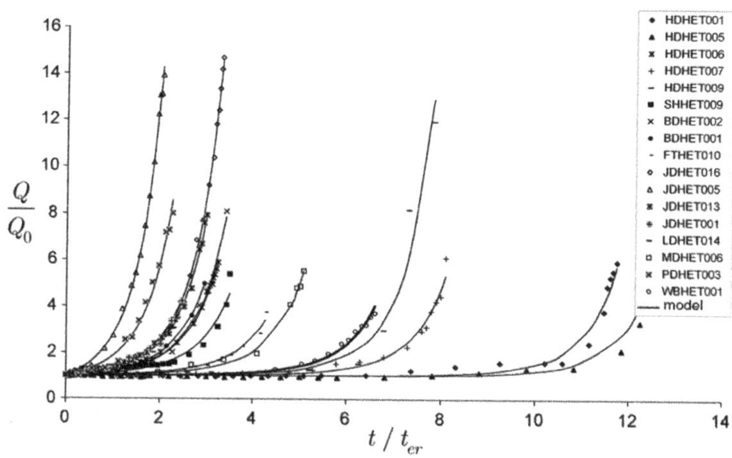

Figure 5.12. *Non-dimensional flow depending on non-dimensional time of 17 HETs with an imposed pressure difference*

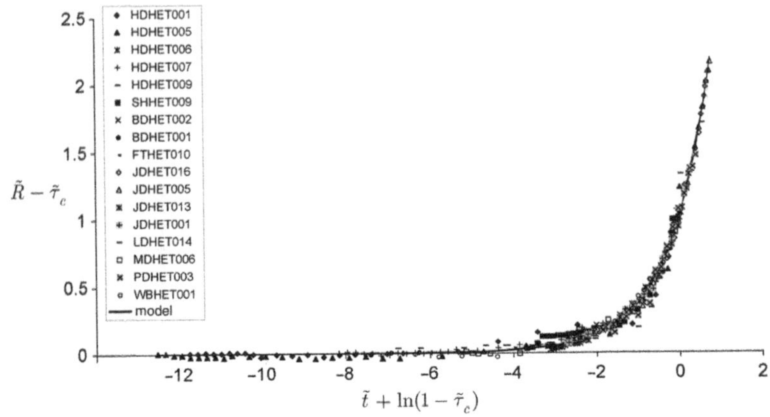

Figure 5.13. *Gathering on the master curve defined by equation [5.15] and the results from 17 HETs with an imposed pressure difference imposed*

5.4.3.2. *Erosion with a constant flow rate*

Let us consider a test of pipe flow with erosion, with a constant flow rate ($Q(t) = Q_0$). From the flow equations of pipes with erosion, we may deduce the following relations (Chapter 6):

$$\tilde{R}(t) = \left[\Delta\tilde{P}(t)\right]^{-1/5} = \left[\tilde{V}(\tilde{t})\right]^{-1/2} \qquad [5.14]$$

$$f(\tilde{\tau}_c^{1/4}\tilde{R}) = f(\tilde{\tau}_c^{1/4}) + \tilde{\tau}_c^{5/4}\tilde{t}, \quad f(x) = \tfrac{1}{2}(\arctan x + \operatorname{arctanh} x) - x \qquad [5.15]$$

Equation [5.14] allows us to estimate the radius from the pressure difference. Adjusting equations [5.14] and [5.15] allows us to identify the erosion parameters (τ_c, C_e).

The analysis of 16 HETs with a constant flow rate demonstrates the validity of this model. These tests were carried out on several types of soil [BEN 12]: white kaolinite (KW), proclay kaolinite (KP), illite, silt, a mixture of KP and Hostun sand, a mixture of KP and Fontainebleau sand. Table 5.3 shows the ranges of geotechnical parameters and the ranges of the test parameters. The critical shear shear stress ranges from 1 to 260 Pa. The erosion coefficient ranges from 2.83×10^{-5} to 2.83×10^{-3} s m^{-1}. The initial Reynolds number (which is at its maximum in the beginning of the test) ranges from 12,000 to 58,000: the outflows are turbulent. The erosion number ranges from 10^{-4} to 5×10^{-3}: the erosion kinetics are small.

		Minimum	Maximum
Porosity	n	0.32	0.56
Initial moisture content	S_r	0.55	0.93
Critical shear stress	τ_c (Pa)	1.1	260.5
Erosion coefficient	C_e (s m^{-1})	2.83×10^{-5}	3.24×10^{-3}
Initial pipe radius	R_0 (mm)	3	
Length of sample	L (mm)	120	155
Initial pressure difference	ΔP_0 (kPa)	2.24	164.52
Initial shear stress	τ_0 (Pa)	8.58	1,970.27
Initial outflow velocity	V_0 (s m^{-1})	1.51	9.66
Initial Reynolds number		12,855	57,951
Initial erosion number	$C_e V_0$	10^{-4}	5×10^{-3}
Characteristic erosion time	t_{er} (s)	45	1,397

Table 5.3. *The range of geotechnical parameters of 16 types of soil and the range of test parameters of 16 HETs with an imposed flow*

Figure 5.14 shows the evolution of the pressure difference over time. Figure 5.15 gathers the 16 results of the tests carried out on the master curve defined by equation [5.15]. This result is a validation element of the simplified modeling.

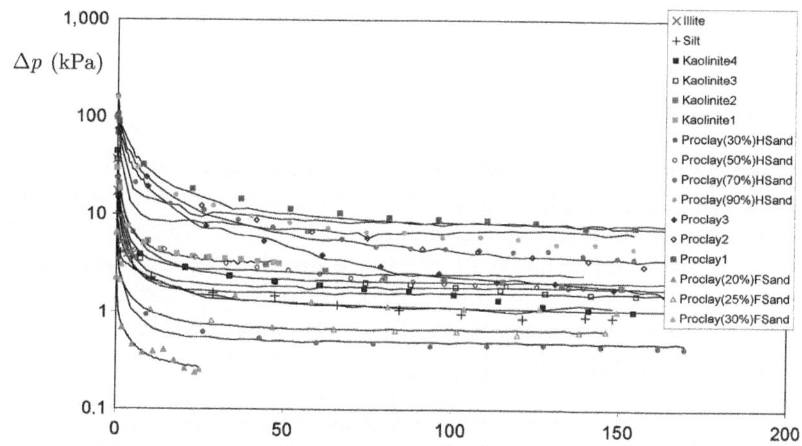

Figure 5.14. *Pressure difference according to time of 16 HETs with imposed flow rate*

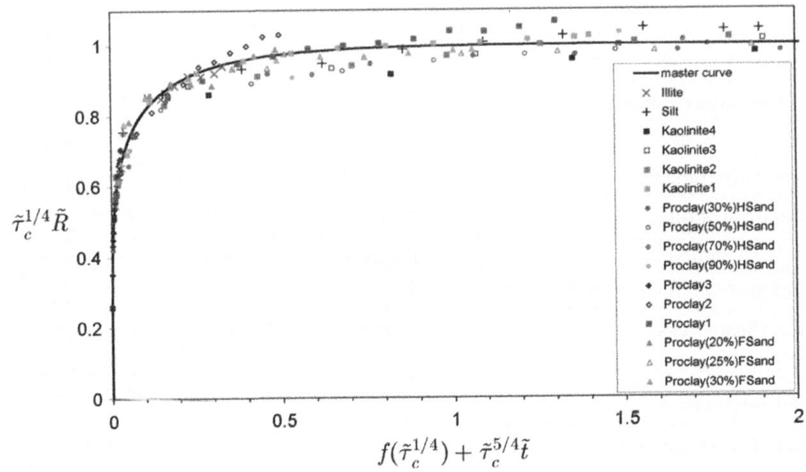

Figure 5.15. *Gathering on the master curve defined by equation [5.17] the results of 16 HETs with imposed flow rate*

5.5. Effect of different soil parameters on erosion

In what follows, we will present the measuring campaigns that were carried out using the previously described devices [PHA 08, BEN 12, HAG 12]. They allow us to obtain several tendencies of the influence that different constitutive soil parameters have on the sensitivity of these soils to erosion.

5.5.1. *Trial campaign*

This section presents the different HETs that were carried out. To use the soils that have a satisfying quantity and quality, as well as being from a wide range of textures, it has been decided to recreate the specimen from two basic materials that represent two different categories of grain size distributions: clay and sand. This allowed for a better control of the properties of the soil and a satisfying repeatability when preparing the samples. The types of clay that were used were KA, KP, KW, illite, and montmorillonite (Mo).

The sands used for the mixtures are fine and uniform sands, namely the Hostun sand HN31 (SH) and the Fontainebleau sand NE34 (SF). The SH is made from angular grains to subangular grains, ranging from 200 to 500 µm with a D_{50} equal to 350 µm. The SF has round-shaped grains, of a size ranging from 100 to 300 µm, with a D_{50} equal to 199 µm. The physicochemical properties of these materials were detailed in references [BEN 12], [BON 09], [PHA 08], and [REI 07].

To better examine the role of each raw material that consists of soil mixtures, several mixtures between clay and sand have been used. A review of the tests carried out with the prepared soils characteristics is given in Table 5.4. The numbers next to the acronyms correspond to the mass percentages of the materials. Thus, for example, *KA*30 *SF*70 corresponds to a mixture of KA type of clay at a height of 30% and Fontainebleau sand at a height of 70%.

Type of soil	Texture	Dry mass fraction (%)		Moisture content (%)	Dry density (t m^{-3})
		Argile	Sable		
Armorican kaolinite	*KA*	100	0	26	1.40
White kaolinite	*WK*100-*a*	100	0	20.5	1.40
	*WK*100-*b*	100	0	23.5	1.40
	*WK*100-*c*	100	0	26.5	1.40
	*WK*100-*d*	100	0	23.5	1.25
	*WK*100-*e*	100	0	23.5	1.32
	*WK*100-*f*	100	0	23.5	1.40
	*WK*100-*g*	100	0	23.5	1.47
Montmorillonite	*Mo*100-*a*	100	0	35	1.75
	*Mo*100-*b*	100	0	35	1.50
Proclay kaolinite + Hostun sand HN31	*KP*90 *SH*10	90	10	21	1.59
	*KP*70 *SH*30	70	30	21	1.62
	*KP*50 *SH*50	50	50	21	1.60
	*KP*30 *SH*70	30	70	21	1.66
Armorican kaolinite + Fontainebleau sand NE34	*KA*20 *SF*80	20	80	7	2.10
	*KA*30 *SF*70	30	70	9	2.00
	*KA*50 *SF*50	50	50	16	1.80
	*KA*70 *SF*30	70	30	20	1.60
Montmorillonite + Fontainebleau sand NE34	*Mo*30 *SF*70	30	70	16	1.75
	*Mo*50 *SF*50	50	50	21	1.50
Illite + Fontainebleau sand NE34	*I*20 *SF*80	20	80	10	1.96
	*I*25 *SF*75	25	75	11	1.93
	*I*30 *SF*70	30	70	13	1.91
	*I*40 *SF*60	40	60	14	1.83
	*I*50 *SF*50	50	50	15	1.73
	*I*70 *SF*30	70	30	19	1.56

Table 5.4. *The characteristics of the tested materials*

5.5.2. *The influence of clay content*

Figure 5.16 shows the data that characterize the erosion law, data that were obtained for soil textures made of KA–Fontainebleau sand with different clay percentages: 30, 50, 70, and 100%. Following up on what was observed in the previous part, regarding the mixture *KA*50 *SF*50 (Figure 5.11), and given the texture of the soil considered, the results obtained for different tests are coherent with one another, regardless of the pressure applied and of the duration of the test. The result is that the sensitivity to erosion of the different tested textures can be characterized by the data of the critical shear stress τ_c and the erosion coefficient C_e.

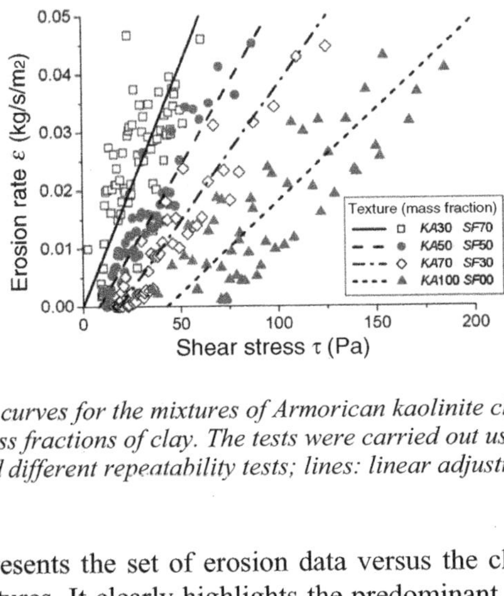

Figure 5.16. *Erosion curves for the mixtures of Armorican kaolinite clay and Fontainebleau sand for different mass fractions of clay. The tests were carried out using different pressure differences and different repeatability tests; lines: linear adjustment of the data*

Figure 5.17 represents the set of erosion data versus the clay content, obtained for the different textures. It clearly highlights the predominant influence of the clay fraction on the critical erosion shear stress and the erosion rate. The typical threshold shear stress values range from 0 to 50 Pa and the erosion rates from 0.3 to 0.9 × 10^{-3} s m^{-1}. The critical shear stress τ_c increases with the increase of the clay percentage, whereas the erosion coefficient C_e decreases: the more a texture is clay-like, the less it is likely that it will develop erosion. This behavior is very similar to the behavior observed in reference [RED 97].

Figure 5.18 represents the results of a series of tests that were carried out, this time on the mixtures of soil made up of KP and SH. The mass clay fraction varied from 30 to 90%. As in the previous case, the critical shear stress and the erosion coefficient were similarly influenced by the clay content. The critical shear stress moves from a τ_c value of 6.3 Pa for a clay percentage equal to 30% to a

τ_c value of 407 Pa for a soil made of 90% clay. However, these values are much more important than the values obtained on the mixtures of Fontainebleau sand and KA clay. This is clearly due not only to the nature of the two types of kaolinite clay (Armorican and proclay) but also to the nature of the types of sand used. SH is angular sand that presents a much more significant friction than in the case of SF, which is finer and more rounded. The influence of the nature of the clays will be analyzed in more detail in the following section.

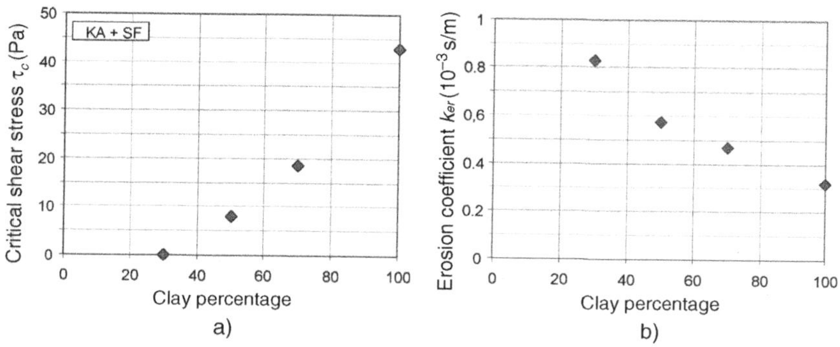

Figure 5.17. *(a) Critical shear stress and (b) erosion coefficient depending on the clay percentage for the mixtures of Armorican kaolinite clay and Fontainebleau sand*

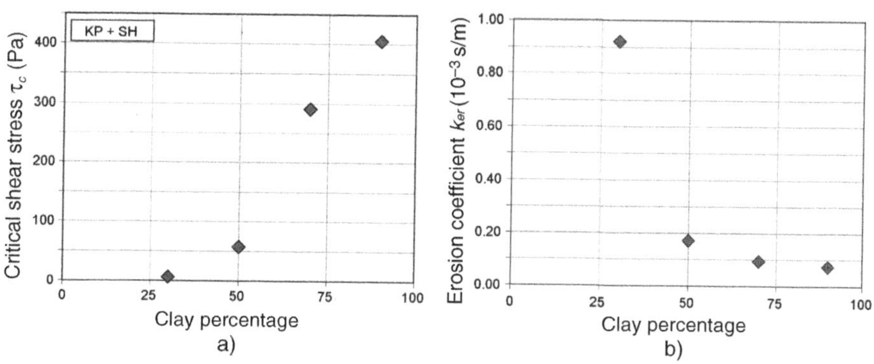

Figure 5.18. *(a) Critical shear stress and (b) erosion coefficient depending on the percentage of clay for the mixtures proclay kaolinite and Hostun sand*

The erosion coefficient decreases with the increase of the clay percentage. However, it is worth noting that in the two cases and regardless of the nature of the sand and the clay used, the erosion coefficient will vary more significantly for a capacity of clay that is lower than 50%.

5.5.3. *The influence of the nature of the clay*

Besides the textures based on kaolinite clay (Armorican and proclay), other mixtures of sand–clay were tested as well: some of them were based on illite clay and others on Mo clay.

The tests carried out on illite-based textures were more difficult to carry out than the tests performed on kaolinite. Indeed, erosion turned out to be much more difficult. In particular, the textures having more than 40% illite clay were not sensitive to the protocol and did not present any significant erosion, regardless of the pressure difference that was applied. Thus, only the tests carried out on textures with 25 and 30% and several tests with 40% could be performed. They are represented in Figure 5.19. These materials are actually much less sensitive to erosion. A moisture content of 25% in illite is thus sufficient for obtaining a critical shear stress τ_c of an order of 150 Pa.

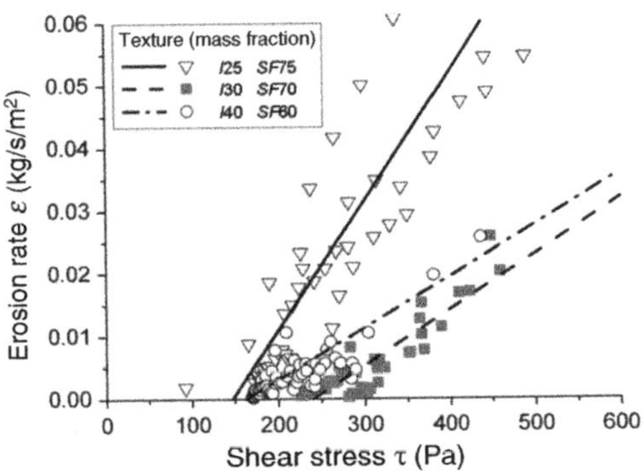

Figure 5.19. *Erosion curves of the mixtures Fontainebleau sand–illite for different mass fractions of clay. Lines: linear adjustment of data*

The behavior is thus very different from the behavior displayed by the mixtures of sand–kaolinite clay, which can have a null threshold shear stress even with 30% kaolinite clay (Table 5.5). The erosion coefficients are equally lower (around one order of magnitude) in the mixtures based on illite clay, compared to the mixtures based on kaolinite clay.

The HET proves to be very decisive in this case, since the different textures, whether kaolinite-based or illite-based, are clearly discernable, i.e. the materials based on illite show a much higher resistance to erosion.

There is no HET carried out on soils made up of Mo clay (30 or 50% clay) that reached significant results. The cause for this is the swelling character of Mo. At the end of each test, the hole pierced in the test tube was indeed clogged, throughout its entire length, with a material that was drenched in water. Sometimes, certain hollows in the upstream and the downstream of the cell could be observed, which were certainly the results of the turbulences caused at its extremities when the test ended.

Texture	Dry mass fraction (%)		Critical shear stress τ_c (Pa)	Erosion coefficient C_e (s m^{-1})
I30 SF70	30	70	230	0.080×10^{-3}
KA30 SF70	30	70	0	0.83×10^{-3}

Table 5.5. *Comparing the erosion parameters τ_c and C_e for the different mixtures of clay and sand, having the same mass percentage of clay*

The swelling phenomenon also manifests itself as an excess of soil that overhangs at the extremities of the test tube (around 5 mm), despite the presence of grids (Figure 5.20(a)). In Figure 5.20(b), we can see the mold of the hole, filled by the muddy soil that is drenched in water. The hole quickly becomes clogged during the test, thus preventing its erosion. When the erosion tests were carried out on test tubes of pure Mo, the soil started swelling after the cell was filled, and then it disintegrated in large aggregates once the test was launched (Figure 5.21). The HET, whether with an imposed pressure difference or an imposed flow rate, is not therefore adapted to types of clay that have a swelling character.

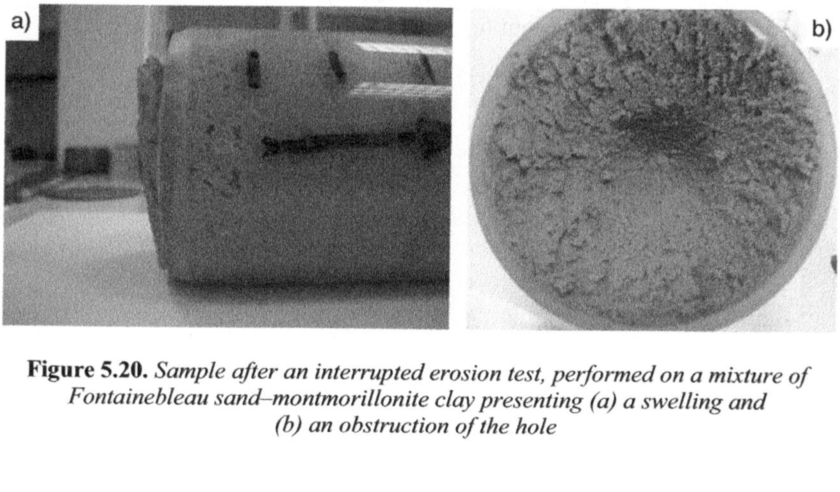

Figure 5.20. *Sample after an interrupted erosion test, performed on a mixture of Fontainebleau sand–montmorillonite clay presenting (a) a swelling and (b) an obstruction of the hole*

Figure 5.21. *Swelling and disintegration of a sample of pure montmorillonite clay during an erosion test*

5.5.4. *The influence of the density and the moisture content*

To examine the influence that the density has on the critical erosion shear stress and the erosion coefficient, a series of tests was carried out on samples of KW clay, having been prepared at compaction energies of about 85, 90, 95 and 100% of Standard Proctor Maximum Dry Density (SMDD), and thus corresponding to the dry densities that range from 1.25 to 1.47 t m^{-3}. The moisture content of the prepared mixture for the entire set of specimens was kept at a constant W_{OPN} of 23.5%.

Figure 5.22 shows that, according to the classical observations, increasing the dry density of the soil increases the critical erosion shear stress. The denser the soil, the better it resists to the erosion. As regards the erosion coefficient, it does not seem to be too affected by the variation in density, and its value (of the order 0.2×10^{-3}) remains almost constant for the densities tested.

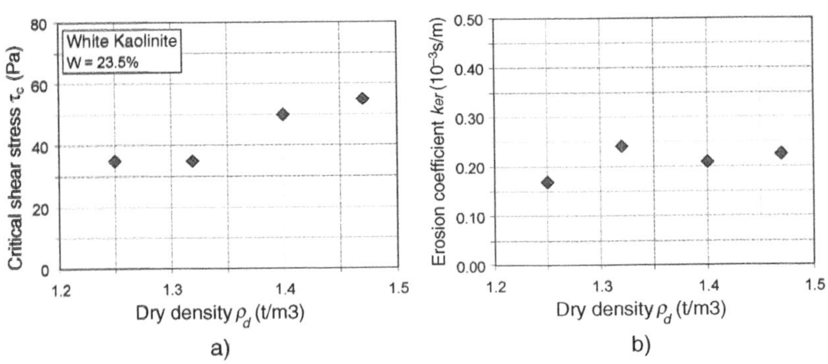

Figure 5.22. *Influence of the density on the critical erosion shear stress and the erosion coefficient of white Kaolinite clay*

Figure 5.23 represents the results of the erosion tests obtained on the specimen of KW compacted at 95% of optimum Proctor energy, corresponding to a dry density of 1.4 t m^{-3}, and prepared at three different moisture contents: 21.5, 23.5, and 26.5%. These results show that there is a clear relation of dependence between the moisture content and the critical erosion shear stress.

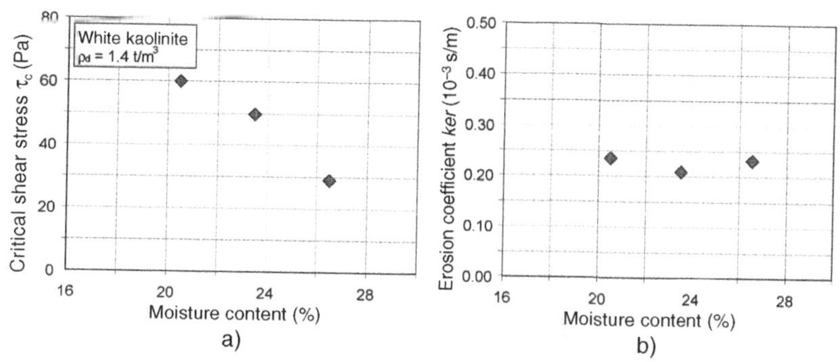

Figure 5.23. *Influence of the moisture content on the critical erosion shear stress and the erosion coefficient of white Kaolinite clay*

The more significant is the moisture content, the lower is the resistance to erosion. It is worth highlighting the fact that this result contradicts the results usually advanced in literature [WAN 02]. The previous research carried out on natural soils indeed showed that the increase in moisture content leads to an improvement of their resistance to erosion. It is thus possible that this behavior is linked to the nature of this clay.

The erosion coefficient is constant and is thus not influenced by the variation of moisture content. It is even worth noting that its value is identical to the value obtained on specimen prepared at different densities. These results show that the erosion coefficient does not vary with the moisture content and the dry density for a given texture of soil.

5.6. Importance of the erosion index for hydraulic structures

How should we assess the failure time once a pipe flow with erosion has been detected on-site? Assessing this failure time is crucial for choosing the type of intervention necessary and for carrying out an adapted procedure of risk management on-site. The erosion kinetic has a bearing on the velocity of the development of pipe flow with erosion, as well as on the duration of the development of a breach, and on the moment where the peak flow appears. This information is crucial for assessing the time that we have at our disposal as well as for choosing the right type of management strategy.

The two main breach parameters are the time of failure and the peak flow. In the case of overflow, an analysis of the existing relations has already been carried out [WAH 98]. These empirical relations are mere regressions that express the breach parameters depending on the information regarding the dam or the water storage. For pipe flow with erosion detected on-site, mechanically based relations were obtained from the same system of equations used for interpreting a HET [BON 09, BON 11]. These relations link the pipe, the time of failure, and the peak flow to parameters relating to the earthfill structure. One of these parameters is the erosion index.

5.6.1. *The case of earthfill dams*

A total of 14 cases of dam failure due to pipe flow with erosion were selected from Wahl's database [WAH 98]. This set of case studies is characterized as follows: the height of the dam ranges from 6 to 93 m, the failure time ranges from 0.5 to 5.25 h, and the peak flow at the time of failure is between 79 and 65,120 $m^3 s^{-1}$. The values of shear stress τ and the water velocity V in the pipe at the time of failure were estimated as follows: τ varies between 262 and 8,051 Pa and V varies between 7 and 40 $m s^{-1}$.

The erosion index I_e was also estimated: it is lower than 3 ($1.6 \leq I_e \leq 3$). In a first stage, we may deduce that:

– the soils of earthfill dams are usually very erodible or extremely erodible ($1 \leq I_e \leq 3$);

– out of these, only a few were subject to failure;

– when failure occurs, it usually happens very fast (within a few hours).

We may also perform another analysis, which seems to better correspond to reality:

– the soils of the earthfill dams are likely to have a highly variable erodibility ($1 \leq I_e$);

– a certain number of these is likely to have been in a failure situation;

– only the earthfill dams that have a very erodible soil or a highly erodible soil ($1 \leq I_e \leq 3$) had enough time to collapse.

5.6.2. *The case of dikes*

Let us now turn towards the flood protection fluvial dikes. These structures are characterized by low height, which eliminates one parameter within the framework of a simplified approach. We may thus deduce the following simplified formula for assessing the failure time of a dike, between the moment where the phenomenon of pipe flow with erosion was detected on-site, and the time when the top of the pipe collapses and a breach forms [BON 09]:

$$\Delta t_u[\text{mn}] \approx 5 \times 10^{I_e - 2} \qquad [5.16]$$

This expression assigns a size order for a dike of 5–10 m in height, subject to a maximum upstream level and to a pipe flow with erosion that has just been detected. The result is that if the soil has an erosion index $I_e \approx 2$, the failure will take place over the next few minutes, and if the soil has an erosion index $I_e \approx 3$, the failure will take place within several hours.

For a soil that has an erosion index $I_e \approx 4$, the failure will take place over the next few days, but we have no knowledge of such cases, i.e. of structures that were left unattended for several days. Therefore, either the fall in the water flow happened before the failure, or the failure took place after several days, and it was not however recorded.

5.7. Conclusion

This chapter dealt with flow erosion as found in a pipe, by means of an experimental approach carried out in the laboratory: the Hole Erosion Test. The developments observed concerned the set-up and the measuring device, as well as the experimental protocol and the test interpretation.

The results that we have obtained show a very good reproducibility that validates the device and the experimental protocol. The systematic tests carried out on the different soil textures that were chosen have allowed us to validate the approach for interpreting the tests. The exploitation of the turbidity signal enables us to directly obtain the relation between the shearing stress τ and the erosion rate $\dot{\varepsilon}$, as well as to deduce the characteristic parameters of erosion (critical shear stress τ_c and erosion coefficient C_e). These parameters are necessary for a numerical modeling (see Chapter 6 of this book).

Systematic tests were carried out on pure clays and on different mixtures of sand and clay. The tests that were carried out on mixtures of sand and clay with different relative concentrations (from 30 to 100% clay mass for kaolinite type and from 25 to 50% for illite) have allowed us to determine the erosion parameters of the different textures and to emphasize the effect of the clay fraction on these parameters. The tests carried out on Montmorillonite clay and the mixtures between sand and this clay did not present any quantitative results, since the swelling of this type of clay at the beginning of the test induces the shutter of the hole. In the given conditions, this particular type of texture cannot be tested using the HET. The parametrical study carried out on KW clay allowed us to emphasize the influence that the moisture content and the dry density had on the characteristics of erosion and the importance of considering them in the assessment of the resistance to erosion.

The HET is part of an array of geotechnical tests that were used during a safety assessment of a hydraulic structure. The critical shear stress and the kinetic erosion coefficient are considered two geotechnical parameters.

5.8. Bibliography

[BEN 12] BENAHMED N., BONELLI S., "Investigating concentrated leak erosion behaviour of cohesive soils by performing Hole Erosion Tests", *European Journal of Environmental and Civil Engineering*, vol. 16, no. 1, pp. 43–58, 2012.

[BON 06] BONELLI S., BRIVOIS O., BORGHI R., BENAHMED N., "On the modelling of piping erosion", *Comptes Rendus de Mécanique*, vol. 8–9, no. 334, pp. 555–559, 2006.

[BON 07] BONELLI S., MAROT D., TERNAT F., BENAHMED N., "Criteria of erosion for cohesive soils", *Assessment of the Risk of Internal Erosion of Water Retaining Structures: Dams, Dykes and Levees*, TUM Edt, pp. 45–59, 2007.

[BON 08] BONELLI S., BRIVOIS O., "The scaling law in the hole erosion test with a constant pressure drop", *International Journal for Numerical and Analytical Methods in Geomechanics*, vol. 32, pp. 1573–1595, 2008.

[BON 09] BONELLI S., BENAHMED N., PHILIPPE P., BERNARD A., GRÉMEAUX Y., NUNES G., "Evaluer le temps de rupture d'une digue en remblai érodée par conduit traversant", *Ingénieries*, no special issue "La prévention des inondations", pp. 85–94, 2009.

[BON 11] BONELLI S., BENAHMED N., "Piping flow erosion in water retaining structures", *International Journal on Hydropower and Dams*, vol. 18, no. 3, pp. 94–99, 2011.

[BRO 02] BROWN G.O., "The history of the Darcy–Weisbach equation for pipe flow resistance", in FREDRICH A., ROGERS J. (eds), *Environmental and Water Resources History*, ASCE, Reston, VA, pp. 34–43, 2002.

[CHE 10] CHEVALIER C., HAGHIGHI I., PHAM T.L., REIFFSTECK P., "Two complementary tests for characterizing the soil erosion", *5th International Conference on Scour and Erosion*, San Francisco, 2010.

[COU 09] COURIVAUD J.-R., FRY J.-J., BONELLI S., BENAHMED N., REGAZZONI P.-L., MAROT D., "Measuring the erodibility of soil materials constituting earth embankments: a key input for dams and levees safety assessment", *HYDRO 2009 International Conference and Exhibition*, 26–28 October, Lyon, France, 2009.

[DEY 99] DEY S., "Sediment threshold", *Applied Mathematical Modelling*, vol. 23, pp. 399–417, 1999.

[FAR 07] FARRAR J.A., TORRES R.L., ERDOGAN Z., "Bureau of reclamation erosion testing for evaluation of piping and internal erosion of dams", *Geotechnical Special Publication*, no. 167, pp. 22–31, 2007.

[FEL 07] FELL R., FRY J.J., (eds), *Internal Erosion of Dams and Their Foundations*, Taylor & Francis, 2007.

[FRY 97] FRY J.J., DEGOUTTE G., GOUBET A., "L'érosion interne: typologie, détection et reparation", *Barrages & Réservoirs* no. 6, p. 126, 1997.

[HAG 12] HAGHIGHI I., CHEVALIER C., DUC M., GUÉDON S., REIFFSTECK P., "Improvement of hole erosion test and results on reference soils", *Journal of Geotechnical and Geoenvironmental Engineering*, in press, 2012. doi: http://dx.doi.org/10.1061/(ASCE)GT.1943-5606.0000747

[LAC 08] LACHOUETTE, D., GOLAY, F., BONELLI, S., "One-dimensional modelling of piping flow erosion", *Comptes Rendus de Mécanique*, vol. 336, pp. 731–736, 2008.

[LEF 85] LEFEBVRE G., KAROL R., DOUVILLE S., "Erosivity of naturel intact structured clay: evaluation", *Canadian Geotechnical Journal*, vol. 22, pp. 508–517, 1985.

[PHA 08] PHAM T.L., Erosion et dispersion des sols argileux par un fluide, Thesis, Ecole Nationale des Ponts et Chaussées, France, 2008.

[POP 00] POPE S.B., *Turbulent Flows*, Cambridge University Press, UK, 2000.

[RED 97] REDDI L.N., BONALA M.V.S., "Critical shear stress and its relationship with cohesion for sand-kaolinite mixtures", *Canadian Geotechnical Journal*, vol. 34, no. 1, pp. 26–33, 1997.

[RED 00] REDDI L.N., LEE I., BONALA M.V.S., "Comparison of internal and surface erosion using flow pump test on a sand-kaolinite mixture", *Geotechnical Testing Journal*, vol. 23, no. 1, pp. 116–122, 2000.

[REI 06] REIFFSTECK P., PHAM T.L., VARGAS R., PAIHUA S., "Comparative study of superficial and internal erosion tests", *3rd International Conference on Scour and Erosion*, Amsterdam, 2006.

[REI 07] REIFFSTECK P., NGUYEN PHAM P.T., ARBAUT J., "Influence de la répartition granulométrique sur le comportement mécanique d'un sol", *Bulletin des Laboratoires des Ponts et Chaussées*, vol. 268–269, pp. 83–104, 2007.

[ROH 86] ROHAN K., LEFEBVRE G., DOUVILLE S., MILETTE J.-P., "A new technique to evaluate erosivity of cohesive material", *Geotechnical Testing Journal*, vol. 9, no. 2, pp. 87–92, 1986.

[SCH 87] SCHLICHTING H., *Boundary Layer Theory*, 7th ed., McGraw-Hill, New York, 1987.

[SHE 76] SHERARD J.-L., DUNNIGAN L.P., DECKER R.S., STEELE E.F., "Pinhole test for identifying dispersive soils", *Journal of the Geotechnical Engineering Division*, vol. 102, no. GT1, pp. 69–84, 1976.

[TER 08] TERNAT F., BOYER P., ANSELMET F., AMIELH M., "Erosion threshold of saturated natural cohesive sediments: modeling and experiments", *Water Resources Research*, vol. 44, no. W11434, pp. 1–18, 2008.

[WAH 98] WAHL T.L., Prediction of embankment dam breach parameters – a literature review and needs assessment, Dam Safety Report No. DSO-98-004, US Department of the Interior, Bureau of Reclamation, Denver, CO, 1998.

[WAN 02] WAN C.F., FELL R., Investigation of internal erosion and piping of soils in embankment dams by the slot erosion test and the hole erosion test, UNICIV Report No. R-412, The University of New South Wales, Sydney, Australia, July 2002.

[WAN 04a] WAN C.F., FELL R., "Investigation of rate of erosion of soils in embankment dams", *Journal of Geotechnical and Geoenvironmental Engineering*, vol. 30, no. 4, pp. 373–380, 2004.

[WAN 04b] WAN C.F., FELL R., "Laboratory tests on the rate of piping erosion of soils in embankment dams", *Journal of Geotechnical Testing Journal*, vol. 27, no. 3, pp. 295–303, 2004.

Chapter 6

Modeling of Interfacial Erosion

6.1. Introduction

The fields of research on river hydraulics and sediment transport have been fruitful in the works on moving bed erosion in an open channel flow [CHA 99, GRA 71, YAL 77]. In a significant number of these works, there is often confusion between sediment transport and erosion: in river hydraulics, erosion is seen as a spatial gradient of transport capacity at equilibrium, that is a record (rather crude) between detachment, transport, and deposition. At the scale at which we place ourselves, which is that of continuous medium and material, we cannot assimilate sediment transport and erosion. It is only recently that a detailed modeling of bedload transport has been developed ([CHA 10, OUR 07, OUR 09a, OUR 09b, PEY 09], see Chapter 8 of this book). This modeling describes the transition between the solid and fluid with a two-phase model. Solid and fluid phases interact through a source term in the equations of motion, which describes the exchange of momentum between phases. The rheology of granular media enables the description of a compact and dense medium or a granular flow. There is no soil/flow interface, but rather an area in which the behavior is intermediate between the solid and fluid. This modeling gives results consistent with the experiment in the case of a laminar flow on a granular bed.

In the oil sector, the question of sand production during the extraction process of the heavy oil is responsible for proposed modeling under the poromechanical context and mixture theory [PAP 01, PAP 05, PAP 06, VAR 96, VAR 01, VAR 04]. This modeling introduces a third phase, the fluidized solid that represents a smooth

---

Chapter written by Stéphane BONELLI, Frédéric GOLAY and Fabienne MERCIER.

transition between the solid and fluid. These three phases interact via the balance equations. A source term in the conservation of mass equations describes the mass exchange between phases and represents the erosion of the solid phase. This approach has been validated mainly on granular soils for laminar flow. It allows us to describe many situations: backward erosion, erosion by tangential flow, contact erosion, and fluidization. This is, to our knowledge, the first mechanically based approach that has been published and validated in the field of geomechanics.

The key point in these approaches is that erosion is controlled by an internal flow. This flow may be caused by a situation of flow from the soil, as an exfiltration, or by an external flow on the soil. The area of soil mobilized by the flow has a thickness. It is located around the fluid/solid interface. In this area, the flow is a Darcy–Brinkman flow. This description is only relevant for high permeability soil as a granular medium. In the case of a fine and cohesive soil, of very low permeability, it is not appropriate to consider an internal flow. Modeling of erosion must be different.

Figure 6.1 shows the erosion of soil saturated by a water flow depending on the type of soil and the direction of the flow in relation to the soil/water interface:

– For a granular soil subjected to tangential flow at the interface, the external flow, which can be laminar or turbulent, induces a Darcy flow in the soil; around the interface a bedload and a Brinkman area develop (transition between an internal Darcy flow and an external boundary layer flow); within these two areas, the granular medium is likely to be fluidized.

– For a granular soil subjected to a normal flow at the interface, the direction of flow must be taken into account:

- If the flow comes from the soil, this is an internal Darcy flow, therefore it is laminar, that induces a fluidization area and an external laminar flow.

- If the flow enters soil, it is a laminar or turbulent flow that is equally susceptible to induce an internal Darcy flow and a fluidization area (but not at the entry point).

– For a fine cohesive soil subjected to a tangential flow at the interface, the external flow, which can be laminar or turbulent, does not induce an internal flow because the permeability is very low.

– For a fine cohesive soil subjected to a normal flow at the interface, the direction of flow must be taken into account:

- If the flow is directed toward the soil, it is a laminar or turbulent flow; the soil being saturated, this impinging jet does not induce any internal flow, but it is likely to induce pressure fluctuations into the ground.

- An outflow is not applicable, the permeability being very low.

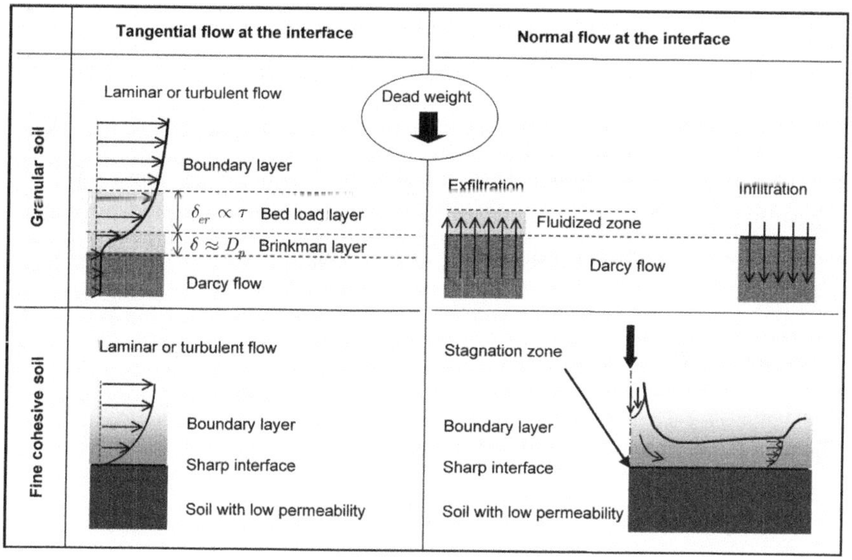

Figure 6.1. *Typology of flow/water interface depending on the type of soil and the direction of the flow*

This chapter deals with the modeling of interfacial erosion of a fine cohesive soil. Our description is original in the sense that we only consider two phases (solid and fluid) separated by a singular interface (of discontinuity) and not by a third phase (fluidized solid). Fields (density, volume fraction, strain, pressure, velocity, etc.) are likely to be discontinuous on the interface. On each side of the interface, the medium is two phases: on the one hand the soil (a compact assembly of particles containing water) and on the other hand the flow (water containing particles in dispersed phase). The key point is the use of jump equations to introduce the law of erosion and simplification of the system of equations by evaluating the orders of magnitudes of phenomena [BON 06, BON 08, BON 11, BRI 07, GOL 10, GOL 11a, GOL 11b, LAC 08].

6.2. Modeling of a two-phase medium

6.2.1. *Definition of volume quantities*

On the microscopic scale, components are distributed in space. We can define for the fluid component – *water* – its characteristic function I_w^μ ($I_w^\mu = 1$ in water, 0 otherwise), its density ρ_w^μ, and its velocity $\mathbf{u}_w^\mu$. Similarly, we can define for the *solid* component (the soil particles) its characteristic function I_s^μ ($I_s^\mu = 1$ in the solid, 0 otherwise), its density ρ_s^μ, and its velocity $\mathbf{u}_s^\mu$. The medium consists of only water and solid, therefore $I_w^\mu + I_s^\mu = 1$.

We consider V, an elementary representative volume of mass M in the two-phase medium: $V = V_w \cup V_s$, where V_w is the volume occupied by water and V_s is the volume occupied by the solid. The volumes of water and solid in V are given by:

$$|V_w| = \int_V I_w^\mu dV, \; |V_s| = \int_V I_s^\mu dV, \; |V| = |V_w| + |V_s| \qquad [6.1]$$

The masses of water and solid in V are given by:

$$M_w = \int_V I_w^\mu \rho_w^\mu dV, \; M_s = \int_V I_s^\mu \rho_s^\mu dV, \; M = M_w + M_s \qquad [6.2]$$

At the macroscopic scale, the two components define by averaging two phases, which are processed by superimposed continuous medium (Figure 6.2).

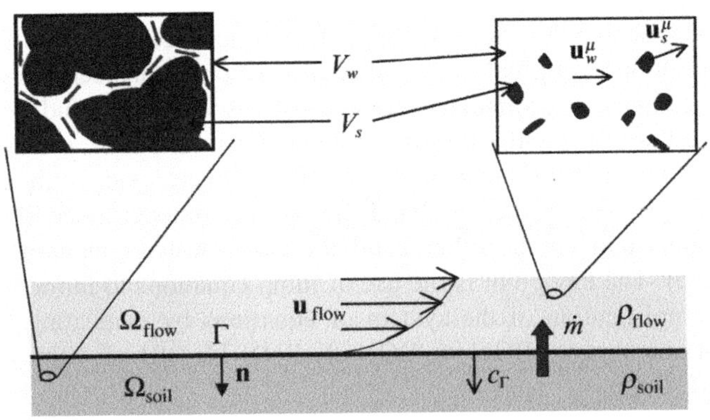

Figure 6.2. *Two-phase flow diagram on two-phase soil, separated by an interface. On the one side water between solid particles in compact arrangement (the soil). On the other side solid particles in water (the flow). The interface is crossed by the eroded mass flux and moves along*

The volume fractions of water and solid are defined as follows:

$$\phi_w = \frac{|V_w|}{|V|}, \; \phi_s = \frac{|V_s|}{|V|}, \; \phi_w + \phi_s = 1 \qquad [6.3]$$

The mass fractions of water and solid are defined as follows:

$$Y_w = \frac{M_w}{M},\ Y_s = \frac{M_s}{M},\ Y_w + Y_s = 1 \qquad [6.4]$$

Volume fractions and mass fractions are linked as follows:

$$\phi_w \rho_w = Y_w \rho,\ \phi_s \rho_s = Y_s \rho,\ \rho = \frac{M}{V} \qquad [6.5]$$

The real average velocities of water phases $\mathbf{u}_w$ and solid phases $\mathbf{u}_s$ are given by:

$$\mathbf{u}_w = \frac{1}{|V_w|}\int_V I_w^\mu \mathbf{u}_w^\mu dV,\ \mathbf{u}_s = \frac{1}{|V_s|}\int_V I_s^\mu \mathbf{u}_s^\mu dV \qquad [6.6]$$

By real average, we mean *average of the actual volume occupied by the component*: if the microscopic quantity is uniform, it is equal to its real average. The components are assumed to be incompressible and homogeneous: the real average of the density is equal to the true density ($\rho_w = \rho_w^\mu$ and $\rho_s = \rho_s^\mu$).

The density of the mixture can be defined in two ways:

$$\rho = \left(\frac{Y_s}{\rho_s} + \frac{Y_w}{\rho_w}\right)^{-1} = \phi_s \rho_s + \phi_w \rho_s \qquad [6.7]$$

The average mass velocity of the mixture, on the one hand, is given by:

$$\mathbf{u} = Y_s \mathbf{u}_s + Y_w \mathbf{u}_w \qquad [6.8]$$

The average volume velocity of the mixture, on the other hand, is given by:

$$\hat{\mathbf{u}} = \phi_s \mathbf{u}_s + \phi_w \mathbf{u}_w \qquad [6.9]$$

These two velocities are linked by the relative velocity of the two phases, which can be expressd by the diffusion flux of mass $\mathbf{j}$:

$$\mathbf{u} = \hat{\mathbf{u}} + \left(\frac{1}{\rho_w} - \frac{1}{\rho_s}\right)\mathbf{j},\ \mathbf{j} = Y_s Y_w \rho(\mathbf{u}_s - \mathbf{u}_w) \qquad [6.10]$$

or by the diffusion flux of volume $\mathbf{q}$:

$$\mathbf{u} = \hat{\mathbf{u}} + \left(\frac{\rho_s - \rho_w}{\rho} \right) \mathbf{q}, \; \mathbf{q} = \phi_s \phi_w (\mathbf{u}_s - \mathbf{u}_w) \qquad [6.11]$$

The diffusion flux of mass $\mathbf{j}$ and the diffusion flux of volume $\mathbf{q}$ are linked as follows:

$$\mathbf{j} = \frac{\rho_s \rho_w}{\rho} \mathbf{q} \qquad [6.12]$$

The equivalent quantities $\mathbf{q}$ and $\mathbf{j}$ do not favor either of the two phases: they are two-phase quantities that represent the fact that the two phases have different velocities. They are adapted to describe a two-phase flow (water + grains). For a porous medium, the description lies within the mechanics of the solids: the solid phase (grains + interstitial field) is preferred. It is what defines the configuration. The fluid phase is described in relation to this configuration. The water diffusion in the soil is defined by the Darcy velocity:

$$\mathbf{q}_D = \phi_w (\mathbf{u}_w - \mathbf{u}_s) \qquad [6.13]$$

6.2.2. *Definition of interfacial quantities*

We have Γ, an interface, separating the two-phase flow of the two-phase soil (Figure 6.2). The mass flux of solids $\dot{m}_s$ and the mass flux of water $\dot{m}_w$ that cross Γ are:

$$\dot{m}_s \underset{def}{=} \phi_s \rho_s (c_\Gamma - \mathbf{u}_s \cdot \mathbf{n}), \; \dot{m}_w \underset{def}{=} \phi_w \rho_w (c_\Gamma - \mathbf{u}_w \cdot \mathbf{n}) \text{ on } \Gamma \qquad [6.14]$$

where c_Γ is the interface celerity and $\mathbf{n}$ the normal unit at Γ directed toward the interior of the soil. The total mass flux $\dot{m} = \dot{m}_s + \dot{m}_w$ that crosses Γ is:

$$\dot{m} \underset{def}{=} \rho (c_\Gamma - \mathbf{u} \cdot \mathbf{n}) \text{ on } \Gamma \qquad [6.15]$$

The erosion process is therefore characterized by the mass flux crossing Γ and the celerity of Γ. Accordingly, Γ is not a material interface as, at two distinct points in time, Γ is not defined by the same particles. We therefore assume that Γ is a singular interface that represents only a geometrical separation with no thickness between the flow of water and the soil. In this sense raising the celerity of the interface is more appropriate than to raise its velocity.

This approach is relevant for fine cohesive soils: the characteristic size of particles is of the order of μm and the interface between clay and surrounding water is better defined than in the case of sand or gravel. The benefit of this approach is to consider the fact that a compact set of cohesive solid particles (fine soil) behaves in a distinctly different way from the same set of solid particles in diluted suspension. Similarly, a phase of water interacting with a dispersed phase of particles has a behavior that differs distinctly from that of a phase of interstitial water, within clay. The disadvantage is the need to consider a singular interface. It is on this point that the effort has been made.

6.2.3. *Conservation of mass equations*

The local form of conservation of solid mass and fluid mass equations is written, in the case of incompressible components:

$$\frac{\partial \phi_s}{\partial t} + \nabla \cdot (\phi_s \mathbf{u}_s) = 0\,, \quad \frac{\partial \phi_w}{\partial t} + \nabla \cdot (\phi_w \mathbf{u}_w) = 0 \ \text{ in } \Omega \tag{6.16}$$

Let us note $[\![a]\!] = a_{\text{soil}} - a_{\text{flow}}$ the jump of any magnitude a through Γ, where a_{soil} is the value of a on Γ on the soil side and a_{flow} is the value of a on Γ on the fluid flow side. On Γ the jump equations are:

$$[\![\dot{m}_s]\!] = 0,\ [\![\dot{m}_w]\!] = 0 \ \text{ on } \Gamma \tag{6.17}$$

Summing up the two equations [6.16], we obtain the conservation of mass equation of the mixture in relation to the average volume velocity. By introducing the result in equation $[6.16]_1$ (the first expression of equation [6.16]), we obtain the diffusion equation of the solid phase within the mixture in relation to the diffusion flux of volume:

$$\nabla \cdot \hat{\mathbf{u}} = 0, \quad \frac{\partial \phi_s}{\partial t} + \nabla \cdot (\phi_s \hat{\mathbf{u}}) = -\nabla \cdot \mathbf{q} \ \text{ in } \Omega \tag{6.18}$$

$$[\![\hat{\mathbf{u}} \cdot \mathbf{n}]\!] = 0,\ [\![\phi_s (c_\Gamma - \hat{\mathbf{u}} \cdot \mathbf{n})]\!] - [\![\mathbf{q} \cdot \mathbf{n}]\!] = 0 \ \text{ on } \Gamma \tag{6.19}$$

By multiplying the equation $[6.16]_1$ by ρ_s and the equation $[6.16]_2$ by ρ_w, then by summing up these two equations, we obtain a conservation of the mass equation of the mixture in relation to the average mass velocity. By introducing the result in equation [6.16]₁, we obtain the diffusion equation of the solid phase within the mixture in relation to the diffusion flux of mass:

$$\frac{\partial \rho}{\partial t} + \nabla \cdot (\rho \mathbf{u}) = 0 \,, \; \frac{\partial (\rho Y_s)}{\partial t} + \nabla \cdot (\rho Y_s \mathbf{u}) = -\nabla \cdot \mathbf{j} \text{ in } \Omega \qquad [6.20]$$

$$[\![\dot{m}]\!] = 0 \,, \; \dot{m}[\![Y]\!] - [\![\mathbf{j} \cdot \mathbf{n}]\!] = 0 \text{ on } \Gamma \qquad [6.21]$$

Equation $[6.21]_1$ reflects the continous total mass flux that crosses Γ. Equation $[6.21]_2$ reflects the fact that part of the mixture is in advection and the other in diffusion, on both sides of the interface, in different proportions. The systems of equations ([6.16], [6.17]), ([6.18], [6.19]), and ([6.20], [6.21]) are equal.

6.2.4. *Equations of motion*

Solid and fluid equations of motion are written as:

$$\rho_s \left[\frac{\partial (\phi_s \mathbf{u}_s)}{\partial t} + \nabla \cdot (\phi_s \mathbf{u}_s \otimes \mathbf{u}_s) \right] = \nabla \cdot \mathbf{T}_s + \phi_s \rho_s \mathbf{g} - \mathbf{f} \text{ in } \Omega \qquad [6.22]$$

$$\rho_w \left[\frac{\partial (\phi_w \mathbf{u}_w)}{\partial t} + \nabla \cdot (\phi_w \mathbf{u}_w \otimes \mathbf{u}_w) \right] = \nabla \cdot \mathbf{T}_w + \phi_w \rho_w \mathbf{g} + \mathbf{f} \text{ in } \Omega \qquad [6.23]$$

Vector $\mathbf{g}$ represents the gravitational forces. Vector $\mathbf{f}$ represents the average value resulting from the balance of forces exerted by the fluid on the particles. This term is essential when the description uses the equations of motion of each phase, both for a laminar flow [MAS 99, MAS 03] and for a turbulent flow [ABU 95]. It is this term that models the erosion seen as an exchange of momentum [CHA 10, OUR 07].

Stress tensors $\mathbf{T}_s$ and $\mathbf{T}_w$ are respectively associated with the solid and fluid phases. On Γ jump equations are:

$$[\![\mathbf{T}_s \cdot \mathbf{n}]\!] = -\dot{m}_s [\![\mathbf{u}_s]\!], \; [\![\mathbf{T}_w \cdot \mathbf{n}]\!] = -\dot{m}_w [\![\mathbf{u}_w]\!] \text{ on } \Gamma \qquad [6.24]$$

By summing up equations [6.22] and [6.23], we obtain the equation of the motion of the mixture in relation to the average mass velocity:

$$\frac{\partial (\rho \mathbf{u})}{\partial t} + \nabla \cdot (\rho \mathbf{u} \otimes \mathbf{u}) = \nabla \cdot \mathbf{T} + \rho \mathbf{g} \text{ in } \Omega \qquad [6.25]$$

$$[\![\mathbf{T}\cdot\mathbf{n}]\!] = -\dot{m}[\![\mathbf{u}]\!] \quad \text{on } \Gamma \tag{6.26}$$

$$\mathbf{T} = \mathbf{T}_s + \mathbf{T}_w - Y_s Y_w \rho(\mathbf{u}_s - \mathbf{u}_w)\otimes(\mathbf{u}_s - \mathbf{u}_w) \tag{6.27}$$

Obviously, the medium being two phases, two equations of motion are necessary. Strictly speaking, equation [6.25] must be completed by one of the two equations, [6.22] or [6.23], or a combination of both [LHU 01]. The other equation provides an expression for vector $\mathbf{f}$, which does not intervene in the equation of motion of the mixture, but which will play a role in the diffusion law of $\mathbf{j}$ (or $\mathbf{q}$). The same applies for jump equations. Nevertheless, in the present model, we will see that concentrations are weak. It is therefore not necessary to clarify the diffusion laws and the interaction vector **f**.

6.3. Modeling of the soil/fluid interface

6.3.1. *Simplifying assumptions*

Taking into account the two-phase nature of the continuous medium studied (equations [6.3]$_3$ and [6.4]$_3$), we note $\phi = \phi_s = 1-\phi_w$ the volume fraction of the solid particles in the mixture. In the soil, ϕ represents the compactness ($\phi_{\text{soil}} = 1-n$, where n is the porosity). Equally, we note $Y = Y_s = 1 - Y_w$ the mass fraction of solid particles in the mixture (in the soil, $Y_{\text{soil}} = (1-n)\rho_s/\rho_{\text{soil}}$).

To simplify the notation, we neglect the deformations of soil and assume that the frame is attached to the soil. It results that $\mathbf{u}_s = 0$ in the soil.

Velocity vectors $\mathbf{u}$ and stress vectors $\mathbf{T}\cdot\mathbf{n}$ are decomposed by normal and tangential components on Γ:

$$\mathbf{u} = u\mathbf{n} + \mathbf{v}, \quad u = \mathbf{u}\cdot\mathbf{n}, \quad \mathbf{v} = \mathbf{u} - (\mathbf{u}\cdot\mathbf{n})\mathbf{n} \tag{6.28}$$

$$\mathbf{T}\cdot\mathbf{n} = \sigma\mathbf{n} + \boldsymbol{\tau}, \quad \sigma = \mathbf{n}\cdot\mathbf{T}\cdot\mathbf{n}, \quad \boldsymbol{\tau} = \mathbf{T}\cdot\mathbf{n} - (\mathbf{n}\cdot\mathbf{T}\cdot\mathbf{n})\mathbf{n} \tag{6.29}$$

We assume that the tangential velocity is continuous through Γ. It is a reasonable classic assumption, but, constitutive, often called *the no-slip condition*. It results that $[\![\mathbf{v}]\!] = 0$. This means that the particles of soil leave those following a normal direction at the interface, which is not intuitive: further research is needed on this point.

6.3.2. *Rewriting of jump equations*

The mass flux that crosses Γ is continuous through Γ (equations [6.17] and $[6.21]_1$). The jump equation $[6.21]_1$ means that the matter coming from the soil crosses through the interface and moves into the flow, which is reflected by:

$$\underbrace{\rho_{\text{soil}}(c_\Gamma - u_{\text{soil}})}_{\substack{\text{Water + Particles}\\ \text{leaving the soil}}} = \underbrace{\dot{m}}_{\substack{\text{Water + Particles}\\ \text{crossing }\Gamma}} = \underbrace{\rho_{\text{flow}}(c_\Gamma - u_{\text{flow}})}_{\substack{\text{Water + Particles}\\ \text{entering the flow}}} \text{ on } \Gamma \qquad [6.30]$$

The mass flux crossing Γ (equation $[6.21]_1$) is the sum of the eroded mass flux (water + particles) and the Darcy flux. In fact, we deduce, from the equality $\rho(c_\Gamma - \mathbf{u}\cdot\mathbf{n}) = \rho(c_\Gamma - \mathbf{u}_s\cdot\mathbf{n}) + \rho(1-Y)(u_s - u_w)$ written on the one side of the soil, that:

$$\underbrace{\dot{m}}_{\text{Mass flux crossing }\Gamma} = \underbrace{\rho_{\text{soil}} c_\Gamma}_{\text{Mass flux of eroded material}} - \underbrace{\rho_w q_D}_{\text{Darcy mass flux}} \text{ on } \Gamma \qquad [6.31]$$

Jump equation $[6.21]_2$ is rewritten, on the other side, as follows:

$$\underbrace{j_{\text{flow}}}_{\substack{\text{Relative motion}\\ \text{of solid particles}\\ \text{in the flow}}} = \underbrace{\rho_{\text{soil}}(Y_{\text{flow}} - Y_{\text{soil}})c_\Gamma}_{\substack{\text{Eroded solid particles leaving the soil}\\ \text{and entering the flow}}} - \underbrace{\rho_w q_D Y_{\text{flow}}}_{\substack{\text{Relative motion}\\ \text{of solid particles due to } q_D}} \text{ on } \Gamma \qquad [6.32]$$

By decomposing the jump equation [6.26] in normal and tangential components, we obtain:

$$\sigma_{\text{flow}} - \sigma_{\text{soil}} = \dot{m}^2 \left(\frac{1}{\rho_{\text{flow}}} - \frac{1}{\rho_{\text{soil}}} \right), \quad \tau_{\text{flow}} - \tau_{\text{soil}} = 0 \text{ on } \Gamma \qquad [6.33]$$

Equation $[6.33]_1$ indicates that the normal stresses are discontinuous when Γ is crossed by a mass flux. This discontinuity is due to the momentum transfer on both sides of the interface and at the difference of the density between the flow and the soil. Equation $[6.33]_2$ indicates that τ is continuous on Γ, because of the assumption of continuity of the tangential velocity of Γ.

6.3.3. *The erosion law*

The erosion law is a constitutive relationship attached to the interface. It is most often a threshold law of the form:

$$\rho_{\text{soil}} c_\Gamma = \begin{cases} k_{\text{er}}(\tau - \tau_c) \text{ if } \tau > \tau_c \\ 0 \text{ else} \end{cases} \text{ on } \Gamma \qquad [6.34]$$

where $\tau = \|\boldsymbol{\tau}\|$ is the norm of the tangential stress on Γ. The threshold τ_c is the critical stress and k_{er} the erosion coefficient. Equally, we can formulate an erosion law in relation to the single solid mass that crosses Γ:

$$\rho_d c_\Gamma = \begin{cases} C_e(\tau - \tau_c) \text{ if } \tau > \tau_c \\ 0 \text{ else} \end{cases} \text{ on } \Gamma \qquad [6.35]$$

A third expression links the normal celerity of the interface to the tangential stress:

$$c_\Gamma = \begin{cases} k_d(\tau - \tau_c) \text{ if } \tau > \tau_c \\ 0 \text{ else} \end{cases} \text{ on } \Gamma \qquad [6.36]$$

Equations [6.34], [6.35], and [6.36] are equal. The relationship between k_{er}, coefficient C_e of Wan and Fell [WAN 04a, WAN 04b], and coefficient k_d of Hanson [HAN 91, HAN 04] is given by:

$$\frac{k_{\text{er}}}{\rho_{\text{soil}}} = k_d = \frac{C_e}{\rho_d} \qquad [6.37]$$

where ρ_d is the density of the dry soil. Concerning units, ρ is expressed in kg m^{-3}, k_{er} and C_e in s m^{-1}, and k_d in m s^{-1} Pa^{-1} or in m^3 s^{-1} N^{-1} (in literature, k_d is often expressed in cm^3 s^{-1} N^{-1}, which corresponds to 10^{-6} m s^{-1} Pa^{-1}). Parameter C_e is the most used in the geotechnical community. These three parameters are equal (to the knowledge of the density of the near soil).

The physical significance of the erosion coefficient is not obvious. Parameters k_{er} and C_e vary on several orders of magnitudes, between 10^{-2} and 10^{-6} s m^{-1}: it is, by no means, the opposite of a velocity. Similarly, k_d, which can be expressed in m^2 s kg^{-1}, cannot be considered as the opposite of a mass flux. Coefficient k_d can be seen as the ratio between a characteristic dimension ℓ_Γ and the surface viscosity μ_{soil} of the eroded soil. Let us assume that ℓ_Γ represents the depth of influence of the flow inside the soil: ℓ_Γ can be seen as a porous characteristic dimension, quantified by the geometric permeability of the soil κ_{soil}, so that $\ell_\Gamma \approx (\kappa_{\text{soil}})^{1/2}$. Therefore, $k_d \propto (\kappa_{\text{soil}})^{1/2}/\mu_{\text{soil}}$. The geometric permeability and the viscosity of a fine soil varies on several magnitudes (between 10^{-13} and 10^{-16} m^2 for κ_{soil}, between 1 and 10^5 Pa s for μ_{soil}). This interpretation is likely to explain why the erosion coefficient varies on several orders of magnitude from one soil to the other. It is only at this stage that an assumption has to be confirmed by research.

6.4. Modeling of flow with erosion

6.4.1. *Two-phase flow with erosion and transport*

The flow problem with erosion is defined by the following equations:

– Conservation equation in Ω

$$\nabla \cdot \mathbf{u} = 0 \text{ in } \Omega \tag{6.38}$$

$$\rho\left[\frac{\partial \mathbf{u}}{\partial t} + (\nabla \mathbf{u}) \cdot \mathbf{u}\right] = \nabla \cdot \mathbf{T} \text{ in } \Omega \tag{6.39}$$

– Conditions on Γ

$$\mathbf{u} = u\mathbf{n},\quad u = \frac{\rho_w}{\rho_{\text{flow}}} q_D - \left(\frac{\rho_{\text{soil}}}{\rho_{\text{flow}}} - 1\right) c_\Gamma,\ \text{on } \Gamma \tag{6.40}$$

$$c_\Gamma = \begin{cases} k_d(\tau - \tau_c) \text{ if } \tau > \tau_c \\ 0 \text{ else} \end{cases},\quad \tau = \| \boldsymbol{\tau} \| = \sqrt{(\mathbf{T} \cdot \mathbf{n})^2 - (\mathbf{n} \cdot \mathbf{T} \cdot \mathbf{n})^2} \tag{6.41}$$

The condition expressed by equation [6.40] is deduced from equations [6.30] and [6.31]. It expresses the fact that the normal velocity at Γ is non-zero, as a result of infiltration or exfiltration of water, on the one hand, and of discontinuity of densities on either side of Γ when there is erosion, on the other hand.

The stress tensor in flow $\mathbf{T}$ consists of three terms:

$$\mathbf{T} = \underbrace{-p\mathbf{I}}_{\text{Static pressure}} + \underbrace{2\mu\mathbf{D}(\hat{\mathbf{u}})}_{\text{Viscous stress}} - \underbrace{\rho\overline{\mathbf{u}' \otimes \mathbf{u}'}}_{\text{Turbulent stress}} \tag{6.42}$$

$$\mathbf{D}(\hat{\mathbf{u}}) = \frac{1}{2}\left[\nabla \hat{\mathbf{u}} + (\nabla \hat{\mathbf{u}})^T\right] \tag{6.43}$$

where $\mathbf{I}$ is the stress identity, p the static pressure, and μ the viscosity of the mixture, which depends on the concentration of solid particles (the more important the concentration is, the more viscous the mixture).

The tensor $\mathbf{D}(\hat{\mathbf{u}})$ is the symmetrical part of the gradient of the average volume velocity, of zero divergence (equation [6.18]$_1$), because the mixture of the two incompressible components is incompressible. This concerns a choice, now well known for concentrated suspensions [LHU 01, ZHA 97].

In the case of a description by the Navier–Stokes average incompressible Reynolds average Navier–Stokes (RANS) equations, the turbulent stress tensor $-\rho_w \overline{\mathbf{u}' \otimes \mathbf{u}'}$ is defined from velocity fluctuations $\mathbf{u}'$ relative to the average mass velocity $\mathbf{u}$. When the flow is laminar, this tensor is negligible relative to the viscous stresses. However, when the flow is turbulent, it is necessary to choose a model of this tensor.

Transport of soil particles can be described by:

$$\frac{\partial(\rho Y)}{\partial t} + \nabla \cdot (\rho Y \mathbf{u}) = -\nabla \cdot \mathbf{j} \qquad [6.44]$$

$$\mathbf{j} \cdot \mathbf{n} = \rho_{\text{soil}}(Y - Y_{\text{soil}})c_\Gamma - \rho_w q_D Y \quad \text{on } \Gamma \qquad [6.45]$$

Equation [6.44] must be completed by the diffusion (and dispersion) law linking $\mathbf{j}$ to the gradient of concentration Y, but this will not be necessary here.

This system describes a two-phase flow (of water containing soil particles), likely to be turbulent, within the field defined by a mobile interface, whose mobility depends on the flow [BRI 07]. This description is very complex. It is essential to study situations for which modeling can be simplified. It requires, for this, the evaluation of the orders of magnitudes that characterize the elementary mechanisms.

6.4.2. *Analysis of the orders of magnitudes*

Let us note ℓ_w is a characteristic dimension of a fluid field, Δp_0 the pressure drop along ℓ_w, V_w the velocity gauge of the flow, ℓ_Γ a characteristic dimension of the area close to the interface, τ_Γ the gauge of the tangential stress on the interface, and k_{soil} the permeability gauge of the soil close to the interface. The velocity gauge of erosion is $V_{\text{er}} = k_{\text{er}}\tau_\Gamma/\rho_{\text{soil}}$. The time characteristic of erosion is $t_{\text{er}} = \ell_\Gamma / V_{\text{er}}$. The density gauge is ρ_w. Let us introduce the following dimensionless quantities:

$$\tilde{t} = \frac{t}{t_{\text{er}}}, \; \tilde{\mathbf{u}} = \frac{1}{V_w}\mathbf{u}, \; \tilde{\hat{\mathbf{u}}} = \frac{1}{V_w}\hat{\mathbf{u}}, \; \tilde{\nabla} = \ell_\Gamma \nabla, \; \widetilde{\overline{\mathbf{u}' \otimes \mathbf{u}'}} = \frac{1}{V_w^2}\overline{\mathbf{u}' \otimes \mathbf{u}'} \qquad [6.46]$$

$$\widetilde{\nabla p} = \frac{\ell_w}{\Delta p_0}\nabla p,\ \tilde{\sigma} = \frac{\sigma}{\tau_\Gamma},\ \tilde{\tau} = \frac{1}{\tau_\Gamma}\tau,\ \tilde{\tau}_c = \frac{\tau_c}{\tau_\Gamma},\ \tilde{c}_\Gamma = \frac{c_\Gamma}{V_{er}} \qquad [6.47]$$

$$\tilde{\rho} = \frac{\rho}{\rho_w},\ \tilde{Y} = \frac{Y}{Y^{\text{ref}}},\ \tilde{q}_D = \frac{q_D}{k_{\text{soil}}},\ \tilde{\mathbf{j}} = \frac{1}{j^{\text{ref}}}\mathbf{j} \qquad [6.48]$$

The gauges are chosen in such a way that the dimensionless quantity $\tilde{a}$ corresponding to a is to a maximum in $O(1)$. The choice of these gauges can only be evaluated afterwards, by checking the orders of magnitudes on a laboratory test or on a numerical modeling.

From equation [6.45], we deduce that $j^{\text{ref}} = \rho_{\text{soil}}(Y^{\text{ref}} - Y_{\text{soil}})V_{\text{er}} - \rho_w k_{\text{soil}} Y^{\text{ref}}$. The concentration gauge Y^{ref} will be deduced from the analysis of transport equations.

The dimensionless system becomes:

$$\tilde{\nabla}\cdot\tilde{\mathbf{u}} = 0 \qquad [6.49]$$

$$A_1\frac{\partial\tilde{\mathbf{u}}}{\partial\tilde{t}} = \widetilde{\nabla p} + A_2\tilde{\nabla}\cdot\left[\mathbf{D}(\tilde{\mathbf{u}})\right] - A_3\tilde{\nabla}\cdot\left[(\tilde{\mathbf{u}}\otimes\tilde{\mathbf{u}}) + \widetilde{\mathbf{u}'\otimes\mathbf{u}'}\right] \qquad [6.50]$$

$$\tilde{\mathbf{u}} = \tilde{u}\mathbf{n},\ \tilde{u} = \frac{\rho_w k}{\rho_{\text{flow}}V_w}\tilde{q}_D - \left(\frac{\rho_{\text{soil}}}{\rho_{\text{flow}}} - 1\right)\frac{V_{\text{er}}}{V_w}\tilde{c}_\Gamma,\ \text{on } \Gamma \qquad [6.51]$$

$$\tilde{c}_\Gamma = \begin{cases}\tilde{\tau} - \tilde{\tau}_c \text{ if } \tilde{\tau} > \tilde{\tau}_c \\ 0 \text{ else}\end{cases} \quad \text{on } \Gamma \qquad [6.52]$$

$$A_1 = \frac{\rho_w}{\rho_{\text{soil}}}\frac{V_w k_{\text{er}}\tau_\Gamma}{\Delta p_0 \ell_\Gamma/\ell_w},\ A_2 = \frac{2\mu_w V_w}{\Delta p_0 \ell_\Gamma^2/\ell_w},\ A_3 = \frac{\rho_w V_w^2}{\Delta p_0 \ell_\Gamma/\ell_w} \qquad [6.53]$$

The dimensionless system describing the transport of particles is:

$$B_1\frac{\partial(\tilde{\rho}\tilde{Y})}{\partial\tilde{t}} + B_2\tilde{\nabla}\cdot(\tilde{\rho}\tilde{Y}\tilde{\mathbf{u}}) + B_3\tilde{\nabla}\cdot\tilde{\mathbf{j}} = 0 \qquad [6.54]$$

$$B_1 = Y^{\text{ref}}\rho_w\frac{V_{\text{er}}}{V_w},\ B_2 = Y^{\text{ref}}\rho_w,\ B_3 = \frac{j^{\text{ref}}}{V_w} \qquad [6.55]$$

From equation [6.54] we deduce the concentration gauge close to the wall:

$$Y^{\text{ref}} = \frac{\rho_{\text{soil}} V_{\text{er}}}{\rho_w (V_w + k_{\text{soil}}) + \rho_{\text{soil}} V_{\text{er}}} Y_{\text{soil}} \qquad [6.56]$$

We define three dimensionless numbers:

– The Reynolds number of the flow, R_e, which is the ratio between the inertial momentum $\rho_w V_w^2$ and the viscous momentum $\mu_w V_w / \ell_\Gamma$:

$$R_e = \frac{\rho_w V_w \ell_\Gamma}{\mu_w} \qquad [6.57]$$

– The erosion kinetics, $\tilde{V}_{\text{er}}$, which is the ratio between the erosion velocity V_{er} and the flow velocity V_w:

$$\tilde{V}_{\text{er}} = \frac{k_{\text{er}} \tau}{\rho_{\text{soil}} V_w} \qquad [6.58]$$

– The erosion number, $\tilde{k}_{\text{er}}$, which is the ratio between the tangential momentum $\rho_w V_{\text{er}} V_w$ due to the mobility of Γ and the stress τ_Γ:

$$\tilde{k}_{\text{er}} = \frac{\rho_w}{\rho_{\text{soil}}} k_{\text{er}} V_w \qquad [6.59]$$

In the case of laminar flows (R_e = 1), therefore A_3 = A_2 and $A_2 = 1$. The orders of magnitudes are:

$$\tau_\Gamma = \frac{2\mu_w V_w}{\ell_\Gamma}, \ \Delta p_0 = \frac{2\mu_w V_w}{\ell_\Gamma^2 / \ell_w}, \ V_{\text{er}} = \frac{2\mu_w k_{\text{er}} V_w}{\ell_\Gamma \rho_{\text{soil}}} \qquad [6.60]$$

If, in addition, the erosion kinetics is small compared to the flow velocity ($\tilde{V}_{\text{er}} \ll 1$), then the number of erosion is small ($\tilde{k}_{\text{er}} \ll 1$):

$$\underbrace{k_{\text{er}} \ll}_{\text{Low kinetics of erosion}} \frac{\ell_\Gamma \rho_{\text{soil}}}{\mu_w} \underbrace{\ll V_w^{-1}}_{\text{Laminar flow}} \qquad [6.61]$$

In the case of turbulent flows ($R_e \gg 1$), then $A_3 \gg A_2$ and $A_3 = 1$, therefore:

$$\tau_\Gamma = \rho_w V_w^2,\ \Delta p_0 = \frac{\rho_w V_w^2}{\ell_\Gamma/\ell_w},\ V_{er} = \frac{\rho_w k_{er} V_w^2}{\rho_{soil}} \qquad [6.62]$$

This time, it comes down to assuming a small number of erosions ($\tilde{k}_{er} \ll 1$) and a small erosion kinetics ($\tilde{V}_{er} \ll 1$). This situation arises when:

$$\underbrace{k_{er} \ll}_{\text{Low kinetics of erosion}} \quad V_w^{-1} \quad \underbrace{\ll \frac{\ell_\Gamma \rho_{soil}}{\mu_w}}_{\text{Turbulent flow}} \qquad [6.63]$$

In any case, whether the flow is laminar or turbulent, when the kinetic erosion and the number of erosions are small ($\tilde{V}_{er} \ll 1$ and $\tilde{k}_{er} \ll 1$), then:

– $A_1 \ll 1$, therefore the flow is stationary (i.e. it is looked at in the time step of erosion).

– $B_1 \ll 1$, therefore the evolution of the concentration can be considered as stationary (it is done primarily by advection).

– $Y^{ref} \ll Y_{soil}$, therefore the concentration in solid particles close to the wall is very small; we can then assume that the erosion is not influenced by the concentration at the wall[1].

– The order of magnitude of the normal velocity on Γ (equation [6.40]$_2$) is $\tilde{u} = \max(k, V_{er})/V_w$; the permeability is weak in the case of a fine cohesive soil ($k/V_w \ll 1$), we can then consider that $u = 0$ on Γ, as ρ_{soil} and ρ_w are of the same order of magnitude (which would not have been the case if the erosive fluid was air).

The order of magnitude of the discontinuity of σ on Γ (equation [6.33]$_1$), of inertial nature, is $\sigma_{flow} - \sigma_{soil} \approx \rho_w c_\Gamma^2$. This information can be necessary to introduce the role of σ in the erosion law. Further progress should be made to decide in which condition we can neglect this discontinuity.

An example of this application possible for a laminar flow is the modeling of the suffusion process [BON 11, GOL 11b]. The assumption of low kinetics of erosion allows modeling, the results of which are consistent with the experimentation.

1 In fact, erosion law depends on particle concentrations.

For a turbulent flow, an example of an application is the modeling of a pipe flow with erosion [BON 08]. The assumptions of small number of erosions, of small erosion kinetics and diluted flow, are verified experimentally. In particular, we note velocities V_w of 1–10 m s^{-1}, erosion coefficients k_{er} between 10^{-2} and 10^{-6} s m^{-1}, and particle concentrations of the order of 0.1% to a maximum, for soils tested the most erodible. These assumptions have also been numerically checked [LAC 08].

The case of the erosion of a fine cohesive soil by a turbulent flow is often encountered. In this case, it comes down to assuming a small number of erosions and small erosion kinetics. The orders of magnitudes of the densities are the same for water and soil. We therefore deduce that a small number of erosion is equal to $k_{er}V_w \ll 1$ or $C_e V_w \ll 1$ or $\rho_w k_d V_w \ll 1$. This is one of the most important results of the analysis of two-phase equations with erosion.

6.4.3. *Modeling of dilute flow with slow erosion*

Let us consider surface erosion of a fine cohesive soil caused by a water flow. The soil is rigid and saturated. This soil is eroded by a water flow, which detaches particles, then transports them. We assume that the number of erosion is small and the erosion kinetics is weak. These assumptions enable us to concentrate on the erosion phenomenon. They may be subsequently lifted, thereby leading to an ever more complex modeling.

If the system is such that the eroded particles do not accumulate in the flow, but are well trained, then the maximum concentration is at the eroded wall. We can then make the assumption of diluted flow. It means, in particular, that the presence of solid particles in the flow does not influence the density ($\rho \approx \rho_w$) or the viscosity ($\mu \approx \mu_w$), nor the turbulence[2].

The problem of flow with erosion is defined by the following equations:

– Conservation equations in Ω:

$$\nabla \cdot \mathbf{u} = 0 \text{ in } \Omega \qquad [6.64]$$

$$\rho_w \left[\frac{\partial \mathbf{u}}{\partial t} + (\nabla \mathbf{u}) \cdot \mathbf{u} \right] = \nabla \cdot \mathbf{T} \text{ in } \Omega \qquad [6.65]$$

2 Describing the turbulence for a flow charged with particles is very complex [CHA 08].

– Behavior law of the flow:

$$\mathbf{T} = -p\mathbf{I} + 2\mu_w \mathbf{D}(\mathbf{u}) - \rho_w \overline{\mathbf{u}' \otimes \mathbf{u}'} \qquad [6.66]$$

– Conditions on the mobile interface Γ of celerity c_Γ:

$$\mathbf{u} = 0, \quad c_\Gamma = \begin{cases} k_d(\tau - \tau_c) \text{ if } \tau > \tau_c \\ 0 \text{ else} \end{cases} \text{ on } \Gamma \qquad [6.67]$$

This system calls for a few comments:

– The particles are not forgotten: it is their influence on the flow and on the erosion that is neglected. The concentration is now a secondary unknown and the mass conservation equation of the particles is no longer necessary: it can be integrated in a second phase, independently.

– Although the transient term of the motion equation is negligible, the methods of solving turbulent flows are often more effective in maintaining this term, whereby its presence in equation [6.65].

– The feedback of erosion on the flow, visible in equation [6.40], has disappeared. We can therefore decouple (at least incrementally) the calculation of the flow from the calculation of the evolution of Γ.

6.4.4. *Turbulence modeling*

Reynolds stress $-\rho_w \overline{\mathbf{u}' \otimes \mathbf{u}'}$ corresponds to a momentum transport by velocity fluctuations. We use models of class $k-\varepsilon$ [LAU 72], which represent a good compromise between simplicity and relevance to the flows that we have considered.

We introduce the kinetic energy of velocity fluctuations[3] $k = \overline{\mathbf{u}' \cdot \mathbf{u}'}/2$, which is proportional to the Reynolds stress trace, and the level of viscous dissipation of turbulent kinetic energy $\varepsilon = -\nu_w \overline{\nabla \mathbf{u}' \cdot \nabla \mathbf{u}'}$ where $\nu_w = \mu_w/\rho_w$ is the molecular kinematic viscosity of water.

The Boussinesq assumption introducing turbulent viscosity μ_t is written as:

$$-\rho_w \overline{\mathbf{u}' \otimes \mathbf{u}'} = 2\mu_t \mathbf{D}(\mathbf{u}) - \frac{2}{3}\rho_w k\mathbf{I} \qquad [6.68]$$

3 Usual notations of models $k-\varepsilon$ are used here. We should not mistake kinetic energy of fluctuation velocity with the erosion coefficient.

Model $k-\varepsilon$ is an empirical turbulence model based on transport equations k and ε :

$$\rho_w\left[\frac{\partial k}{\partial t}+(\nabla k)\cdot\mathbf{u}\right]=\nabla\cdot\left[\left(\mu_w+\frac{\mu_t}{\sigma_k}\right)\nabla k\right]+P_k-Y_k \quad [6.69]$$

$$\rho_w\left[\frac{\partial \varepsilon}{\partial t}+(\nabla \varepsilon)\cdot\mathbf{u}\right]=\nabla\cdot\left[\left(\mu_w+\frac{\mu_t}{\sigma_\varepsilon}\right)\nabla \varepsilon\right]+P_\varepsilon-Y_\varepsilon \quad [6.70]$$

where P_k (respectively P_ε) is the production term for k (respectively ε) caused by the average velocity gradient and where Y_k (respectively Y_ε) is the dissipation term for k (respectively ε) caused by the turbulence. The turbulent viscosity is defined as follows:

$$\mu_t=\alpha^*\rho_w\frac{k^2}{\varepsilon} \quad [6.71]$$

The standard value of the model parameters and the definition of the quantities (P_k, Y_k, P_ε, Y_ε, and α^*) are well known [FLU 06].

Reynolds stress $-\rho_w\overline{\mathbf{u}'\otimes\mathbf{u}'}$ thus modelized comprises a spherical part and a deviatoric part. The spherical part on the one hand corresponds to the kinetic energy: $\mathrm{tr}(-\rho_w\overline{\mathbf{u}'\otimes\mathbf{u}'})=-2\rho_w k$. The deviatoric part is on the other hand proportional to the average velocity stress, which is one of the limitations of the model[4]: $\mathrm{dev}(-\rho_w\overline{\mathbf{u}'\otimes\mathbf{u}'})=2\mu_t\mathbf{D}(\mathbf{u})$.

6.5. Numerical modeling

Within the numerical simulation of flows in the presence of interface, there are two main approaches: interface-capturing and interface-tracking. The first, called Eulerian approach, consists of defining the medium (water–soil) in a given field (fixed mesh) and to determine its evolution. The second, called Lagrangian approach, consists of modeling only the fluid part, then to displace the boundary over time (mobile mesh).

4 To overcome this limitation, we need a tensor equation and not a scalar equation instead of [6.69]: it is the Reynolds stress model (RSM), whose calibration is extremely sensitive.

6.5.1. *Interface capturing: the fictitious domain method and the level set method*

The fictitious domain method enables us to modelize a flow and the obstacles of a complex form with a Cartesian mesh, fixed and independent of the shape of the obstacles. This enables us to limit volumes of calculations, namely in three-dimensional (3D) cases. That is $\Omega = \Omega_w \cup \Omega_{soil}$ the studied field, of boundary $\partial\Omega$. Formally omitting Reynolds stresses, a new term is introduced in the motion equation [6.65], which leads to:

$$\rho_w\left[\frac{\partial \mathbf{u}}{\partial t}+(\nabla\mathbf{u})\cdot\mathbf{u}\right]-\nabla\cdot\left[2\mu_w\mathbf{D}(\mathbf{u})\right]=-\nabla p-\frac{\mu_w H_{soil}}{\lambda_{soil}}\mathbf{u} \text{ in } \Omega \qquad [6.72]$$

where H_{soil} is the characteristic function of soil and where λ_{soil} is a very small number comparable to a penalty coefficient. In Ω_w, $H_{soil}=0$ and equation [6.72] is the Navier–Stokes equation. In Ω_{soil}, $H_{soil}=1$ and the right-hand side term of equation [6.72] is dominant before other terms, which leads to a very small velocity $\mathbf{u}$ in Ω_{soil}. The fictitious domain method has been developed and studied in a theoretical framework[5] [ANG 99a, ANG 99b, DIA 03, KHA 00].

Once the position of the interfaces is taken into account, we choose to describe the evolution of the interfaces by the level set method. It was introduced by Osher and Sethian [OSH 88] and is very often used for multifluid flows and fictitious domains. That is ψ the signed distance to the interface, such that $\psi<0$ in Ω_w, $\psi=0$ on Γ, and $\psi>0$ in Ω_{soil}. Knowing ψ, the position of Γ is given by $H_{soil}(\mathbf{x})=\{1 \text{ if } \psi(\mathbf{x})>0,\ 0 \text{ if } \psi(\mathbf{x})<0\}$. Its evolution is described by transport equation:

$$\frac{\partial \psi}{\partial t}+\mathbf{c}\cdot\nabla\psi=0 \text{ in } \Omega \qquad [6.73]$$

where $\mathbf{c}$ is a velocity defined in Ω such that $\mathbf{c}=c_\Gamma\mathbf{n}$ on Γ. This velocity is calculated by an extension following the norms Γ in Ω_w on the one hand and Ω_{soil} on the other hand [GOL10, GOL 11a]. The norm Γ is given by $\mathbf{n}=\nabla\psi/\|\nabla\psi\|$.

5 Equation [6.72] is formally comparable to the Navier–Stokes/Darcy/Brinkman model, which has rightly been dismissed in the introduction for fine cohesive soils, due to their very weak permeability. It is here only a penalty, that is of a numerical model. This comment calls nonetheless for an analysis, which is beyond the scope of this chapter.

In a finite volume method, the velocity stress is calculated at the centre of $\mathbf{X}$ of the cell. The evaluation of zero order of the stress consists of using $\tau(\mathbf{X})$ to calculate the velocity of the interface from the law of erosion, which does not lead to good results. Knowing that $\mathbf{X}_\Gamma = \mathbf{X} - \psi(\mathbf{X})\nabla\psi(\mathbf{X})$ is located on the interface, we can improve this with the help of a limited development around $\mathbf{X}$, which leads to $\tau(\mathbf{X}_\Gamma) \approx \tau(\mathbf{X}) - \psi(\mathbf{X})\nabla\psi(\mathbf{X})\cdot\nabla\tau(\mathbf{X})$.

With the assumption of weak erosion kinetics, the coupling of flow/erosion becomes weak and a decoupled sequential solution is possible (Figure 6.3).

The benefit of modeling fluid/solid interactions with the fictitious domain method and the level set method is to be able to consider mesh independent of the interfaces: it does not need to follow the mobile interface and can be Cartesian and fixed. It is therefore possible to consider 2D or 3D complex geometries. The drawback of this modeling rests with the difficulty of precisely modelizing the mechanical magnitudes – whose stress – on the interface: specific developments are necessary to be able to modelize turbulent flows. The results subsequently showed only concern laminar flows for which the nonlinear term $(\nabla\mathbf{u})\cdot\mathbf{u}$ is not considered.

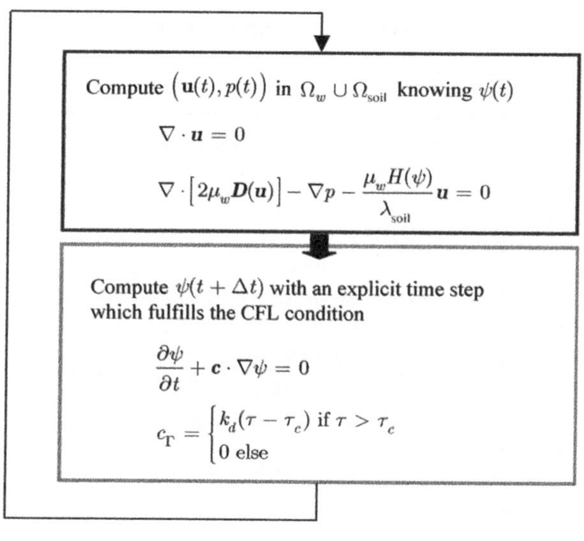

Figure 6.3. *Decoupled sequential scheme with the fictitious domain method and the level set method*

6.5.2. *Interface capturing: Lagrangian approach and remeshing*

Remeshing enables us to modelize a flow and obstacles of complex form, by precisely describing the phenomena close to the walls. This can be interesting when the simulation needs to take into account a very precise way of different flow regimes between the viscous boundary layer at the wall and the turbulent flow far from the wall.

The domain studied is Ω_w and the system equation to be solved constitutes conservation equations and behavior laws, as usual. As a consequence of soil erosion, the wall moves. With the assumption of low erosion kinetics, the flow/erosion coupling is weak and an explicit decoupled sequential solution is possible (Figure 6.4). The geometry is updated at the end of each time step by an explicit Euler scheme:

$$\mathbf{x}(t+\Delta t,\mathbf{X}) = \mathbf{x}(t,\mathbf{X}) + \begin{cases} \Delta t k_d(\tau - \tau_c)\mathbf{n} \text{ if } \tau > \tau_c \\ 0 \text{ else} \end{cases} \qquad [6.74]$$

where $\tau = \tau\left[t, \mathbf{x}(t,\mathbf{X})\right]$ is the tangential stress, $\mathbf{n} = \mathbf{n}\left[t, \mathbf{x}(t,\mathbf{X})\right]$ the normal unit directed by the fluid toward the soil, and $\mathbf{x}(t,\mathbf{X})$ the position at time t of point $\mathbf{X}$. Once the position of the walls is updated, the field is remeshed near the interface to obtain a discretization adapted to this new configuration.

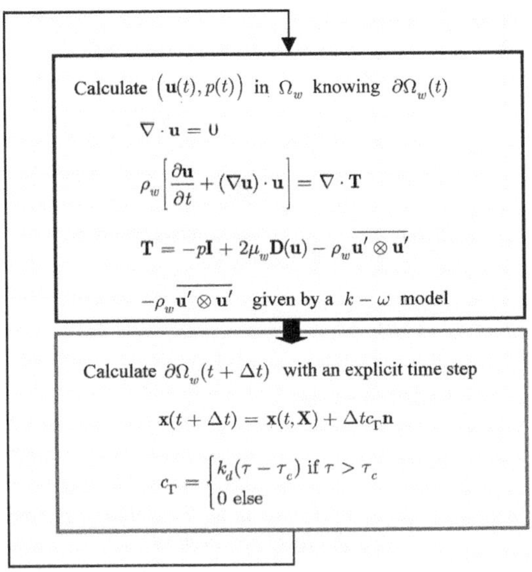

Figure 6.4. *Decoupled sequential scheme with Lagrangian displacing of the interface and remesh*

The benefit of this modeling by remeshing is to be able to precisely modelize magnitudes of the fluid/solid interface, enabling the use of the law of the wall and of the complex models of turbulence. The disadvantage rests with the difficulty to take into account the evolution of complex geometries, namely topological singularity locations (e.g. a corner). It is equally very sensitive to carry out remeshing on 3D geometries. The results subsequently showed concern turbulent flows with model type $k-\varepsilon$.

6.6. Validation of numerical models

6.6.1. *Simplified modeling and exact solutions in the case of pipe flows*

The advantage in simple modeling rests with the possibility of calculating exact solutions of flow problems with erosion. We consider here a circular pipe (Figure 6.5), of constant length L and variable radius $R(t)$ (initial value R_0). Parameters of the erosion law are the yield stress τ_c and the erosion coefficient k_d. We make the assumption of small erosion kinetics. We note time t, the water flow Q, the average velocity of water V such that $Q = \pi R^2 V$, $\dot{m}$ the mass flux of eroded matter, and τ the tangential stress exerted by the flow on the interface. The driving pressure is $P(t) = R_0 \Delta p(t)/(2L)$ where $\Delta p(t)$ is the pressure drop in the pipe. The reference erosion velocity is $V_{\text{er}} = k_d P_0$. The characteristic time of erosion is $t_{\text{er}} = R_0 / V_{\text{er}}$.

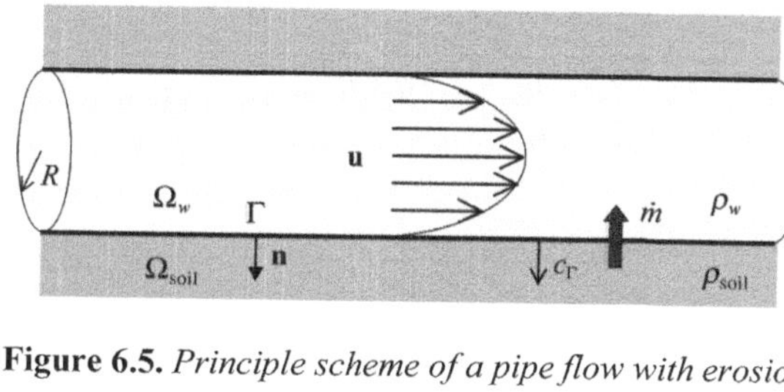

Figure 6.5. *Principle scheme of a pipe flow with erosion*

We introduce the following dimensionless variables:

$$\tilde{t} = \frac{t}{t_{\text{er}}}, \ \tilde{R} = \frac{R}{R_0}; \ \tilde{\dot{m}} = \frac{\dot{m}}{\rho_{\text{soil}} V_{\text{er}}}, \ \tilde{Q} = \frac{Q}{Q_0} \qquad [6.75]$$

$$\tilde{V} = \frac{V}{V_0}, \ \tilde{P} = \frac{P}{P_0}, \ \tilde{\tau} = \frac{\tau}{P_0}, \ \tilde{\tau}_c = \frac{\tau_c}{P_0} \qquad [6.76]$$

When $\tilde{\tau}_c < 1$, the pipe expansion generated by the erosion is described by the following equations [BON 08]:

$$\underbrace{\frac{d\tilde{R}}{d\tilde{t}} = \dot{\tilde{m}}}_{\text{Jump mass equation}}, \quad \underbrace{\tilde{R}\tilde{P} - \tilde{\tau} = 0}_{\text{Motion equation}}, \ \underbrace{\tilde{\tau} = \tilde{R}^{n-2}\tilde{V}^n}_{\text{Flow behavior law}}, \quad \underbrace{\dot{\tilde{m}} = \tilde{\tau} - \tilde{\tau}_c}_{\text{Erosion behavior law}} \qquad [6.77]$$

where $n = 1$ if the flow is laminar and $n = 2$ if the flow is turbulent. The flow–pressure relationship is given by:

$$\tilde{Q} = \tilde{R}^{1+3/n}\tilde{P}^{1/n} \qquad [6.78]$$

In the case of a constant pressure drop $\Delta p(t) = \Delta p_0$, the evolution of the radius is exponential and does not depend on the type of flow (laminar or turbulent):

$$\tilde{R}(\tilde{t}) = \tilde{\tau}_c + (1 - \tilde{\tau}_c)\exp(\tilde{t}) \qquad [6.79]$$

In the case of a constant flow $Q(t) = Q_0$, the exact solution, this time, depends on the type of flow:

$$\tilde{t} = \frac{f(\tilde{\tau}_c^{1/m}\tilde{R}) - f(\tilde{\tau}_c^{1/m})}{\tilde{\tau}_c^{1+1/m}} = \frac{\tilde{R}^{1+m} - 1}{1+m} + \sum_{k=2}^{\infty} \frac{\tilde{\tau}_c^{k-1}}{1+mk}(\tilde{R}^{1+mk} - 1) \qquad [6.80]$$

– laminar flow ($n = 1$):

$$f(\eta) = \frac{1}{\sqrt{3}}\operatorname{atan}\left(\frac{1+2\eta}{\sqrt{3}}\right) + \frac{1}{6}\ln\left(\frac{1+\eta+\eta^2}{(1-\eta)^2}\right) - \eta \text{ and } m = 3 \qquad [6.81]$$

– turbulent flow ($n = 2$):

$$f(\eta) = \frac{1}{2}\left(\operatorname{atan}\eta + \operatorname{atanh}\eta\right) - \eta \text{ and } m = 4 \qquad [6.82]$$

Equation [6.80] shows that, in this case, the evolution of the radius is limited if $\tilde{\tau}_c > 0$, by the asymptotic value:

$$\lim_{\tilde{t}\to\infty} \tilde{R} = \tilde{\tau}_c^{-1/m} \qquad [6.83]$$

These exact solutions show a new number: the characteristic erosion time $t_{er} = 2L/(k_d \Delta p)$. This number appears on flow problems with interfacial erosion.

It explains the crucial role of the erosion coefficient: if the coefficient k_d is 10 times greater, t_{er} is 10 times smaller and the erosion is 10 times faster. Similarly, if the global pressure stress $\Delta p/L$ is 10 times greater, t_{er} is 10 times smaller and the erosion is 10 times faster.

6.6.2. *Plane Poiseuille flow with erosion*

To validate the numerical approach proposed in section 6.5.1 by a level set and fixed Cartesian mesh, we consider a 2D pipe of height $2h$ subjected to a laminar flow, whose walls have been eroded. The representation is similar to Figure 6.5, but this time, this is a plane flow. The exact solution of this problem is obtained in section 6.6.1. If we note the horizontal velocity of fluid u, its viscosity μ_w, its average velocity U, and the pressure p, we obtain for an evolution of constant pressure difference Δp:

$$\frac{h}{h_0} = \frac{\tau_c L}{h_0 \Delta p} + \left(1 - \frac{\tau_c L}{h_0 \Delta p}\right) e^{\frac{t}{t_{er}}}, \quad t_{er} = \frac{\rho_{soil} L}{k_{er} \Delta p} \qquad [6.84]$$

If on the other hand it is the maximum velocity U_{max} which is constant, in the simplified case where $\tau_c = 0$, we obtain:

$$h = h_0 \sqrt{2\frac{t}{t_{er}} + 1}, \quad t_{er} = \frac{\rho_{soil} h_0^2}{2\mu_d k_{er} U_{max}} \qquad [6.85]$$

The computation field is of size $L = H =$ 1 cm, with a channel of initial half-height $h_0 = 1$ mm. The fluid crossing the channel is of density $\rho_w = 10^3$ kg m^{-3} and of viscosity μ_w 10^{-3} Pa s. The soil, of density $\rho_{soil} =$ 2.10^3 kg m^{-3}, is assumed erodible such that $\tau_c =$ 0 Pa and $k_{er} = 10^{-3}$ s m^{-1}. The penalty coefficient of the fictitious domain method used hereafter is $\lambda_{soil} = 10^{-9}$ m^2.

In the first instance, we try to numerically find the equation result [6.84] in the case where the pressure is imposed: $\Delta p = 10^{-2}$ Pa. In Figure 6.6, we check the excellent agreement between the numerical simulation and the analytical solution. Second, we try to numerically find the equation result [6.85] in the case where the maximum velocity is imposed: $U_{max} = 10^{-3}$ m s^{-1}. In Figure 6.7, we check once more the excellent agreement between the numerical simulation and the analytical solution. This validation was equally led with the numerical method by remeshing described in section 6.5.2.

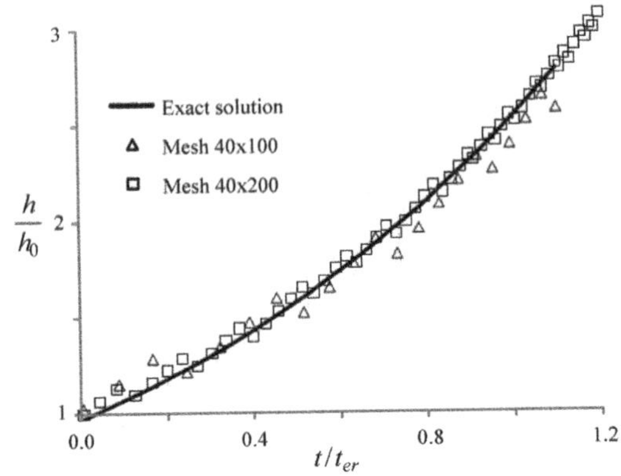

Figure 6.6. *Half-height of the channel in the case of a flow of constant stress pressure in relation to a dimensional time for two mesh and analytical solutions*

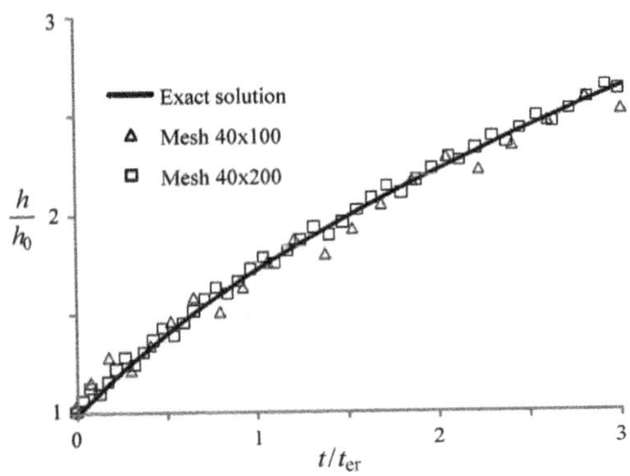

Figure 6.7. *Half-height of the channel in the case of a flow at maximum constant velocity in relation to a dimensional time for two mesh and analytical solutions*

6.6.3. *Axisymmetric Poiseuille flow with erosion*

In the 3D configuration, to validate the proposed numerical approach in section 6.5.1 by a level set and fixed Cartesian mesh, we consider a straight cylinder of constant length L and of variable radius $R(t)$ (initial value R_0). The walls are eroded by a laminar flow.

The pressure difference Δp between the entry and the exit is constant. The exact solution of this problem is given by equations [6.78] and [6.79]. The simulations are realized with the following values: $\Delta p = 0.1$ Pa, $L = 2$ m, $R_0 = 12$ cm, $\tau_c = 0$ Pa, $k_d = 3.7 \times 10^{-7}$ m Pa^{-1} s^{-1}. The characteristic erosion time is $t_{er} = 1.08 \times 10^8$ s. The penalty coefficient of the fictitious domain method is $\lambda_{soil} = 10^{-7}$ m^2.

The influence of the fixed Cartesian mesh on a 3D modeling (Figure 6.8) is shown here. The cases represented are the following:

– m050 10-9: mesh 50 × 50 and zero-order stress;

– m100 10-9: mesh 100 × 100 and zero-order stress;

– m050 10-9++: mesh 50 × 50 and order 1 stress;

– m100 10-9++: mesh 100 × 100 and order 1 stress.

The comparison between the exact solution and the results obtained by numerical simulations in the four cases studied is depicted in Figure 6.9 for the radius in relation to time and in Figure 6.10 for the flow in relation to the radius. The error decreases when the mesh is smaller, or when the the stress computation on the interface is more accurate. The importance of the accuracy of the computation of the mechanical magnitudes on the interface is well represented here.

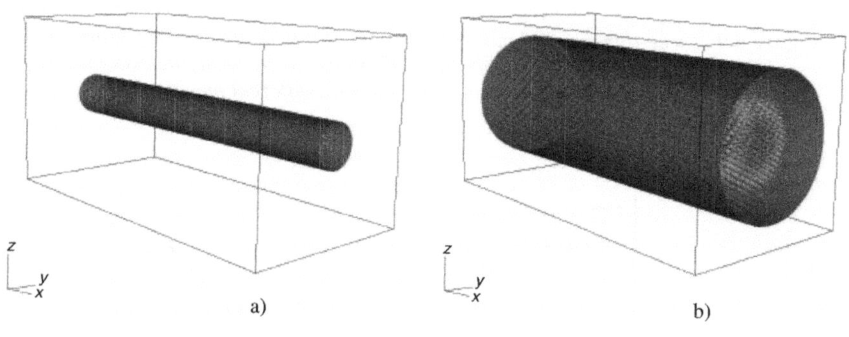

Figure 6.8. *Three-dimensional modeling of pipe flow with erosion. Initial pipe and pipe at* $t/t_{er} = 1.52$

6.7. Illustrative examples

6.7.1. *Scouring of a vertical cylinder*

Soil erosion around a vertical cylinder is a 3D problem. We consider a vertical non-erodible cylinder of radius 25 cm (Figure 6.11), rooted in a layer of erodible soil

($\tau_c = 0$ Pa and $k_d = 3.7 \times 10^{-7}$ m Pa^{-1} s^{-1}). The sizes of the field are 2 m × 1 m × 1 m. The numerical simulation is carried out by Cartesian mesh and level set (section 6.5.1). The penalty coefficient of the fictitious domain method is $\lambda_{soil} = 10^{-9}$ m^2. Two level set functions are used: one for the bridge pier and the other for the soil. The mesh is coarse (150 × 50 × 50) because the simulation is exploratory and designed to establish the feasibility of the approach in a 3D case as shown in Figure 6.11. To be in a situation corresponding to reality (e.g. a bridge pier), it would be necessary to take into account the free surface flow and the turbulence, which exceeds the present framework.

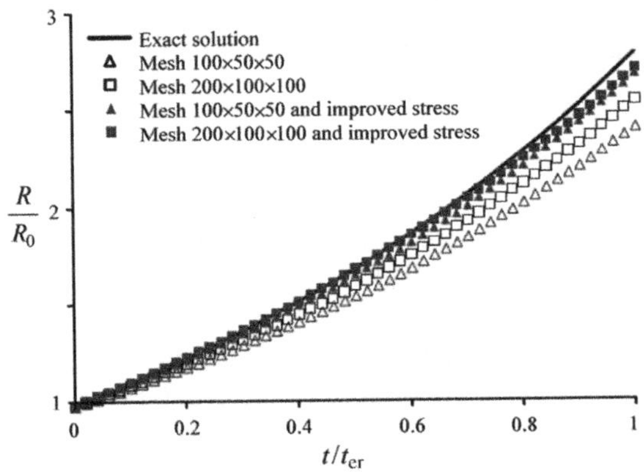

Figure 6.9. *Three-dimensional modeling of pipe flow with erosion with constant pressure drop. Evolution of radius with time. Comparison between the exact solution and the numerical results obtained on different mesh*

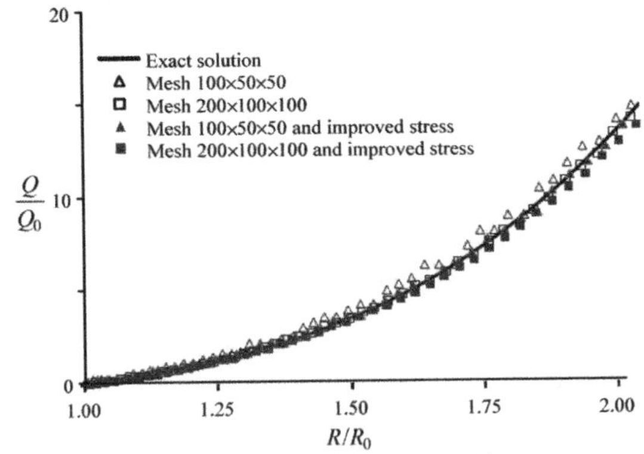

Figure 6.10. *Three-dimensional modeling of pipe flow with erosion with constant pressure drop. Evolution of flow rate with time. Comparison between the exact solution and the numerical results obtained on different mesh*

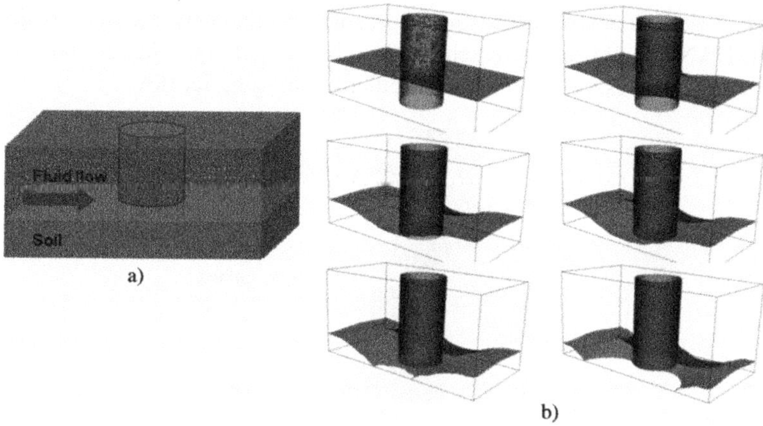

Figure 6.11. *(a) Scouring of a vertical cylinder; (b) three-dimensional modeling and position of interface at times t = 0, 19, 42, 55, 66, 95 (×10⁶ s)*

6.7.2. *Modeling of Jet Erosion Test*

We seek to modelize a Jet Erosion Test whose device enables us to study the resistance of erosion soils in the laboratory, but equally in the field [HAN 91, HAN 04, PIN 11]. The geometry is axisymmetric (Figure 6.12).

Water enters the cylinder inlet at controlled pressure, crosses the nozzle, and will impact on the surface of the material (Figure 6.12). This water then comes out by the laterals openings respecting the asymmetry of the geometry. The critical stress and the erosion coefficent of soil are τ_c=11 Pa and k_d= 10.26 × 10^{-6} m Pa^{-1} s^{-1}. We impose a differential pressure Δp of 30 kPa between the entry and the pressure of the free surface. The order of magnitude of the velocity at the exit of the nozzle is $u_0 = (2\Delta p/\rho_w)^{1/2}$, that is 7 m s^{-1}.

The numerical simulations are carried out with the *Fluent* computational fluid dynamics (CFD) code [FLU 06], with the numerical method and mesh adaptation proposed in section 6.5.2. The mesh was developed to satisfy: (1) a global independence of results under mesh refinement; (2) the geometrical conditions inherent to wall laws of the turbulent model. Several numerical simulations with several meshes are necessary for this. The initial mesh contains approximately 50,000 cells. After 150 s and 30 remesh, it contains 160,000 cells.

Figure 6.13 shows the velocity field after 5 s (a), then after 160 s (b). These results highlight the present modeling for this type of situation. In fact, the central peak visible in Figure 6.13, due to the fact that the tangential stress is zero at the center, is in reality not observed on the results obtained by the Jet Erosion Test [HAN 91, HAN 04]. Further research needs to be carried out to better understand the erosion of a fine cohesive soil under a turbulent impinging jet flow.

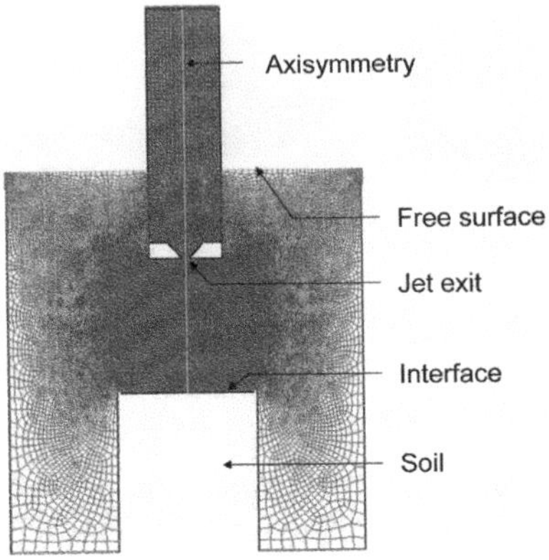

Figure 6.12. *Geometry and mesh of Jet Erosion Test*

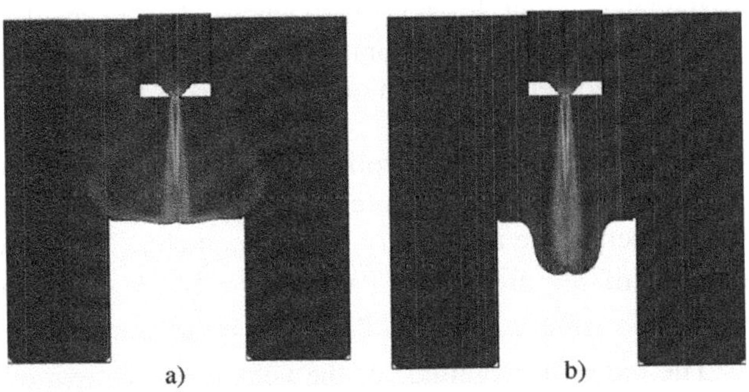

Figure 6.13. *Jet Erosion Test, results of the numerical modeling, visualization of the velocity and the interface after 5 s (a) and after 160 s (b)*

6.7.3. *Modeling of Hole Erosion Test*

The Hole Erosion Test is a laboratory test that enables us to determine the erosion law of a soil (see Chapter 5). It relates to a turbulent pipe flow with erosion. We consider an axisymmetric representation of the Hole Erosion Test (Figure 6.14). The central pipe has an initial radius $R_0 = 2.5$ mm and a length $L = 150$ mm. Three sets of erosion parameters have been considered: Table 6.1 contains data characteristic of three corresponding simulations C1, C2, and C3. The numerical simulations are carried out with the *Fluent* CFD code [FLU 06], with the numerical method and mesh adaptation proposed in section 6.5.2.

The flow is imposed as a constant, between 0.27 and 0.55 $m^3 h^{-1}$. The difference of the initial pressure is between 17.75 and 65.10 kPa. From the simplified modeling in section 6.4.1, we deduce that the average shear stress at the wall is between 51 and 169 Pa and that the average velocity of the flow is between 3.95 and 7.78 m s^{-1}. The Reynolds number is between 19,364 and 38,787.

Computation cases		C1	C3	C2
Uniform inlet velocity	u_{in} (m/s)	3×10^{-2}	2×10^{-2}	1.5×10^{-2}
Critical stress	t_c (Pa)	25.81	43.86	5.00
Hanson erosion coefficient	k_d (cm^3/(N.s))	0.14	0.18	1.00
Outlet	Q_0 (m^3/h)	0.55	0.38	0.27
Initial pressure drop	DP_0 (kPa)	65.10	34.02	17.75
Initial tangential stress	t_0 (Pa)	169	93	51
Initial average velocity	V_0 (m/s)	7.78	5.55	3.95
Initial Reynolds number	Re_0	38,787	27,455	19,364
Characteristic time of erosion	t_{er} (s)	106.54	145.96	47.88

Table 6.1. *Hole Erosion Test, parameters of the numerical modeling*

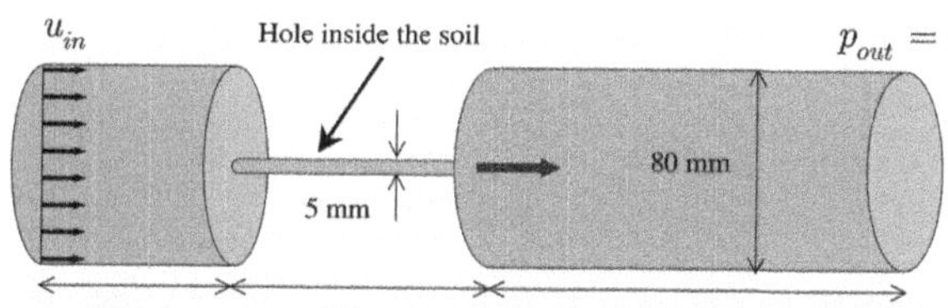

Figure 6.14. *Hole Erosion Test, principle scheme*

Figure 6.15 represents the pressure difference in relation to time and compares the results obtained by numerical simulation and those given by simplified modeling in section 6.4.1. Figure 6.16 shows that the three results obtained by numerical simulation regroup themselves by the index contour line of the simplified modeling equations [6.80] and [6.82]. These results consist of an element of mutual validation of the numerical modeling and of the simplified modeling, which is furthermore used to interpret Hole Erosion laboratory tests ([BON 08] and Chapter 5). Figure 6.17 shows an example of the velocity field.

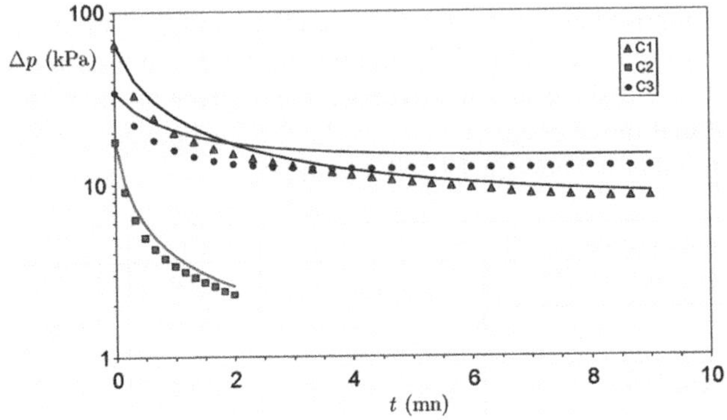

Figure 6.15. *Hole Erosion Test at constant flow, pressure drop in relation to time, comparison between the numerical modeling (solid lines) and the simplified modeling (symbols)*

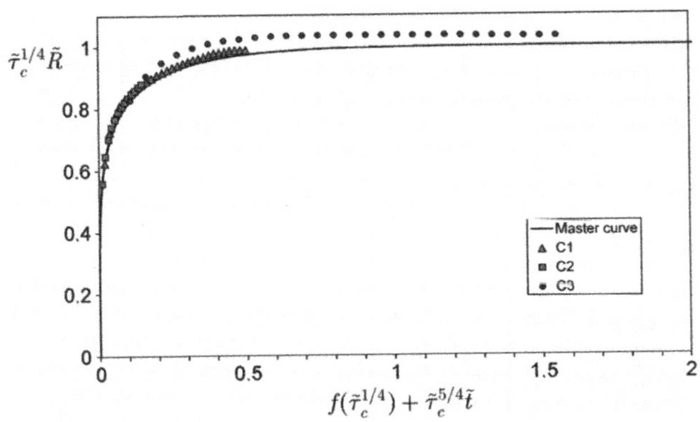

Figure 6.16. *Hole Erosion Test at constant flow, the index contour line, comparison between the numerical modeling (solid lines) and the simplified modeling (symbols)*

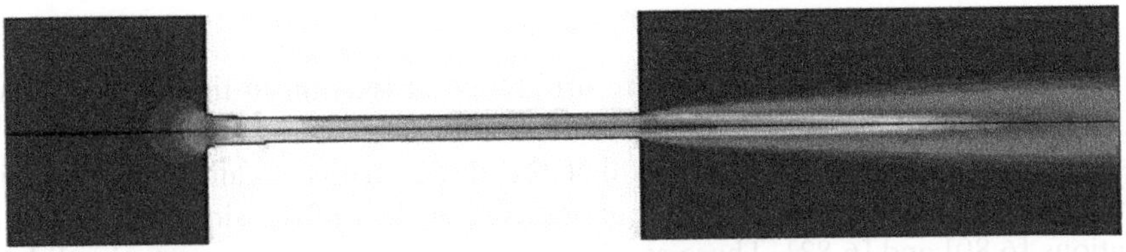

Figure 6.17. *Hole Erosion Test, results of the numerical modeling, representation of the water velocity*

6.8. Conclusion

This chapter concerns the numerical modeling of interfacial erosion of a soil by a laminar or turbulent flow. The erosion is characterized by the fact that the interface between the flow and the fine soil is crossed by the flux of eroded matter and that the interface is mobile. The celerity of this interface is linked to the stress exerted by the flow via the erosion law. This description is well suited to the erosion of a fine cohesive soil.

The orders of magnitude that led on the two-phase transitory flow equations with erosion shows that the product of the erosion coefficient and the flow velocity is an important dimensionless number: it is the erosion number. When this number is small, the description can therefore be simplified by a system of stationary flow equations, diluted with slow erosion.

Two numerical models are proposed. The first, with an Eulerian interface approach, consists of defining the medium (water–soil) in a discretized field by fixed mesh, defining the interface by the fictitious domain method, and determining the evolution by the level set method. The second, with a Langrangian interface approach, consists of modelizing only the fluid part, displacing the interface over time, and then proceeding to mesh adaptation.

The numerical results are compared with the exact solutions of plan and axisymmetric Poiseuille flows with erosion. Illustrative examples are then shown. The Jet Erosion Test modeling shows that further research is necessary to better understand the erosion of a fine cohesive soil under a turbulent impinging jet flow. The results obtained by the numerical modeling of the Hole Erosion Test are consistent with the simplified modeling, used furthermore for the interpretation of laboratory tests in Chapter 5 of this book.

6.9. Bibliography

[ABU 95] ABU-ZAID S., AHMADI G., "A thermodynamically consistent rate-dependent model for turbulent two-phase flows", *International Journal of Non-Linear Mechanics*, vol. 30, no. 4, pp. 509–529, 1995.

[ANG 99a] ANGOT P., BRUNEAU C.-H., FABRIE P., "A penalization method to take into account obstacles in incompressible viscous flows", *Numerische Mathematik*, vol. 81, no. 16, pp. 497–520, 1999.

[ANG 99b] ANGOT P., "Analysis of singular perturbations on the Brinkman problem for fictitious domain models of viscous flows", *Mathematical Methods in the Applied Sciences*, vol. 22, pp. 1395–1412, 1999.

[BON 06] BONELLI S., BRIVOIS O., BORGHI R., BENAHMED N., "On the modelling of piping erosion", *Comptes Rendus de Mécanique*, vol. 8–9, no. 334, pp. 555–559, 2006.

[BON 08] BONELLI S., BRIVOIS O., "The scaling law in the hole erosion test with a constant pressure drop", *International Journal for Numerical and Analytical Methods in Geomechanics*, vol. 32, no. 13, pp. 1573–1595, 2008.

[BON 11] BONELLI S., MAROT D., "Micromechanical modeling of internal erosion", *European Journal of Environmental and Civil Engineering*, vol. 18, no. 8, pp. 1207–1224, 2011.

[BRI 07] BRIVOIS O., BONELLI S., BORGHI R., "Soil erosion in the boundary layer flow along a slope : a theoretical study", *European Journal of Mechanics/B Fluids*, vol. 26, pp. 707–719, 2007.

[CHA 99] CHANSON H., *The Hydraulics of Open Channel Flows: An Introduction*, Butterworth-Heinemann, Oxford, UK, 1999.

[CHA 08] CHAUCHAT J., GUILLOU S., "On turbulence closure for two-phase sediment-laden flow models", *Journal of Geophysical Research*, vol. 113, C11017, 2008.

[CHA 10] CHAUCHAT J., MÉDALE M., "A three-dimensional numerical model for incompressible two-phase flow of a granular bed submitted to a laminar shearing flow", *Computer Methods in Applied Mechanics and Engineering*, vol. 199, nos. 9–12, pp. 439–449, 2010.

[DIA 03] DIAZ-GOANO C., MINEV P.D., NANDAKUMAR K., "A fictitious domain/finite element method for particulate flows", *Journal of Computational Physics*, vol. 192, no. 1, pp. 105–123, 2003.

[FLU 06] Fluent user's guide, vol. 3, Technical report, Fluent Inc., 2006.

[GOL 10] GOLAY F., LACHOUETTE D., BONELLI S., SEPPECHER P., "Interfacial erosion: a three-dimensional numerical model", *Comptes Rendus de Mécanique*, vol. 338, no. 6, pp. 333–337, 2010.

[GOL 11a] GOLAY F., LACHOUETTE D., BONELLI S., SEPPECHER P., "Numerical modelling of interfacial soil erosion with viscous incompressible flows", *Computer Methods in Applied Mechanics and Engineering*, vol. 200, nos. 1–4, pp. 383–391, 2011.

[GOL 11b] GOLAY F., BONELLI S., "Numerical modeling of suffusion as an interfacial erosion process", *European Journal of Environmental and Civil Engineering*, vol. 15, no. 8, pp. 1225–1241, 2011.

[GRA 71] GRAF W.H., *Hydraulics of Sediment Transport*, McGraw-Hill, 1971.

[HAN 91] HANSON G.J., "Development of a jet index to characterize erosion resistance of soils in earthen spillways", *Transactions of the ASAE*, vol. 34, no. 5, pp. 2015–2020, 1991.

[HAN 04] HANSON G.J., COOK K.R., "Apparatus, test procedures, and analytical methods to measure soil erodibility in situ", *ASAE Applied Engineering in Agriculture*, vol. 20, no. 4, pp. 455–462, 2004.

[KHA 00] KHADRA K., ANGOT P., PARNEIX S., CALTAGIRONE J.P., "Fictitious domain approach for numerical modelling of Navier-Stokes equations", *International Journal for Numerical Methods in Fluids*, vol. 34, no. 8, pp. 651–684, 2000.

[LAC 08] LACHOUETTE D., GOLAY F., BONELLI S., "One-dimensional modelling of piping flow erosion", *Comptes Rendus de Mécanique*, vol. 336, no. 6, pp. 731–736, 2008.

[LAU 72] LAUNDER B.E., SPALDING D.B., *Lectures in Mathematical Models of Turbulence*, Academic Press, London, 1972.

[LHU 01] LHUILLIER D., "Internal variables and the non-equilibrium thermodynamics of colloidal suspensions", *Journal of Non-Newtonian Fluid Mechanics*, vol. 96, nos. 1–2, pp. 19–30, 2001.

[MAS 99] MASSOUDI M., RAJAGOPAL K.R., PHUOC T.X., "On the fully developed flow of a dense particulate mixture in a pipe", *Powder Technology*, vol. 104, pp. 258–268, 1998.

[MAS 03] MASSOUDI M., "Constitutive relations for the interaction force in multicomponent particulate flows", *International Journal of Non-Linear Mechanics*, vol. 38, pp. 313–336, 2003.

[OSH 88] OSHER S., SETHIAN J.A., "Fronts propagating with curvature-dependent speed: algorithms based on Hamilton-Jacobi formulations", *Journal of Computational Physics*, vol. 79, no. 1, pp. 12–49, 1988.

[OUR 07] OURIEMI M., AUSSILLOUS P., MEDALE M., PEYSSON Y., GUAZZELLI E., "Determination of the critical Shields number for particle erosion in laminar flow", *Physics of Fluids*, vol. 19, no. 6, 061706, 2007.

[OUR 09a] OURIEMI M., AUSSILLOUS P., GUAZZELLI E., "Sediment dynamics. Part 2. Dune formation in pipe flow", *Journal of Fluid Mechanics*, vol. 636, pp. 321–336, 2009.

[OUR 09b] OURIEMI M., AUSSILLOUS P., GUAZZELLI E., "Sediment dynamics. Part 1. Bed-load transport by laminar shearing flows", *Journal of Fluid Mechanics*, vol. 636, pp. 295–319, 2009.

[PAP 01] PAPAMICHOS E., VARDOULAKIS I., TRONVOLL J., SKJAERSTEIN A., "Volumetric sand production model and experiment", *International Journal for Numerical and Analytical Methods in Geomechanics*, vol. 25, no. 8, pp. 789–808, 2001.

[PAP 05] PAPAMICHOS E., VARDOULAKIS I., "Sand erosion with a porosity diffusion law", *Computers and Geotechnics*, vol. 32, no. 1, pp. 47–58, 2005.

[PAP 06] PAPAMICHOS E., DESMORAT R., VARDOULAKIS I., "Geomechanical and structural issues in energy production", Special issue of *European Journal of Civil Engineering*, Hermes, vol. 10, nos. 6–7, pp. 640–902, 2006.

[PEY 09] PEYSSON Y., OURIEMI M., MEDALE M., AUSSILLOUS P., GUAZZELLI E., "Threshold for sediment erosion in pipe flow", *International Journal of Multiphase Flow*, vol. 35, no. 6, pp. 597–600, 2009.

[VAR 96] VARDOULAKIS I., STAVROPOULOU M., PAPANASTASIOU P., "Hydromechanical aspects of sand production problem", *Transport in Porous Media*, vol. 22, no. 2, pp. 225–244, 1996.

[VAR 01] VARDOULAKIS I., PAPANASTASIOU P., STAVROPOULOU M., "Sand erosion in axial flow conditions", *Transport in Porous Media*, vol. 45, no. 2, pp. 267–281, 2001.

[VAR 04] VARDOULAKIS I., "Fluidization in artesian flow conditions: II. Hydro-mechanically unstable granular media", *Géotechnique*, vol. 54, no. 3, pp. 165–177, 2004.

[WAN 04a] WAN C.F., FELL R., "Investigation of rate of erosion of soils in embankment dams", *Journal of Geotechnical and Geoenvironmental Engineering*, vol. 30, no. 4, pp. 373–380, 2004.

[WAN 04b] WAN C.F., FELL R., "Laboratory tests on the rate of piping erosion of soils in embankment dams", *Journal of Geotechnical Testing*, vol. 27, no. 3, pp. 295–303, 2004.

[YAL 77] YALIN M.S., *Mechanics of Sediment Transport*, 3rd ed., McGraw-Hill, London, 1977.

[ZHA 97] ZHANG D.Z., PROSPERETTI A., "Momentum and energy equations for disperse two-phase flows and their closure for dilute suspensions", *International Journal of Multiphase Flow*, vol. 23, no. 3, pp. 425–453, 1997.

Chapter 7

Physics of Sediment and Aeolian Transport

7.1. Introduction

It is usual to distinguish different modes of sediment transport as a function of the type of forces dominating the dynamics involved (Figure 7.1). Soil particles are subject to three types of forces: hydrodynamic forces, gravity, and interparticle contact forces. When hydrodynamic forces dominate, we refer to transport in suspension. In general, this is the case of the transport of fine sediments. When gravity is great enough to confine the transport in a layer at the surface of a bed of grains, we refer to bedload. This type of transport concerns the coarsest sediments (sand, pebbles, and boulders). Therefore, different types of trajectories are distinguished. When the grains move in a succession of jumps, the term saltation is used. The dominant forces are therefore gravity and hydrodynamic forces. On the contrary, when the flow rolls grains to the surface of the bed, with long-term contacts between the grains, the terms bedload and traction are used. The three forces therefore act simultaneously. Finally, when the impact of grains in saltation on the granular bed is high enough, particles of the bed are displaced by what is termed reptation [BAG 41]. Reptation is therefore a mode of transport driven by contact forces and gravity.

The aim of this chapter is to show how the complexity of sediment transport mentioned earlier in numerous modes of transport can be summarized into three (or four) key quantities: (1) entrainment threshold, (2) equilibrium (or saturated) flux, and (3) length (and time) of saturation – more details can be found in the review [DUR 11]. In particular, we show how this length linked to grain transport

Chapter written by Bruno ANDREOTTI and Philippe CLAUDIN.

properties controls the scale to which a flat erodible bed is destabilized, leading to the formation of dunes. The outline of this chapter follows this progression precisely. We have focused most of the text on aeolian transport (saltation); that is to say, where the density of transported particles ρ_p is considerably greater than that of the fluid ρ_f. Moreover, we consider turbulent flows. Nevertheless, the formalism and the methods we develop here are very general and we discuss how they can be applied to other situations in section 7.6.

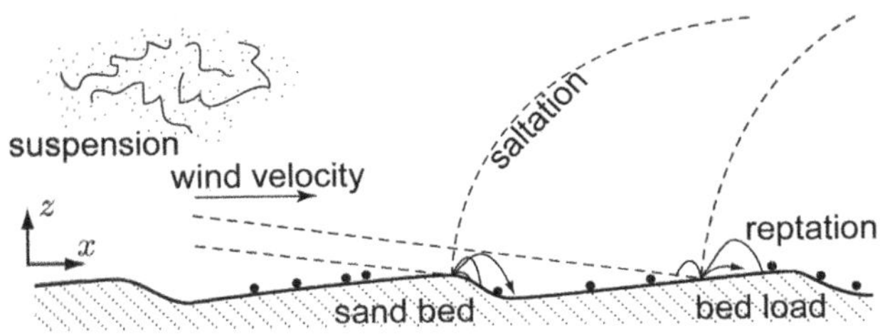

Figure 7.1. *Scheme representing different modes of transport*

7.2. Static transport threshold

A fluid flow cannot set grains in motion at the surface of a granular bed unless it is strong enough. In this section, we introduce the Shields number that is the dimensionless number traditionally used to describe the transport threshold. We then describe the influence of the slope and the presence of adhesion between grains on this threshold.

7.2.1. *Shields number*

Let us consider a spherical grain trapped between two of its fixed neighbors and subjected to a fluid flux above it (see Figure 7.2(b)). In the first instance, we assume that the grains interact only through friction and that there is no cohesion. The grain loses static equilibrium when the driving force F_{drag} is equal to the weight of the grain shifted by the buoyant force $P \sim \pi/6\,(\rho_p - \rho_f)gd^3$ multiplied by an effective friction coefficient μ. We thus obtain a quantitative motion criterion on the basis of the ratio $F_{\text{drag}}/(\rho_p - \rho_f)gd^3$.

To obtain the threshold value, it is necessary to link the force exerted on the particle to hydrodynamic control parameters. Dimensionally, the hydrodynamic force exerted by the fluid on a flat surface of the size of a grain is proportional to

τd^2, where τ is the shear stress at the grain/fluid interface. We therefore deduce that the relevant number is the Shields number, defined by:

$$\Theta = \frac{\tau}{(\rho_p - \rho_f)gd} \qquad [7.1]$$

This analysis at grain scale suggests that the motion threshold is controlled by a Shields threshold, $\Theta_{th} \propto \mu$, independent of the size of the grain, its density, and the nature of the surrounding fluid. Figure 7.2(a) shows Shields threshold measurements in liquids of different viscosities ν and for natural sand of different sizes in laminar and turbulent regimes. To compare these different situations, we introduce the viscous diameter $d_\nu \equiv (\rho_p/\rho_f - 1)^{-1/3}\nu^{2/3}g^{-1/3}$ and observe that the critical Shields number is approximately constant for large diameters but systematically increases when d/d_ν reduces. This dependence shows that the relationship between force F_{drag} and basal stress τ is not trivial.

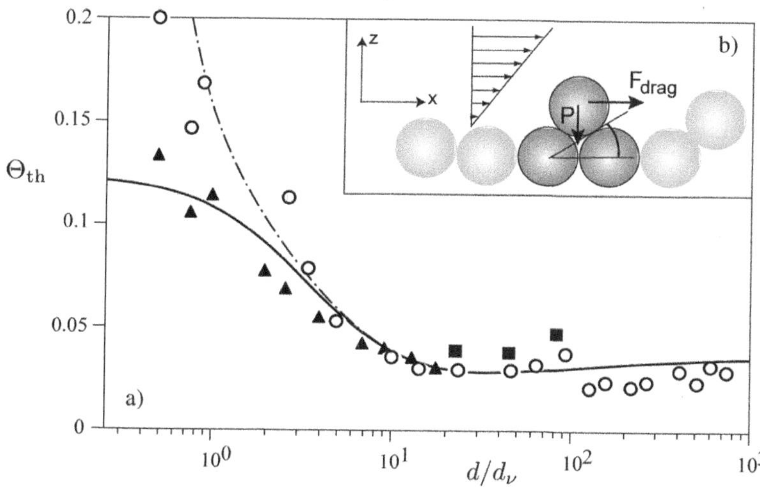

Figure 7.2. *(a) Shields threshold Θ_{th} as a function of the size of grains d made dimensionless by viscous diameter d_ν. The symbols correspond to measurements carried out by [FER 76] (■) and those collected by [YAL 79] in water (○) and in more viscous liquids (▲). Solid line: prediction of the model proposed here. Dotted line: account taken of correction due to cohesion. (b) Diagram showing the mechanical source of the transport threshold at grain scale*

We can write this driving force in the form $\pi d^2 C_d \rho_f U^2/8$, where the phenomenological drag coefficient is written as $C_d = (C_\infty^{1/2} + s\mathcal{R}_d^{-1/2})^2$, and where $\mathcal{R}_d = Ud/\nu$ is the Reynolds number on the basis of the size of the grain. For natural sand, these constants apply $C_\infty \simeq 1$ and $s \simeq 5$ [FER 04]. The difficulty is to correctly calculate the velocity of the fluid at the level of grain U (e.g. at altitude $z = d/2$) to connect it to shear τ. This can be done by decomposing the shear in a viscous term and with a contribution of the turbulent Prandtl mixture. Within purely viscous and turbulent limits, we obtain the following thresholds, respectively:

$$\Theta_{th} = \frac{8\mu}{3s^2} \qquad [7.2]$$

$$\Theta_{th} = \frac{4\mu\kappa^2}{3C_\infty \ln^2[1 + d/(2z_0)]} \qquad [7.3]$$

In the second term, $\kappa = 0.4$ is the von Kármán constant and z_0 the hydrodynamic roughness. This model efficiently replicates the experimental measurements taken under water (Figure 7.2(a)). The wind tunnel measurements of [RAS 96] are also in good agreement with these predictions.

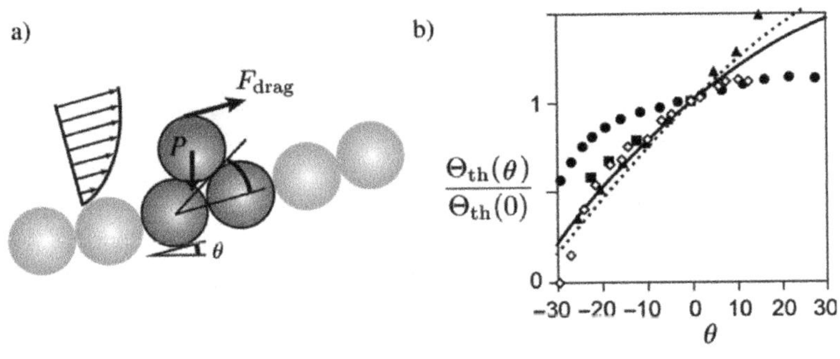

Figure 7.3. *(a) Diagram showing the dependence of the Shields threshold Θ_{th} with angle θ of inclination of the sandy bed. (b) Variation of the threshold with the bed slope. The black symbols correspond to measurements taken in water by [FER 76] (■) and [DEY 03] (▲) for grains of natural sand, to those of [LOI 05] for glass beads (●) in a narrow channel filled with a water–glycerol mixture. The white symbols (◊) correspond to measurements taken in the air by [HAR 88]. The dotted curve shows function $\cos\theta + \sin\theta / \mu$, with $\mu = \tan 35°$. The solid line curve shows the model prediction for natural sand grains, taking into account the change of the Reynolds number*

7.2.2. *The influence of longitudinal slope*

Let us consider the case of a sandy bed inclined at angle θ according to the flow direction (Figure 7.3(a)). If the slope is positive, the flow velocity must be stronger to drive the grains. If the slope is negative, the threshold is lower [FER 76, IVE 94, RAS 96]. By rewriting the force balance on a grain, we obtain a modified tangential force in $F_{\text{drag}} - P\sin\theta$ and a normal force in $P\cos\theta$. Consequently, the threshold is reached when $F_{\text{drag}} - P\sin\theta = \mu P\cos\theta$. This results in the following equation:

$$\Theta_{\text{th}}(\theta) = \Theta_{\text{th}}(0)\left(\cos\theta + \frac{\sin\theta}{\mu}\right) \qquad [7.4]$$

This law is validated correctly, but with an effective friction coefficient slightly higher than the avalanche slope (Figure 7.3(b)). The effect is therefore slightly weaker than expected.

7.2.3. *Influence of soil cohesion*

The threshold is increased if a supplementary cohesion force exists between grains. The smaller the d is, the higher the threshold. A realistic calculation of adhesion between two spheres can be performed by assuming that, in reality, the "Hertz" contact occurs via several microcontacts. Whether the latter are in an elastic or plastic regime, the scaling laws remain globally the same and the transport threshold is written in the following form:

$$\Theta_{\text{th}} = \Theta_{\text{th}}^{\infty}\left[1 + \frac{3}{2}\left(\frac{d_m}{d}\right)^{5/3}\right] \qquad [7.5]$$

with $d_m \propto (\gamma / M)^{3/5}[E / (\rho_s g)]^{2/5}$, where E is the Young's modulus of the grains, M the Young's modulus (elastic regime) or hardness modulus (plastic regime), and γ the surface tension of the material composing them [CLA 06]. It is this cohesive term that links the threshold curve to the small values of d (Figure 7.2(a)).

7.3. Aeolian transport

In this section, we provide a qualitative description of the properties of the aeolian transport of grains. We then show how these properties can be expressed by a dynamic threshold and a *focal point* for the wind velocity profile.

7.3.1. *Qualitative description*

When a grain is torn from the bed by wind, it starts to accelerate by bouncing on the bed and making jumps of increasing height. This amplification is caused by surface irregularities: part of the horizontal momentum of the grains can be redirected upward during bouncing. At each impact with the bed, the grain ejects new grains, which can then be accelerated by the wind (Figure 7.4(a)). This amplification of the number of grains transported becomes saturated due to the feedback effect of transport on the flow. The higher the number of grains transported, the weaker the velocity of the wind in the transport layer. Indeed, each time the flow accelerates a grain, the latter exerts a force on the flow in return. The fluid stress exerted on the bed is therefore reduced in comparison to the fluid stress far from the bed, part of the momentum having been transferred to the grains [OWE 64].

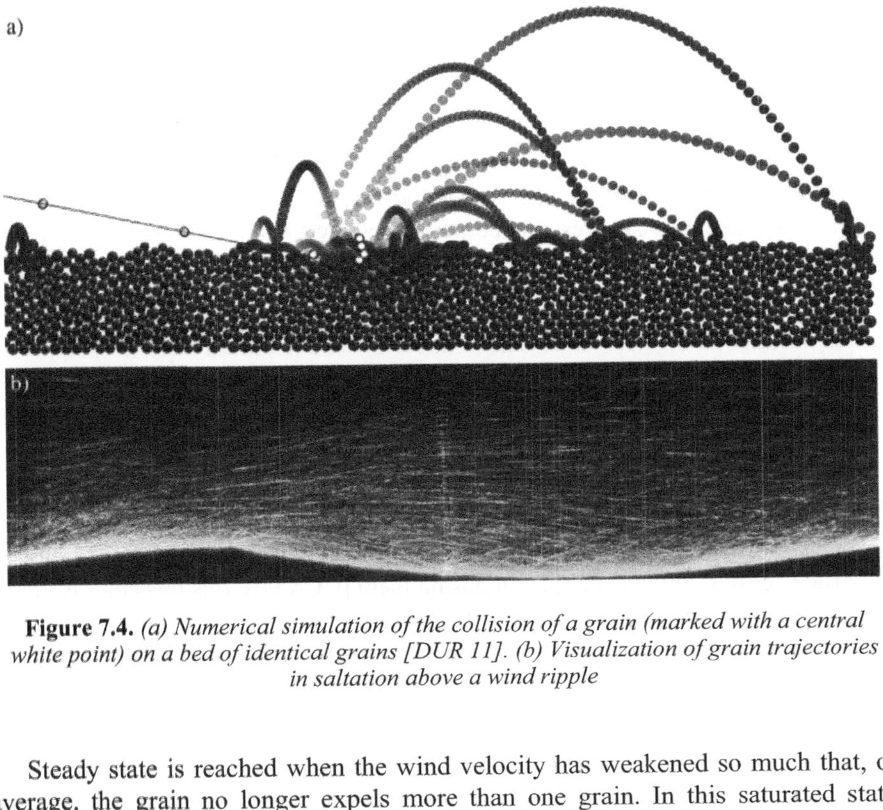

Figure 7.4. *(a) Numerical simulation of the collision of a grain (marked with a central white point) on a bed of identical grains [DUR 11]. (b) Visualization of grain trajectories in saltation above a wind ripple*

Steady state is reached when the wind velocity has weakened so much that, on average, the grain no longer expels more than one grain. In this saturated state, aeolian transport is characterized by a diffuse layer above the bed, as illustrated in Figure 7.4(b), which shows the typical particle trajectories above a wind ripple. Quantitatively, the distribution of grains in this layer has been measured for different wind intensities [CRE 09]. The vertical profiles of the volume fraction $\phi(z)$ decrease exponentially with altitude (Figure 7.5). This corresponds to a Maxwell distribution of vertical velocity emerging from the bed.

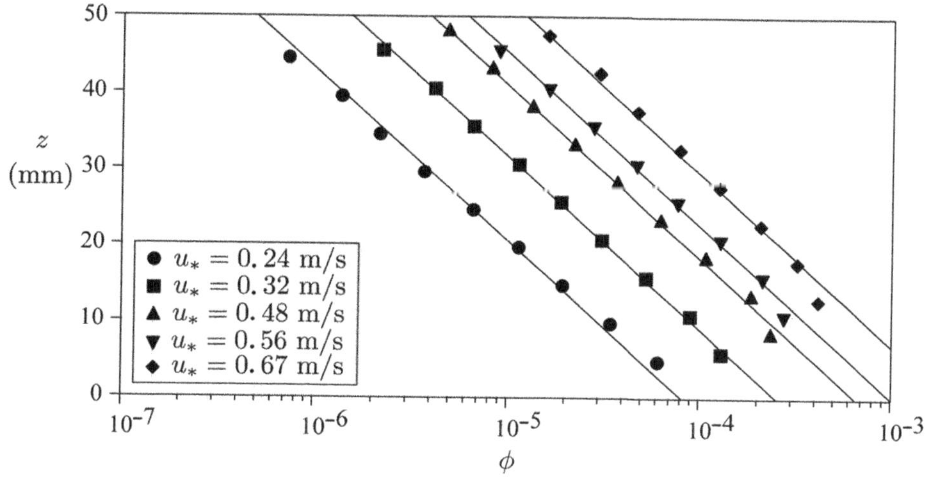

Figure 7.5. *Vertical profiles of volume fraction* $\phi(z)$ *of grains transported in saltation at different wind velocities* u_*. *In this range of velocities, the volume fraction decreases exponentially with altitude, with a characteristic length in the region of 10 mm, independent of shear velocity* u_* *[CRE 09]*

This entire process in which grains bounce, is accelerated by wind and expels other grains on impact, is called saltation and the associated grains are called saltons. It is the main mode of transport of material by wind. However, there is a second mode of transport called reptation that stems from the motion of the grains of the bed under the effect of salton collisions. The flux of grains in reptation is proportional to the flux of grains in saltation. This mode, which participates modestly in global transport (from 1/6 to 1/3 of total transport), is responsible for the formation of aeolian ripples.

7.3.2. *Dynamic transport threshold*

Collisions by grains in saltation comprise mechanisms other than hydrodynamic drag to set the grains of the bed in motion. The transport threshold therefore presents hysteresis: once the saltation process has been triggered, transport can be maintained for wind velocities lower than the static threshold. Transport is in fact stationary if each collision produces, on average, a new grain in saltation.

We can estimate the dynamic transport threshold in the following way. Below the static threshold, the force required to free the grains from the bed is proportional to $gd^3(\rho_p - \rho_f)(1 - \Theta/\Theta_{th})$. Thus, it is necessary to compare the kinetic energy of the colliding grains $\rho_p d^3 u_*^2$ to the effort of this force to ensure a displacement in the region of d. This gives a Shields threshold number Θ_{dyn} given by:

$$\Theta_{\mathrm{dyn}} = b\frac{\rho_f}{\rho_p}\left(1-\frac{\Theta_{\mathrm{dyn}}}{\Theta_{\mathrm{th}}}\right) \quad [7.6]$$

where b is a numerical constant. By solving Θ_{dyn}, we finally obtain:

$$\Theta_{\mathrm{dyn}} = \frac{\Theta_{\mathrm{th}}}{1+b\frac{\rho_p}{\rho_f}\Theta_{\mathrm{th}}} \quad [7.7]$$

For $\rho_f \ll \rho_p$ (the aeolian case), $\Theta_{\mathrm{dyn}} \to b\rho_f/\rho_p$ is a lot lower than the static threshold. This dynamic threshold presents the same slope and cohesion corrections as the static threshold. To obtain a more accurate determination of the threshold, it is necessary to calculate the trajectory of the saltons. This requires a more accurate formula of the grain rebound laws that can be either stochastic or non-stochastic that in turn requires the numerical integration of the motion equations. Readers should refer to [AND 04] and [CLA 06] for detailed calculations. Figure 7.6 shows the relationship between the dynamic threshold and grain size in the case of saltation transport in air and in the Martian atmosphere.

7.3.3. *Focal point of the velocity profiles*

As we have seen in the earlier sections, the saturation of aeolian transport is due to the feedback effect of transport on the flow, since the number of particles transported is adjusted to guarantee that the basal stress remains equal to the stress threshold τ_{dyn}. In an initial approximation, the velocity profile $u(z)$ in the saltation layer (Figure 7.7(a)) remains in the form:

$$u = \frac{u_{\mathrm{dyn}}}{\kappa}\ln\left(\frac{z}{z_0}\right) \quad \text{with} \quad \tau_{\mathrm{dyn}} \equiv \rho_f u_{\mathrm{dyn}}^2 \quad [7.8]$$

Since this velocity profile is independent of shear velocity u_*, the same applies to the typical trajectory of the grains. The height of the transport layer, noted H_f, is therefore also independent of u_*. This property is highlighted directly by the curves in Figure 7.5.

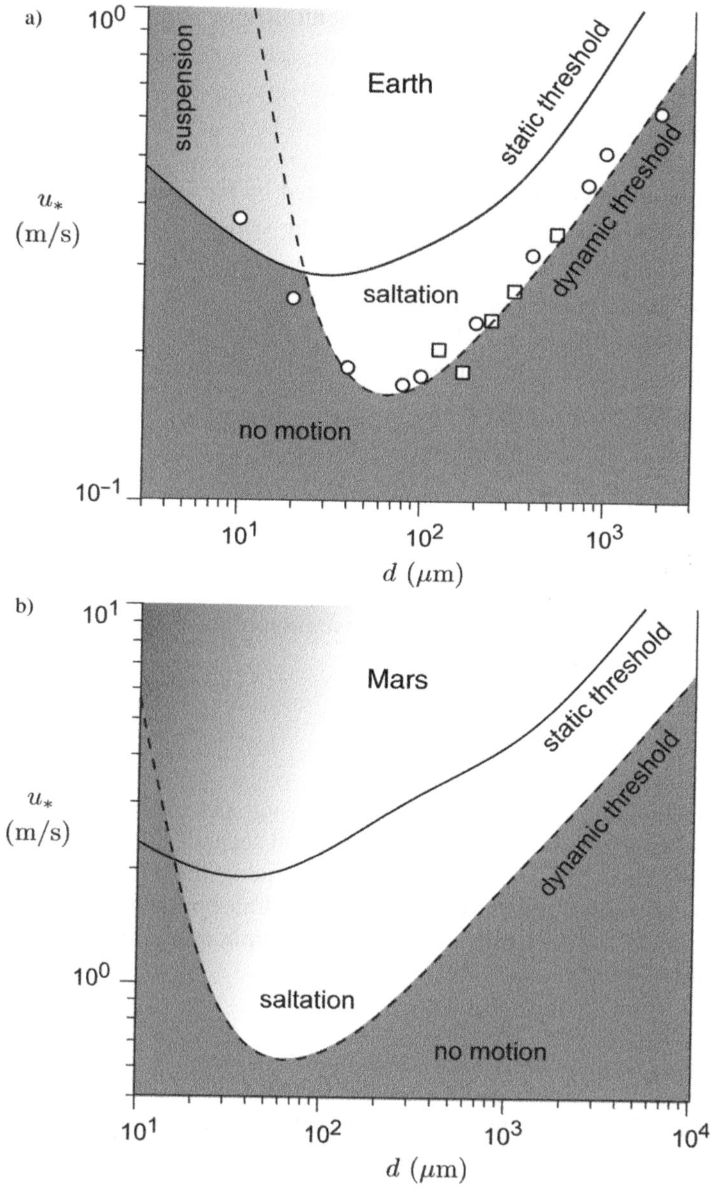

Figure 7.6. *Shear velocity above which the atmospheric flow can maintain the transport of grains of a bed of natural sand of size d (dynamic threshold, u_{dyn}, dotted line). The static threshold is also shown (u_{th}, solid line). (a) Saltation on Earth. The symbols correspond to the measures performed by [CHE 45] (○) and [RAS 96] (□). The dotted line curve shows the model prediction proposed by [CLA 06]. (b) Predictions in the Martian case*

Above the saltation layer, we find a flow controlled by shear velocity u_*:

$$u = \frac{u_*}{\kappa} \ln\left(\frac{z}{z_s}\right) \qquad [7.9]$$

where z_s is the aerodynamic roughness generated by the layer of transport. At the upper limit of the saltation layer, in $z = H_f$, the same velocity, $u = U_f$, must be obtained whatever the value of u_*. The velocity profiles obtained for different winds thus converge at the same focal point (Figure 7.7(a)), a phenomenon that has been observed experimentally [BAG 41, IVE 99]. The characteristics of this focal point can be obtained quantitatively by measuring aerodynamic roughness z_s. By equaling the two expressions of U_f, we obtain:

$$z_s = H_f \exp\left(-\frac{\kappa U_f}{u_*}\right) \qquad [7.10]$$

Figure 7.7(b) shows that z_s is an increasing function of u_* and therefore of the sand flux. This curve is a convincing demonstration of the existence of the feedback effect of transport on the flow.

7.4. Quantitative description of transport

In this section, we show how to describe erosion and transport in a quantitative and unified way. First, we define the interface that separates the granular bed of the fluid, then the fluxes of materials, thereby describing the exchanges through and along it. Second, we address the concept of steady transport and the saturated flux of grains associated with it. Finally, we propose a simple description of the saturation transients of transport.

7.4.1. *Interface between the sediment bed and the fluid*

Two approaches are possible when defining the interface between a sediment bed and the fluid. The first corresponds to interface ξ_s between the static and mobile grains (Figure 7.8(a)). In an initial approximation, we can consider that the velocity of the fluid vanishes at ξ_s. This interface can be defined either by dynamic fields (partition of stresses into granular and fluid stresses) or kinematics (interface between the static and mobile grains). A second definition, less immediate, but useful for expressing the conservation of the material, consists of defining the dummy interface

ξ that the sediment bed would have had if all the grains in motion were redeposited at the bottom (Figure 7.8(b)). In the case of a continuous description by an area of study ϕ, the effective position of the interface is written as:

$$\xi = \int_{-\infty}^{\infty} \frac{\phi}{\phi_b} \mathrm{d}z \qquad [7.11]$$

where ϕ_b is the volume fraction of the bed.

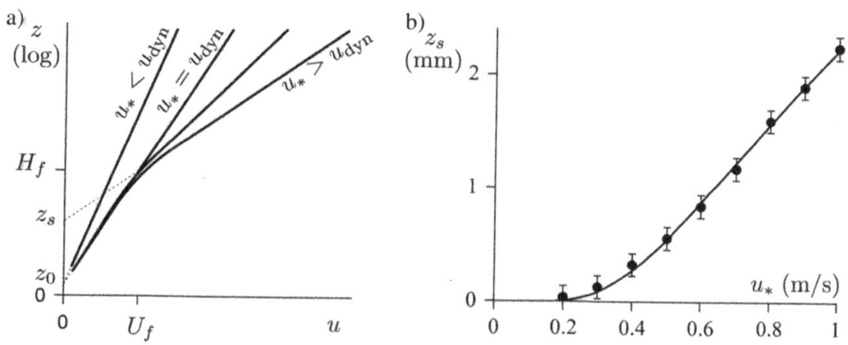

Figure 7.7. *(a) Diagram of the velocity profile $u(z)$ through a layer of saltation (NB: logarithmic scale in z). The profiles above the transport threshold converge at the same "focal point" at height H_f and speed U_f (see Figure 7.10). (b) Measurement of aerodynamic roughness z_s seen from above the layer of transport [IVE 99] for $d = 242$ μm. Solid line curve: adjustment of equation [7.10] ($H_f = 9$ mm and $U_f = 3.5$ m/s)*

In the case of a thick flux of grains entrained by a carrier fluid, ξ_s is the position of the interface between the static and rolling phases, and ξ the interface between the rolling phase and the fluid. In most interesting situations, the gap between interfaces ξ_s and ξ is small. Subsequently, we assume that the two interfaces defined in Figure 7.8 coincide and specify which interface is involved only when it is important. We will also consider that the sediment bed has a well-defined volume fraction of grains, ϕ_b.

7.4.2. *Flux and conservation of matter*

To quantify transport, we use two types of flux that should be distinguished. The first, noted as $\vec{q}$, quantifies the flux of grains crossing through a surface of unit width by unit of time in a direction transversal to the direction of the transport, and which extends vertically from the soil to infinity (Figure 7.9(a)). In the case of a continuous description by an area ϕ and a velocity field $\vec{u}^p$ for particles, the following is written as:

$$\vec{q} = \int_{-\infty}^{\infty} \frac{\phi}{\phi_b} \vec{u}^p \mathrm{d}z \qquad [7.12]$$

Thus defined, $\vec{q}$ is a *volumetric* flux (at the volume fraction of the bed): it is a volume by unit width and by unit time. It has therefore the dimension of a diffusion coefficient (L^2T^{-1}). Alternatively, we can define the mass flux $\vec{q}_m = \rho_p \phi_b \vec{q}$.

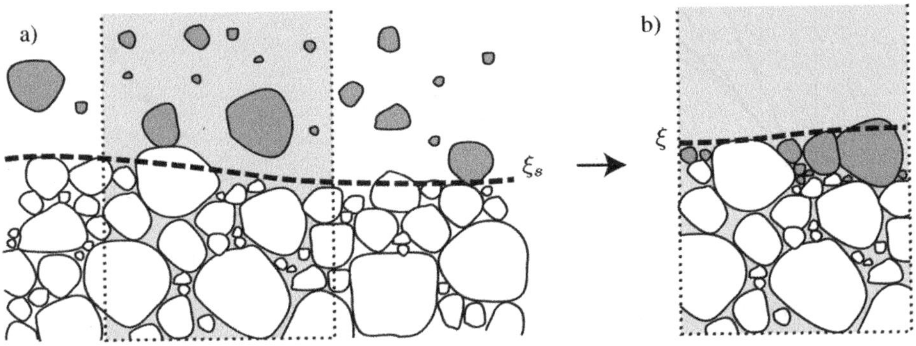

Figure 7.8. *(a) Diagram defining the interface between the static bed and the phase consisting of fluid and grains in motion. (b) By virtually reducing the grains transported on the surface, the effective interface ξ is defined by reconstituting a bed with a homogeneous volume fraction*

Using the effective position ξ of the sand bed, the conservation of material, known in geomorphology as the Exner equation, is therefore written as:

$$\frac{\partial \xi}{\partial t} = -\vec{\nabla} \cdot \vec{q} \qquad [7.13]$$

Furthermore, we define the upward flux $\varphi_\uparrow(z)$ and downward flux $\varphi_\downarrow(z)$ as the volume of grains (with the same volume fraction ϕ_b) that crosses a horizontal unit surface at altitude z by unit of time, from the bottom to the top, and from the top to bottom, respectively (Figure 7.9(b)). These fluxes have the dimension of velocities (LT^{-1}). This time, the equilibrium between erosion and the deposition governing the evolution of the height of the bed is written as:

$$\frac{\partial \xi_s}{\partial t} = \varphi_\downarrow - \varphi_\uparrow \qquad [7.14]$$

We observe that the difference $\varphi_\downarrow - \varphi_\uparrow$ between the accretion rate and the erosion rate is simply the velocity at the surface of the sand. It is measured simply with a dipstick. This velocity is not material since, by definition, the velocity of grains at the surface of the static bed is nil.

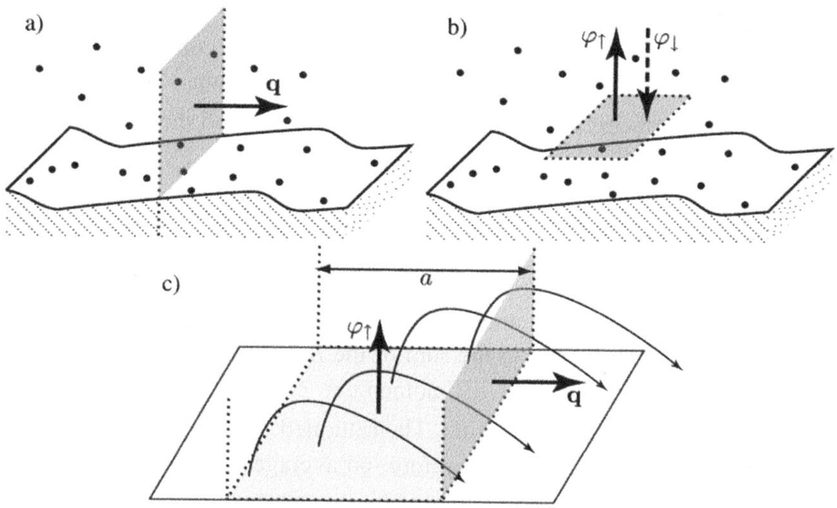

Figure 7.9. *Diagram defining (a) the horizontal flux $\vec{q}$ and (b) the ascending flux $\varphi_\uparrow$ and descending flux $\varphi_\downarrow$. (c) In stationary regime, the horizontal and vertical fluxes are linked by an average length a of the particle trajectories: $q = a\varphi$*

7.4.3. *Saturated flux*

Let us consider an infinite flat sediment bed subjected to a stationary flow. Equilibrium is established between particle transport and the flow characterized by flux $q = q_{\text{sat}}$, called saturated flux. As this flux is spatially homogeneous, there is globally neither erosion nor accretion of particles. Likewise, as many particles are deposited as are torn from the bed: $\varphi = \varphi_\downarrow = \varphi_\uparrow$. If the grains travel an average distance a between the time when they leave the soil and the time where they return to it, the horizontal and vertical fluxes are linked by equality $q = a\varphi$.

The stronger the flow, the more grains it transports at equilibrium: the saturated flux is therefore an increasing function of shear velocity u_* and vanishes below the shear velocity threshold. This dynamic threshold is measured by extrapolating curve $q_{sat}(u_*)$ to zero. The experimental calibration of the relationship between saturated flux q_{sat} and velocity u_* is essential to understand and predict sediment morphodynamics. The appropriate measurements are performed in a wind tunnel [IVE 94, RAS 96] by integrating the measurement of the local flux or by calculating integral mass balances.

To obtain the scaling law on q_{sat}, only the average trajectory of grains is considered. Saltons leave the soil at an average horizontal velocity $u_\uparrow^p$, and, after having made a jump of size a and accelerated by the flow, collide with the soil at horizontal velocity $u_\downarrow^p$. By writing the mechanical equilibrium of a slice contained between two unit surfaces, one located just above the bed and the other far from it, we show that:

$$\tau \equiv \rho_f u_*^2 = \tau_b + \rho_p \phi_b \varphi (u_\downarrow^p - u_\uparrow^p) = \tau_b + \rho_p \phi_b \frac{(u_\downarrow^p - u_\uparrow^p)}{a} q \qquad [7.15]$$

The left term τ is the stress applied on the layer far from the bed, i.e. the non-disturbed fluid stress. It is equal to the sum of the fluid stress exerted at the base of bed τ_b and the horizontal flux of momentum caused by the ascending and descending motion of grains in saltation. The saturated flux is reached when τ_b has decreased to threshold value τ_{dyn}. Therefore, on average, each grain ejects a single grain on collision. The saturated flux becomes:

$$q_{sat} = \frac{a(\tau - \tau_{dyn})}{\rho_p \phi_b (u_\downarrow^p - u_\uparrow^p)} \qquad [7.16]$$

However, it is still necessary to explain the average jump length a and the two velocities $u_\downarrow^p$ and $u_\uparrow^p$. The important point is that, at saturation, the wind velocity in the transport layer is reduced to the value it has at the transport threshold, whatever be the force of the wind above this layer. Consequently, the trajectories of grains in saltation are independent of wind velocity. In particular, in equation [7.16], a, $u_\downarrow^p$, and $u_\uparrow^p$ are independent of τ. Thus, dimensionally, we have $u_\downarrow^p - u_\uparrow^p \propto \sqrt{gd}$ and $a \propto d$ and therefore find:

$$q_{sat} \sim \frac{\rho_p - \rho_f}{\rho_p} \sqrt{gd^3}\,(\Theta - \Theta_{dyn}) \qquad [7.17]$$

The saturated flux is thus proportional to the distance to the basal stress threshold [UNG 87, AND 04] as observed by the measurements taken in the wind tunnel (Figure 7.10(a) and references [IVE 99] and [CRE 09]).

7.4.4. *Length of saturation*

Let us now consider a situation in which the flux or the relief is no longer homogeneous but vary in space or time, like on the top of a dune. The saturation process of transport described does not occur instantaneously: the flux of grains q readapts to reach saturated flux q_{sat} corresponding to the local value of the stress, but with a delay in time and space [BAG 41, AND 88, AND 91, SAU 01, AND 02, AND 04, VAL 05, CHA 06]. By linearizing the problem around the homogeneous and steady state, these delays can be taken into account by a simple relaxation equation. For one space dimension, we obtain the following first-order differential linear equation:

$$T_{sat}\frac{\partial q}{\partial t}+L_{sat}\frac{\partial q}{\partial x}=q_{sat}-q \qquad [7.18]$$

where T_{sat} and L_{sat} are the time and length of saturation of the flux.

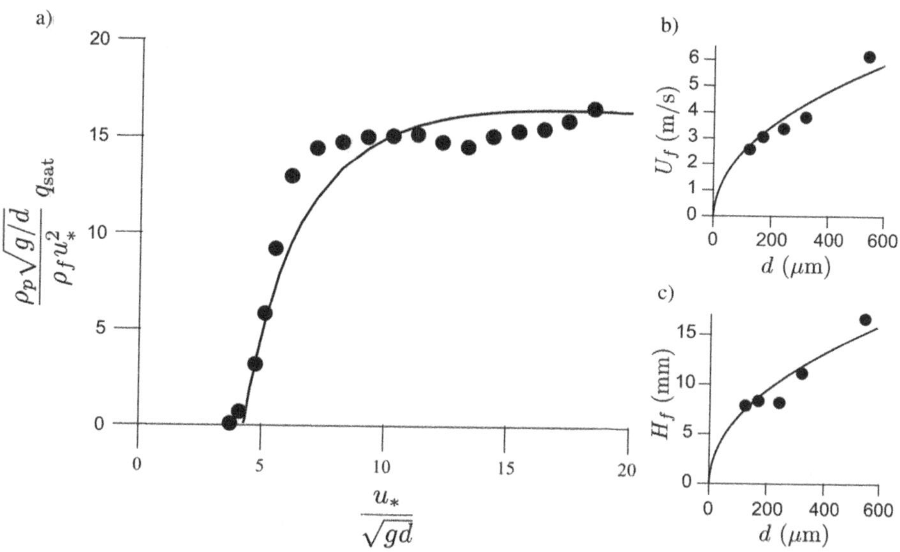

Figure 7.10. *(a) Saturated flux measurements in the case of aeolian transport for $d = 242\ \mu m$ [IVE 99]. The flux is normalized by u_*^2 to show the asymptomatic behavior. (b) Velocity at the focal point in relation to the size of grains. (c) Height of this focal point in relation to d*

To measure these quantities, we must consider two pure situations separately. The first is the case of a homogeneous sand bed, also subjected to a homogeneous flow, but whose intensity increases by increments. The flux is therefore spatially uniform but relaxes exponentially in time toward saturated flux $q_{sat}(u_*)$. The second fundamental configuration consists of having a homogeneous flow on a sandy bed, which only extends in the half-space $x > 0$. Upwind of this bed ($x < 0$) we consider that the soil cannot be eroded and that it can possess the same hydrodynamic roughness as sand. The flux of sand q at the entry ($x = 0$) of the sandy bed is nil; it increases then relaxes exponentially toward its saturation value q_{sat} over a characteristic length L_{sat} (Figure 7.11). The relaxation time T_{sat} is generally shorter than the time during which the ground evolves so that it can be neglected in geomorphological problems, see the following section. The hydrodynamics and transport can therefore be considered as stationary.

This formalism allows us to account for the fact that neither flux q nor the erosion rate is a function of shear stress τ. Indeed, when the flux is locally higher than the saturated flux, it decreases in space, leaving a deposit. If the flux is locally lower than the saturated flux, it increases, leading to the erosion of the bed. The saturation equation thus clearly expresses the possibility of erosion or deposition for very large wind velocities.

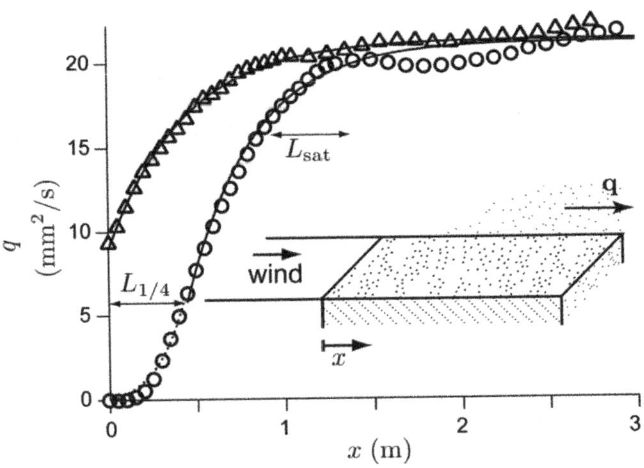

Figure 7.11. *Measurement of sand flux as a function of position, in the case of aeolian transport, without (○) and with (Δ) an incoming flux. x corresponds to the direction of the wind. $L_{1/4}$ is the length necessary so that the flux reaches a quarter of its saturated value. L_{sat} is the length over which the flux relaxes in the vicinity of the saturation*

Note should be taken, however, of two interesting limit conditions. The first is that for which L_{sat} is nil. In this case, the flux is saturated everywhere ($q = q_{sat}$) and is thus a function of τ. The second limit is that where the length of saturation is considerable in comparison to that of the system. In this case, the flux is still very small in comparison to that of the saturated flux and, when neglecting T_{sat}, equation [7.18] becomes $L_{sat}\partial q/\partial x \simeq q_{sat}$. By combining the latter with the mass conservation equation [7.13], it is the erosion rate $-\partial h/\partial t = q_{sat}/L_{sat}$ that this time becomes a function of τ. This limit condition can describe the erosion of very cohesive materials, in particular rocks, for which there is no redeposition of eroded sediments.

To understand the origin of the length of saturation in the case of aeolian transport, it is noteworthy that the flux is written as the product of a grain density and velocity. As a consequence, this process of adapting q to a given wind can be dominated by two mechanisms. On the one hand, the grains must be accelerated at the velocity of the flow. On the other hand, the erosion of the bed results in the number of grains transported tending toward its saturated value. To describe the first mechanism, let us simply consider the horizontal motion of a grain entrained by wind, by writing that the acceleration is equal to the turbulent drag force as follows:

$$\frac{\pi}{6}\rho_p d^3 \frac{du^p}{dt} = \frac{\pi}{8} C_d \rho_f (u - u^p)^2 d^2 \qquad [7.19]$$

where u^p is the velocity of the grain and u the wind velocity.

This equation can be integrated for an essentially constant coefficient C_d, and it can be shown that the relaxation of the velocity of the particle to that of the fluid velocity occurs over a saturation length that scales as:

$$L_{sat} \sim \frac{\rho_p}{\rho_f} d \qquad [7.20]$$

with a pre-factor of the order of 2 [AND 02, AND 04, AND 10]. It must be noted that this length is independent of shear velocity u_*, confirming the direct and indirect measurements of L_{sat} (Figure 7.12). The direct measurements are carried out by adjusting an exponential law on the part of the data close to saturation (Figure 7.11).

On the curve without input flux in Figure 7.11 ($\circ$), an initial flux growth regime can be observed close to the start of the bed, with a typical spatial extension $L_{1/4}$, which is the length necessary for q to reach a quarter of its saturated value.

It corresponds to the second mechanism where the number of grains transported increases. This length increases as a function of the closeness of the shear velocity to its threshold value u_{th}, which means that the total length for which the flux grows from 0 to q_{sat} can be significantly larger than L_{sat}. However, regarding the process of formation of a dune from a flat homogeneous bed (see the following section), only length L_{sat} is relevant as the flux is always close to its saturated value.

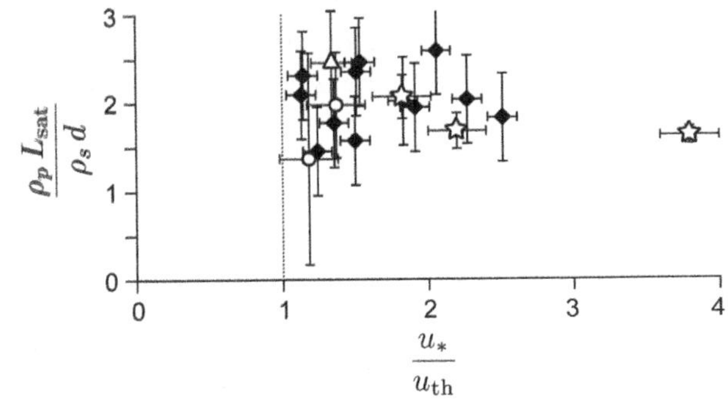

Figure 7.12. *Saturation length, rescaled by the drag length $d\rho_p/\rho_f$ as a function of the shear velocity u_* divided by its threshold value u_{th}. Symbols: (♦) direct measurements in a wind tunnel; direct measurements in the field (Δ); values deduced from the measurement of the wavelength at which the flanks of a Barchan dune is destabilized after a change of wind direction, or of dome size (○), see the following section and [AND 10]*

7.5. Linear stability analysis of a flat erodible bed

By combining the different ingredients developed in the previous sections, in this section we proceed with the linear stability analysis of a flat erodible bed and show how the length of saturation controls the wavelength at which aeolian dunes (on Earth and on Mars) and aquatic ripples occur.

7.5.1. *Qualitative description of instability*

A flat sand bed subjected to a flow whose velocity is above the transport threshold is destabilized and forms periodic ranges of transverse dunes. A nice example is shown in Figure 7.13, where undulations appear on the flank of a large Barchan dune, which in this case plays the role of the flat bed. Qualitatively, this instability can be understood as follows. Let us consider a sinusoidal bed of weak amplitude $\xi = \zeta \exp ikx$, where $k = 2\pi/\lambda$ is the wave number (Figure 7.14). In the first order, the effect of the relief is to increase the velocity and stress at the crests by narrowing the streamlines, and reduce it in the hollows. The transport flux q thus increases along the upwind (stoss) slope, giving rise to erosion, and decreases along the downwind (stoss) side, giving rise to deposition (see the mass conservation equation [7.13]). The result is a propagation of the undulations in the direction of the flow. To explain the amplification, it is necessary to simultaneously take into account shear stresses and the inertial effects of the fluid. In this case, there is a phase shift between the disturbance of the relief and the basal stress, with the maximum stress situated upwind of the crests. This shift therefore leads to an accumulation of sand at the crest and thus amplification [KEN 63].

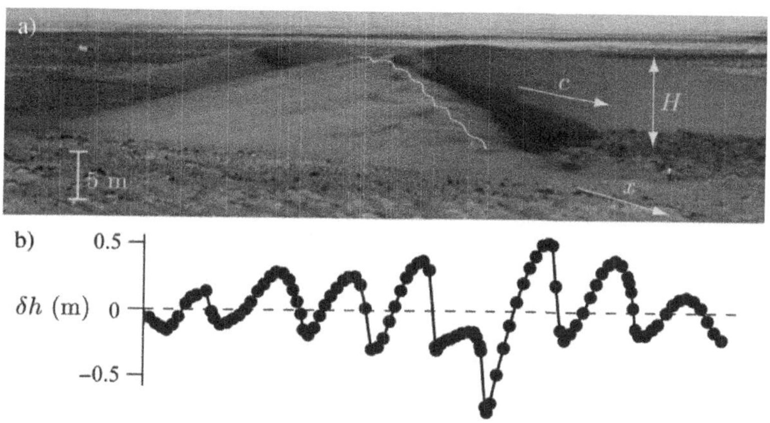

Figure 7.13. *(a) Linear instability on the flank of a large Barchan dune. (b) Longitudinal profile of the disturbance of the dune surface*

This destabilizing hydrodynamic mechanism is the source of dune formation in both turbulent flows [ENG 70, RIC 80, AND 02] and laminar flows [CHA 00, LAG 03, VAL 05, CHA 05]. There are two stabilizing mechanisms of different natures. On the one hand, the effect of a slope induced by gravity tends to bring grains at the crest toward the troughs and thus leads to diffuse stabilizing transport. This mechanism is not associated with any intrinsic length scale.

On the other hand, as discussed in the previous section, the flux follows the basal stress with a delay. This second stabilizing mechanism leads to the occurrence of a length scale, the length of saturation L_{sat}. This is, however, not the only length of the problem: the hydrodynamics and geometry of the problem can lead to others. Examples are the thickness of the viscous boundary sublayer [SUM 84], the geometrical size of the rigid walls confining the flow [OUR 09], and the depth of water H in the case of a free surface flow [FOU 10]. We show next the linear stability analysis performed for a turbulent flow in a semi-infinite medium, a situation for which none of the latter three length scales is relevant.

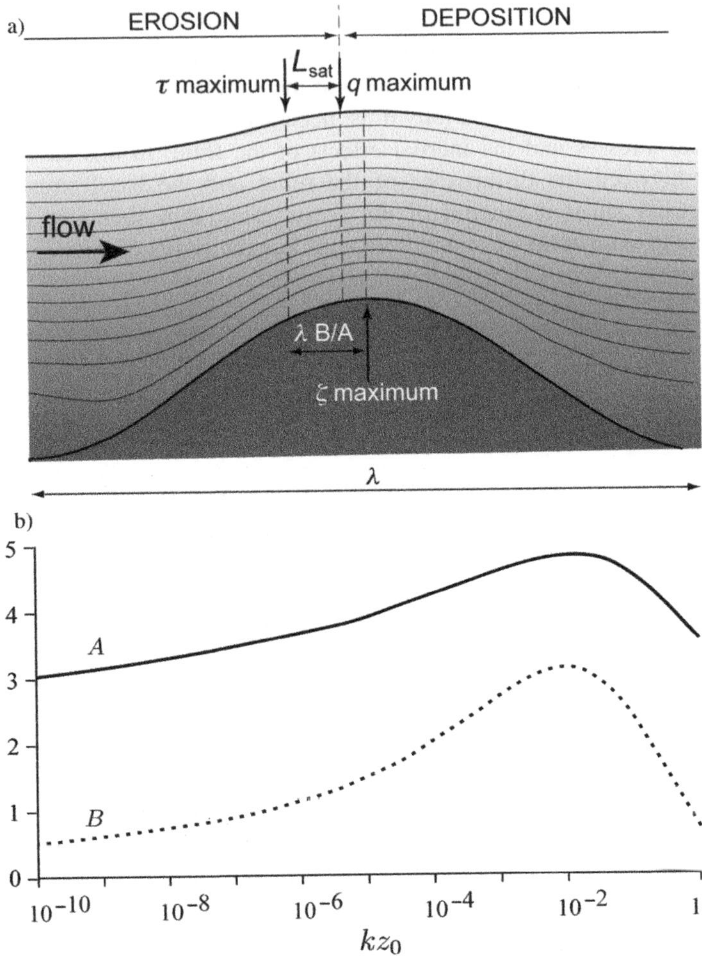

Figure 7.14. *(a) Scheme showing the origin of dune instability. (b) Coefficients A and B as functions of the wave number k made dimensionless by the aerodynamic roughness*

7.5.2. *Linear stability analysis*

Let us consider small disturbances around a flat bed of sediments subjected to a one-directional flow characterized by a stress far from the bed τ_0. The evolution of the sediment bed is controlled by a mass conservation equation:

$$\frac{\partial \xi}{\partial t} = -\frac{\partial q}{\partial x} \quad [7.21]$$

However, we have seen that there is a delay between the actual flux and its saturated value that can be described by the following relaxation equation:

$$L_{\text{sat}} \frac{\partial q}{\partial x} = q_{\text{sat}} - q \tag{7.22}$$

where L_{sat} is the length of saturation of the flux. It is then necessary to consider the law $q_{\text{sat}}(\tau)$ between the saturated flux of sediments and fluid stress that can be chosen in the generic form:

$$q_{\text{sat}} = \chi(\tau - \tau_{\text{th}})^{\gamma} \tag{7.23}$$

where τ_{th} is the stress threshold. As we have seen in the previous section, the aeolian case is described appropriately by $\gamma = 1$ (equation [7.17]), whereas the aqueous case is described better by $\gamma = 3/2$ [MEY 48]. We have also seen that the transport threshold in stress depends on the local slope θ of the bed ($\tan\theta = \partial\xi/\partial x$) according to:

$$\tau_{\text{th}}(\theta) \simeq \tau_{\text{th}}^{0}\left(\cos\theta + \frac{\sin\theta}{\mu}\right) \tag{7.24}$$

where τ_{th}^{0} is the stress threshold on the flat bed and μ a friction coefficient. The final closing relationship required is the stress term τ exerted by the flow on the deformed bed. Insofar as the evolution of the bed is very slow, we generally assume that the flow of the fluid can be determined by considering the bottom as stationary. Under these assumptions, the dimensional analysis enables us, in the case of small sinusoidal disturbances, to write the stress in the generic form:

$$\tau = \tau_0 + \tau_0(A + \mathrm{i}B)k\xi \tag{7.25}$$

where A denotes the stress component in phase with the relief and B the component in quadrature. The exact expressions of A and B depend on the flow regime considered. In a laminar regime, a calculation shows that A and B are functions of the product kl_v, where $l_v = (\eta^2/\rho_f k\tau_0)^{1/3}$ is a length of penetration involving viscosity η with the fluid and its density ρ_f [BEN 59, CHA 00]. In a turbulent regime, A and B depend only slightly (logarithmically) on the hydrodynamic roughness by kz_0 [JAC 75, HUN 88, AND 02, KRO 02a, KRO 02b, FOU 10] and can be considered as constants in practice (Figure 7.14(b)). Whatever the case, A and B are positive. This means that the stress above a hump is greater than the stress without hump ($A > 0$), and that the maximum stress is located upwind of the hump ($B > 0$).

The following stage consists of disturbing the system around the base state according to the Fourier modes in $\exp[\sigma t + \mathrm{i}k(x - ct)]$, which we indicate by a hat ($\hat{\ }$). c is the phase velocity and σ the increasing rate of the ripples on the bed. When linearized equations [7.26]–[7.30] become:

$$(\sigma - \mathrm{i}kc)\zeta = -\mathrm{i}k\hat{q} \tag{7.26}$$

$$\mathrm{i}kL_{\mathrm{sat}}\hat{q} = \hat{q}_{\mathrm{sat}} - \hat{q} \tag{7.27}$$

$$\hat{q}_{\mathrm{sat}} = \frac{Q}{\tau_0}(\hat{\tau} - \hat{\tau}_{\mathrm{th}}) \tag{7.28}$$

$$\hat{\tau}_{\mathrm{th}} = \frac{\tau_{\mathrm{th}}^0}{\mu}\mathrm{i}k\zeta \tag{7.29}$$

$$\hat{\tau} = \tau_0(A + \mathrm{i}B)k\zeta \tag{7.30}$$

where we have defined $Q = \tau_0\gamma\chi(\tau_0 - \tau_{\mathrm{th}}^0)^{\gamma-1}$, which has the dimension of a flux. We note that in linear order, the saturated flux stems from the modulation of wind by the relief and from the slope effect:

$$\hat{q}_{\mathrm{sat}} = Q\left[(A + \mathrm{i}B)k\zeta - \frac{\tau_{\mathrm{th}}^0}{\mu\tau_0}\mathrm{i}k\zeta\right] \tag{7.31}$$

When the stress of the base flow is large in comparison to its threshold value, the slope effect is negligible so that the equation is simply reduced to $\hat{q}_{\mathrm{sat}} = Q(A + \mathrm{i}B)k\hat{\xi}$. In the general case, the effect of the slope can be incorporated in the saturated flux through the component in quadrature with the relief by posing $B_\mu = B - (\tau_{\mathrm{th}}^0/\mu\tau_0)$.

The linear system formed by equations [7.26]–[7.30] has non-zero amplitude solutions only if k, c, and σ are related by the following dispersion relationship:

$$\sigma - \mathrm{i}kc = -\frac{\mathrm{i}Q(A + \mathrm{i}B_\mu)k^2}{1 + \mathrm{i}kL_{\mathrm{sat}}} \tag{7.32}$$

that is

$$\frac{L_{\text{sat}}^2 \sigma}{Q} = \frac{(kL_{\text{sat}})^2 (B_\mu - AkL_{\text{sat}})}{1 + (kL_{\text{sat}})^2} \quad [7.33]$$

$$\frac{L_{\text{sat}} c}{Q} = \frac{(kL_{\text{sat}})(A + B_\mu kL_{\text{sat}})}{1 + (kL_{\text{sat}})^2} \quad [7.34]$$

We note that the length of saturation L_{sat} appears as the characteristic length of the problem, as the characteristic time is ratio L_{sat}^2/Q. These functions are shown in Figure 7.15.

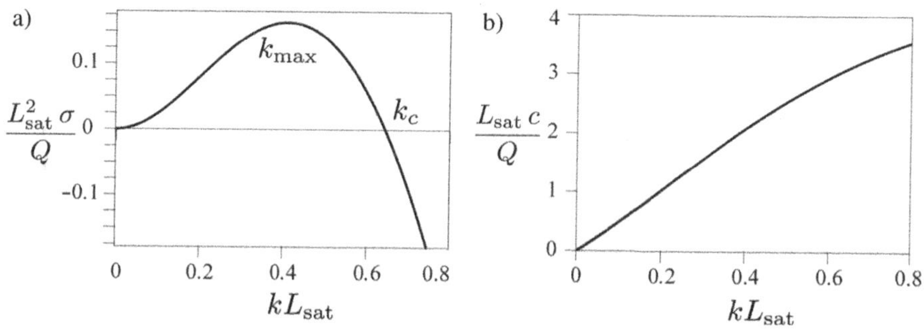

Figure 7.15. *Typical dispersion relation for the present model of an aeolian dune and aquatic ripple formation. (a) Dimensionless growth rate and (b) dimensionless phase velocity*

To understand the properties of this dispersion relationship, we can assume that A and B_μ are independent of wave number k. The flat bed is unstable when growth rate σ is positive for at least a wave number k, that is to say for $B_\mu > 0$. Near the transport threshold, for $\tau_0 \simeq \tau_{\text{th}}^0$, B_μ approaches $B - \mu^{-1}$.

If μB is smaller than 1, then B_μ is negative and there is no instability. In other words, the instability threshold is distinct from the transport threshold. Above the stability threshold, there is a cut-off wave number $k_c = B_\mu / A$ beyond which the disturbances are stable, as shown in Figure 7.15(a).

We are dealing with a large wavelength instability. The stabilization of small wavelengths can be simply interpreted in terms of shifts. The maximum hydrodynamic stress is located at distance $B/(kA)$ upwind of the peak of the sinusoid considered (equation [7.25]). By including the slope effects, we obtain the position of the maximum saturated flux, also located upwind of the peak, at a distance $B_\mu/(kA)$. The response of the actual flux to stress variations is delayed. The maximum flux is thus located at a distance $B_\mu/(kA) - L_{sat}$ from the crest. In fact, this maximum separates the erosion area upwind from the deposition zone downwind. Knowing that a disturbance increases only if the crest is located in the deposit area, we deduce that amplification occurs only for wave numbers verifying $B_\mu/(kA) > L_{sat}$ (Figure 7.14(a)).

We can also calculate the most unstable mode k_{max} from equation [7.33], which verifies:

$$k_{max} L_{sat} \simeq X^{-1/3} - X^{1/3} \quad \text{with} \quad X = -\frac{B_\mu}{A} + \sqrt{1 + \left(\frac{B_\mu}{A}\right)^2} \qquad [7.35]$$

7.5.3. *A scaling law for elementary dunes*

The main result is that, above the stability threshold (with $\tau_0 > \tau_{th}^0$ given), the most unstable wavelength of the bed is proportional to the length of saturation L_{sat}. We saw in the previous section that the length of saturation stems from the inertia of the grains (equation [7.20]). By assuming $B_\mu \ll A$, we therefore find that, in the first approximation, the selected wavelength validates the following scaling law:

$$\lambda_{max} \sim \frac{A}{B_\mu} \frac{\rho_p}{\rho_f} d \qquad [7.36]$$

This law expresses the essence of the instability: the source of the destabilizing mechanism (associated with term B_μ/A) is hydrodynamic, while the stabilizing term (associated with term $(\rho_p/\rho_f)d$) stems from the grain transport. How does this prediction compare to the observations? Figure 7.16 shows measurements carried out in very different environments with grains of different sizes [CLA 06]. Wavelengths in natural situations have been measured on emerging transverse dune areas, recognizable by their morphologies. The wavelength is understood as being the average distance from peak-to-peak, which does not exclude a relatively wide distribution. We note that the scaling law equation [7.36] applies to a wide range of sizes and environments. The dunes of the Atlantic Sahara are formed with a size of about 20 m, for a grain size d of about 180 μm [ELB 05]. Snow dunes are formed during storms during which powdery and light flakes of about 1 cm [FAH 00] are deposited [FRE 02] on the frozen soil (in the Antarctic and Baltic region). In this case, the wavelength is also about tens of meters, which is explained by the fact that product $\rho_p d$ is almost the same as for sand.

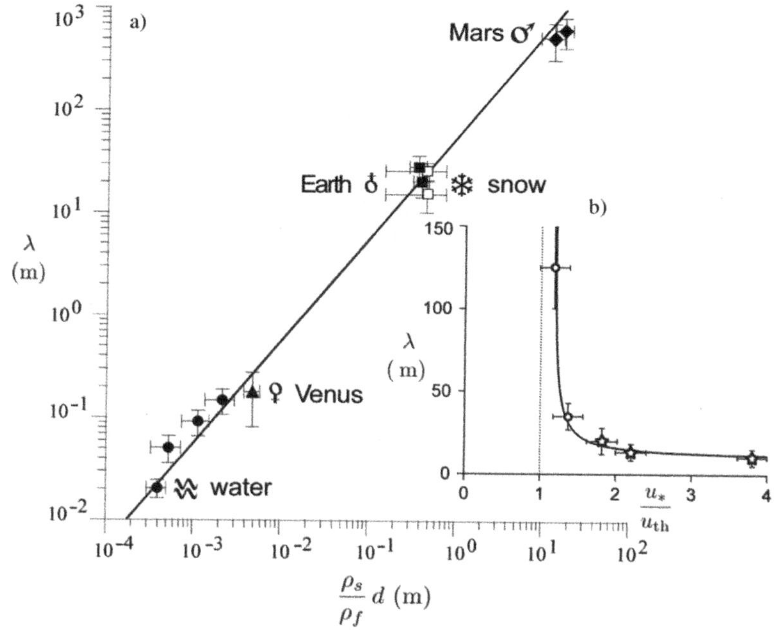

Figure 7.16. *(a) Measurements of the wavelength of elementary dunes formed by linear instability, as a function of the grain to fluid density ratio, multiplied by the size of the grains [CLA 06]. Solid line: angle of equation $\lambda = 53d\rho_p/\rho_f$. (b) The most unstable wavelength as a function of u_*/u_{th}*

Martian dunes are formed in a much diluted CO_2 atmosphere, with grains slightly smaller than those on Earth. It is therefore normal that the elementary dunes are a lot larger, in the order of 600 m. Experiments performed in CO_2 near the critical point have simulated conditions on the surface of Venus. In this atmosphere, which is 100 times denser than air, the elementary dunes are formed with a wavelength of around 10 cm [GRE 84]. Finally, in water, a number of experiments have been performed in turbulent regimes, giving ripples with wavelengths of at the centimeter scale.

It should be emphasized that this scaling law only takes into account the dominating factor and does not detail the instability. In particular, slope effect is expressed by the dependence of coefficients B_μ on flow velocity. Since this is a stabilizing effect, the selected wavelength increases when reaching the transport threshold (Figure 7.16(b)). Also, the hydrodynamic coefficients A and B depend on wavelength k made dimensionless by roughness z_0 (Figure 7.14(b)), which is itself a function of the wind velocity.

7.6. Conclusion

We have shown the scaling laws governing the sediment transport threshold (static and dynamic), the saturation of the sand flux (equilibrium value and length of relaxation to a saturated flux), and finally the wavelength at which a flat erodible bed is destabilized. These results and this formalism developed in the context of aeolian transport can be generalized and adapted to other modes of transport.

In the case of bedload (bed transport in water), the impact of grains in motion on the bed are considerably damped by the viscous flow at the level of contact between grains, so that the coupling saltation/reptation and the existence of a dynamic transport threshold are not relevant. However, saturated flux and the length of saturation still make sense, and their expression depends on the flow regime. For a Shields number close to the threshold, when the feedback of the grains in motion on the fluid is negligible, it can be assumed that saturation is limited by the erosion of grains at the surface of the bed. Thus, the length of saturation corresponds to the typical distance between two traps of sufficient depth to retain a grain at the imposed value Θ. However, saturation is limited by the feedback of the grains in motion on the flow and the analysis is close to that shown in the aeolian case, so that the behavior of L_{sat} is also dominated by the length of the drag $(\rho_p/\rho_f)d$.

In the case of turbulent suspensions, we have shown that the length of saturation is proportional to the length of deposition $(u_*/u_{fall})H$, where u_{fall} is the velocity of the particle sedimentation [CLA 11]. On the one hand, this length depends explicitly on the velocity of the flow, while, on the other hand, it is four or five orders of magnitude greater than the length of saturation of the bedload – typically from 10 to 100 H. It can therefore be understood that this mode of transport has a stabilizing effect for patterns whose wavelength is smaller than the water height, and can be the source of larger structures (anti-dunes, bars, and meanders).

When focusing on the instability mechanism triggering natural structures, it can be noted that the naturalist classification, in particular by resemblance and size, is not necessarily relevant from the point of view of physics. Aeolian dunes, the gigantic Martian dunes, and aquatic ripples are of the same nature since they are formed by the same linear instability involving the interaction between relief and sediment transport. Their size is distinguished only by the density of the fluid surrounding them. On the contrary, aeolian ripples, whose size, like aquatic ripples, is centimetric, are of another nature and have no counterpart in water: they are formed by unstable screening linked to the impact on the relief of grains in saltation. Finally, river dunes result from the interaction with free surface flows, i.e. from an additional mechanism that is totally negligible in the case of aquatic ripples [FOU 10]. Although no strict analogy can be made, giant aeolian dunes also result from an effect of finite size. In this case, hydrodynamic modulation is exerted on the thickness of the atmospheric boundary layer [AND 09].

Finally, to go beyond these scaling laws, greater rigor is required to solve hydrodynamics, the equation that expresses the equilibrium between erosion and deposition, and the trajectories of particles. This work can only be done numerically and leads to numerous variants according to the parameters taken into account. The discussion regarding these issues is beyond the scope of this chapter.

7.7. Acknowledgments

These works were performed in particular for the research project ANR *Zephyr – Morphodynamics Of Rivers And Dunes* (no. ERCS07_18). We thank Y. Forterre and O. Pouliquen for the discussions that helped in the writing of this chapter.

7.8. Bibliography

[AND 88] ANDERSON R.S., HAFF P.K., "Simulation of aeolian saltation", *Science*, vol. 241, pp. 820–823, 1988.

[AND 91] ANDERSON R.S., HAFF P.K., "Wind modification and bed response during saltation of sand in air", *Acta Mechanica Supplement*, vol. 1, pp. 21–51, 1991.

[AND 04] ANDREOTTI B., "A two species model of aeolian sand transport", *Journal of Fluid Mechanics*, vol. 510, pp. 47–70, 2004.

[AND 02] ANDREOTTI B., CLAUDIN P., DOUADY S., "Selection of dune shapes and velocities (2 Parts) ", *European Physical Journal B*, vol. 28, pp. 321–339, 341–352, 2002.

[AND 10] ANDREOTTI B., CLAUDIN P., POULIQUEN O., "Measurements of the aeolian sand transport saturation length", *Geomorphology*, vol. 123, nos. 3–4, pp. 343–348, 2010.

[AND 09] ANDREOTTI B., FOURRIÈRE A., OULD-KADDOUR F., MURRAY A.B., CLAUDIN P., "Giant aeolian dune size determined by the averaged depth of the atmospheric boundary layer", *Nature*, vol. 457, pp. 1120–1123, 2009.

[BAG 41] BAGNOLD R.A., *The Physics of Blown Sand and Desert Dunes*, Methuen, London, 1941.

[BEN 59] BENJAMIN T.B., "Shearing flow over a wavy boundary", *Journal of Fluid Mechanics*, vol. 6, pp. 161–205, 1959.

[CHA 06] CHARRU F., "Selection of the ripple length on a granular bed sheared by a liquid flow", *Physics of Fluids*, vol. 18, 121508, 2006.

[CHA 00] CHARRU F., HINCH E.J., "'Phase diagram' of interfacial instabilities in a two-layer Couette flow and mechanism for the long-wave instability", *Journal of Fluid Mechanics*, vol. 414, pp. 195–223, 2000.

[CHA 05] CHARRU F., HINCH E.J., "Ripple formation on a particle bed sheared by a viscous liquid. Part 1: steady flow", *Journal of Fluid Mechanics*, vol. 550, pp. 111–121, 2005.

[CHE 45] CHEPIL W.S., "Dynamics of wind erosion: I. Nature of movement of soil by wind", *Soil Science*, vol. 60, pp. 305–320, 1945.

[CLA 06] CLAUDIN P., ANDREOTTI B., "A scaling law for aeolian dunes on Mars, Venus, Earth, and for sub-aqueous ripples", *Earth and Planetary Science Letters*, vol. 252, pp. 30–44, 2006.

[CLA 11] CLAUDIN P., CHARRU F., ANDREOTTI B., "Transport relaxation time and length scales in turbulent suspensions", *Journal of Fluid Mechanics*, vol. 671, pp. 491–506, 2011.

[CRE 09] CREYSSELS M., DUPONT P., OULD EL MOCTAR A., VALANCE A., CANTAT I., JENKINS J.T., PASINI J.M., RASMUSSEN K.R., "Saltating particles in a turbulent boundary layer: experiment and theory", *Journal of Fluid Mechanics*, vol. 625, pp. 47–74, 2009.

[DEY 03] DEY S., "Threshold of sediment motion on combined transverse and longitudinal sloping beds", *Journal of Hydraulic Research*, vol. 41, pp. 405–415, 2003.

[DUR 11] DURÁN O., CLAUDIN P., ANDREOTTI B., "On aeolian transport: grain-scale interactions, dynamical mechanisms and scaling laws", *Aeolian Research*, vol. 3, no. 3, pp. 243–270, 2011.

[ELB 05] ELBELRHITI H., CLAUDIN C., ANDREOTTI B., "Field evidence for surface wave induced instability of sand dunes", *Nature*, vol. 437, pp. 720–723, 2005.

[ENG 70] ENGELUND F., "Instability of erodible beds", *Journal of Fluid Mechanics*, vol. 42, pp. 225–244, 1970.

[FAH 00] FAHNESTOCK M.A., SCAMBOS T.A., SHUMAN C.A., ARTHERN R.J., WINEBRENNER D.P., KWOK R., "Snow megadune fields on the East Antarctic Plateau: extreme atmosphere-ice interaction", *Geophysical Research Letters*, vol. 27, pp. 3719–3722, 2000.

[FER 04] FERGUSON R.I., CHURCH M., "A simple universal equation for grain settling velocity", *Journal of Sedimentary Research*, vol. 74, pp. 933–937, 2004.

[FER 76] FERNANDEZ LUQUE R., VAN BEEK R., "Erosion and transport of bed-load sediment", *Journal of Hydraulic Research*, vol. 14, pp. 127–144, 1976.

[FOU 10] FOURRIÈRE A., CLAUDIN P., ANDREOTTI B., "Bedforms in a turbulent stream: formation of ripples by primary linear instability and of dunes by non-linear pattern coarsening", *Journal of Fluid Mechanics*, vol. 649, pp. 287–328, 2010.

[FRE 02] FREZZOTTI M., GANDOLFI S., LA MARCA F., URBINI S., "Snow dunes and glazed surfaces in Antarctica: new field and remote-sensing data", *Annals of Glaciology*, vol. 34, pp. 81–88, 2002.

[GRE 84] GREELEY R., MARSHALL J.R., LEACH R.N., "Microdunes and other aeolian bedlorms on Venus: wind tunnel simulations", *Icarus*, vol. 60, pp. 152–160, 1984.

[HAR 88] HARDISTY J., WHITEHOUSE R.J.S., "Effect of bedslope on desert sand transport", *Nature*, vol. 334, pp. 532–534, 1988.

[HUN 88] HUNT J.C.R., LEIBOVICH S., RICHARDS K.J., "Turbulent shear flows over low hills", *Quarterly Journal of the Royal Meteorological Society*, vol. 114, pp. 1435–1470, 1988.

[IVE 94] IVERSEN J.D., RASMUSSEN K.R., "The effect of surface slope on saltation threshold", *Sedimentology*, vol. 41, pp. 721–728, 1994.

[IVE 99] IVERSEN J.D., RASMUSSEN K.R., "The effect of wind speed and bed slope on sand transport", *Sedimentology*, vol. 46, pp. 723–731, 1999.

[JAC 75] JACKSON P.S., HUNT J.C.R., "Turbulent wind flow over a low hill", *Quarterly Journal of the Royal Meteorological Society*, vol. 101, pp. 929–955, 1975.

[KEN 63] KENNEDY J.F., "The mechanics of dunes and antidunes in erodible bed channels", *Journal of Fluid Mechanics*, vol. 16, pp. 521–544, 1963.

[KRO 02a] KROY K., SAUERMANN G., HERRMANN H.J., "Minimal model for sand dunes", *Physical Review Letters*, vol. 88, 054301, 2002.

[KRO 02b] KROY K., SAUERMANN G., HERRMANN H.J., "Minimal model for aeolian sand dunes", *Physical Review E*, vol. 66, 031302, 2002.

[LAG 03] LAGRÉE P.-Y., "A triple deck model of ripple formation and evolution", *Physics of Fluids*, vol. 15, pp. 2355–2368, 2003.

[LOI 05] LOISELEUX T., GONDRET P., RABAUD M., DOPPLER D., "Onset of erosion and avalanche for an inclined granular bed sheared by a continuous laminar flow", *Physics Fluids*, vol. 17, 103304, 2005.

[MEY 48] MEYER-PETER E., MÜLLER R., "Formulas for bedload transport", *IAHSR 2nd Meeting*, Stockholm, Appendix 2, IAHR, TU Delft, pp. 39–64, 1948.

[OUR 09] OURIEMI M., AUSSILLOUS P., GUAZZELLI E., "Sediment dynamics. Part 1. Bed-load transport by laminar shearing flows; Part 2. Dune formation in pipe flow", *Journal of Fluid Mechanics*, vol. 636, pp. 321–336, 2009.

[OWE 64] OWEN P.R., "Saltation of uniform grains in air", *Journal of Fluid Mechanics*, vol. 20, pp. 225–242, 1964.

[RAS 96] RASMUSSEN K.R., IVERSEN J.D., RAUTAHEIMO P., "Saltation and wind flow interaction in a variable slope wind tunnel", *Geomorphology*, vol. 17, pp. 19–28, 1996.

[RIC 80] RICHARDS K.J., "The formation of ripples and dunes on an erodible bed", *Journal of Fluid Mechanics*, vol. 99, pp. 597–618, 1980.

[SAU 01] SAUERMANN G., KROY K., HERRMANN H.J., "Continuum saltation model for sand dunes", *Physical Review E*, vol. 64, 031305, 2001.

[SUM 84] SUMER B.M., BAKIOGLU M., "On the formation of ripples on an erodible bed", *Journal of Fluid Mechanics*, vol. 144, pp. 177–190, 1984.

[UNG 87] UNGAR J.E., HAFF P.K., "Steady state saltation in air", *Sedimentology*, vol. 34, pp. 289–299, 1987.

[VAL 05] VALANCE A., LANGLOIS V., "Ripple formation over a sand bed submitted to a laminar shear flow", *European Physical Journal B*, vol. 43, pp. 283–294, 2005.

[YAL 79] YALIN M.S., KARAHAN E.J., "Inception of sediment transport", *Journal of the Hydraulics Division*, vol. 105, pp. 1433–1443, 1979.

Chapter 8

Two-Phase Modeling of Bedload Transport

8.1. Introduction

The evolution of a bed of particles subjected to shearing flows is a problem that has been largely studied and discussed for over a century. This problem is central to the understanding of a great variety of natural phenomena, such as sediment transport in rivers and estuaries, erosion and deposit leading to the evolution of streams and formation of dunes in the desert. This is also a fundamental problem in many industrial processes, such as transport of granular matter in the pharmaceutical or food processing industry and hydrate formation in pipelines related to oil production.

In most industrial applications, flows are confined in tubes and the issue is to release the granular matter by a controlled bedload flux. In these problems, the liquid flow rate is generally imposed and it is essential to know the minimum critical flow rate ensuring the grain motion above the threshold. The industrial world has been interested for some 50 years in flow properties of solid phases dispersed for problems related, in particular, to the excavation of mine rocks by pipelines or the dredging of rivers.

Today, technical documentation provides a fairly detailed description of flow regimes (stratified, dispersed, etc.), changes in pressure drop, changes in rates of filling, etc. [SHO 91, MAT 05]. However, the understanding of these phenomena is often empirical and many questions remain open to describe these complex systems.

Chapter written by Pascale AUSSILLOUS, Elisabeth GUAZZELLI and Yannick PEYSSON.

Natural phenomena such as sediment transport in rivers have the specificity of having a free surface that can interact with the granular bed if the thickness of the fluid layer is small enough such as in the case of bedload of sediment matter in streams of torrents type. Thus, many work-related sediment transports were made in channels of rectangular sections having a free surface with a gravity flux of liquid, essentially water.

In all cases (free surface or confined flows), we can mainly distinguish three modes of sediment transport by a fluid (Figure 8.1): suspension, saltation, and bedload transport. We refer to suspension when the particles are transported by the fluid without continuous contact with the sediment bed. The suspension may be of viscous origin [FRA 73, LEI 86] or turbulent [BAG 56] depending on the nature of the flow. It results from a balance between the apparent weight of the particles and the diffusion caused by the shear or turbulent fluctuations of the fluid. Particles of larger sizes may fall and collide with the surface. The energy stored during the collision may enable them to rebound. They then perform a saltation motion. During the collision, particles on the surface of the bed can be dislodged and begin to move in saltation or roll on the surface of the bed. This is referred to as reptation. Saltation and reptation are the main modes of transport in air (see Chapter 6 and [AND 04]). In a liquid, during the collision, energy is dissipated by the fluid, and saltation and reputation are generally negligible [GON 02]. Finally, we refer to the bedload transport when the particles are driven by the fluid shearing and roll and slide over the surface of the sediment bed with which they remain in constant contact. Bedload transport exists independently of the nature of the flow. This is the main mode of transport in a liquid when there is no suspension phenomenon.

In this chapter, we focus only on bedload transport of non-cohesive sediments. We describe the physical mechanisms that govern this type of motion. However, we focus on confined flows (to be more precise pipe flows) for two reasons: the fluid flow is well controlled and, as explained above, it corresponds to many industrial applications for which we observe that the mobile particles remain in contact with the sediment bed. Several types of modeling can be considered to describe bedload transport. In this chapter, we favor a two-phase continuum modeling. We discuss two fundamental aspects for this type of sediment transport: incipient motion of the bed and bedload transport by focusing on the determination of the particle flux.

8.2. Incipient motion

An important issue in bedload transport is to predict the threshold for the onset of motion. In fact, this threshold velocity affects, for example, the useable liquid flow rate range enabling the release of the solid phase for the transport of grains in pipes. In the field of river hydraulics, determining the threshold allows us to define the zones of accumulation or the release of solid particles along rivers and streams.

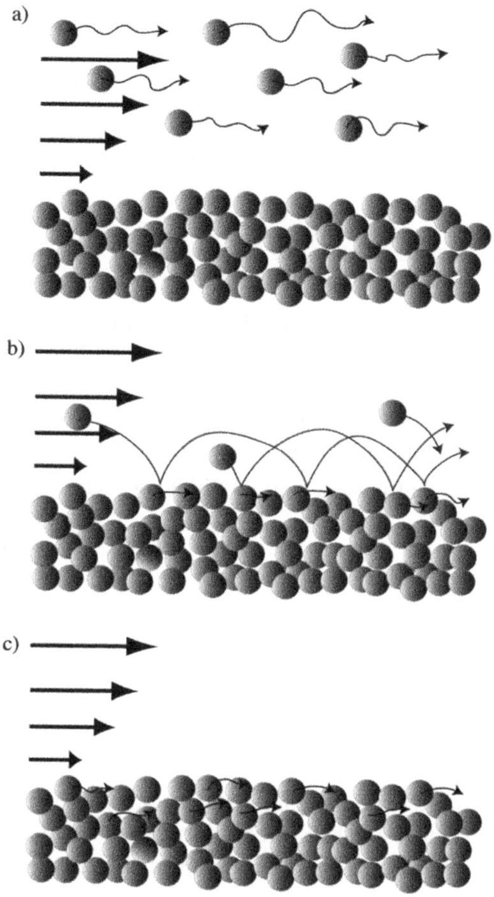

Figure 8.1. *Different modes of sediment transport by a fluid: (a) suspension; (b) saltation and reptation; (c) bedload transport*

In the case of offshore oil production, conditions of low temperature and high pressure can trigger the formation of solid particles called *gas hydrates* which consist of water molecule crystals stabilized by gas molecules, and form solid grains that can exist in dispersed form in the pipe. For these peculiar particles, the presence of a threshold for onset of motion leads to risks of accumulation of particles and clogging of the pipes.

The definition of a criterion enabling us to define the threshold has been the subject of numerous studies, but there is now some consensus around the Shields criterion defined by Alfred Shields in 1936, during his thesis work on sediment transport in rivers [BUF 99].

8.2.1. *Shields number*

The Shields number θ is defined as the hydrodynamic stress τ_b applied at the interface between the fluid and the bed of particles normalized by the effective hydrostatic pressure across a grain of diameter d:

$$\theta = \frac{\tau_b}{(\rho_p - \rho_f)gd} \qquad [8.1]$$

where ρ_f and ρ_p are the respective densities of the fluid and the grains, and g the acceleration of gravity.

The Shields criterion stipulates that the bed of particles is set into motion when the hydrodynamic stress becomes larger than a fraction of the apparent hydrostatic pressure of the particles, that is for a critical Shields number θ^c.

8.2.2. *Force balance on a grain*

The critical Shields number θ^c can be defined from a local force balance on a grain located at the surface of the bed of particles. This approach was proposed early on, namely by White in 1940 [WHI 40], but also, by several authors [VAN 66, FRA 73, MAR 09].

In the very simplified case of three-packed spheres (Figure 8.2), the critical Shields value can be calculated. The force balance involves four forces: the apparent weight of the grain $\pi d^3(\rho_p - \rho_f)g/6$, the contact forces with the first and second particle (N_1, N_2), and the hydrodynamic drag force which is consider to apply on all the surface of the sphere $\pi d^2 \tau_b$. We could take only a fraction of the surface of the sphere as the particle lies on other grains which means that a small part of its surface is in contact with the bed, but the simplification described above is the most often used.

Another important point concerns the definition of the applied stress on the bed, τ_b, which depends on the local flow regime. This point will be discussed in more detail in section 8.2.4.

Before threshold, the particle is immobile and the force balance is zero. This allows us to evaluate two relationships between the forces by writing the projections on the horizontal and vertical axes:

$$\tau_b \pi d^2 + \sin(\alpha)(N_1 - N_2) = 0 \qquad [8.2]$$

$$\frac{\pi}{6} d^3(\rho_p - \rho_f)g - \cos(\alpha)(N_1 + N_2) = 0 \qquad [8.3]$$

where α is defined in Figure 8.2. At the onset of motion, we consider that the particle is loosing contact with the first particle, which enables us to write $N_2 \sin(\alpha) = \tau_b \pi d^2$ and $N_2 \cos(\alpha) = \pi d^3(\rho_p - \rho_f)g/6$. The relationship between these two equations gives the critical Shields value:

$$\frac{\tau_b}{(\rho_p - \rho_f)gd} = \frac{1}{6}\tan(\alpha) \qquad [8.4]$$

For an angle of 30°, corresponding to a cubic packing (Figure 8.2), we obtain a critical Shields value of 0.1.

We have only considered the apparent weight, the contact forces, and the drag force in this force balance. There is another force, the lift force, which can occur due to the difference in speed of the fluid between the top and bottom of the particle. This inertial effect of the flow produces a vertical force directed upwards, which remains quite negligible in the laminar and weak turbulent cases. In fact, most studies show that highly turbulent regimes must be achieved to have significant lift forces [KIN 97, MOL 96].

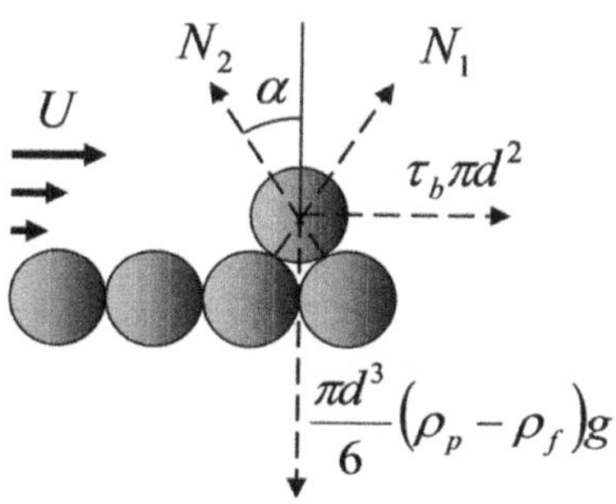

Figure 8.2. *Force balance on a simplified case involving three spheres*

8.2.3. *Force balance on a bed of particles*

In another approach, which we will qualify as continuum approach and which was in particular discussed by Bagnold [BAG 56, BAG 66], the particles are not individualized and the bed is considered as a continuum medium. The onset of motion is then linked to the bed friction coefficient and the volume fraction of the solid grains in the layer.

For a small volume of elementary surface dS, and thickness δ (Figure 8.3), we can write the horizontal force balance. The fluid flow applies a drag force $F_b = \tau_b dS$ on the bed. On the lower layer, a friction of a Coulomb type gives rise to a resisting force $F_f = \mu F_N$ where F_N is the normal force; $F_N = dS\delta(\rho_p - \rho_f)g\phi$ with μ the friction coefficient between layers and ϕ the volume fraction of the layer. At the onset of motion, the volume element is about to move, its thickness is at the minimum of the grain diameter $\delta = d$ and the drag force is equal to the resisting force $\tau_b dS = dS\delta(\rho_p - \rho_f)g\phi\mu$, which leads to:

$$\frac{\tau_b}{(\rho_p - \rho_f)gd} = \mu\phi \qquad [8.5]$$

The Shields value at threshold is $\theta^c = \mu\phi$. For a friction coefficient of magnitude 0.5 and a packing volume fraction of 0.6, we evaluate the critical Shields at 0.3.

It is interesting to note that a friction coefficient is often expressed in a function of the critical friction angle $\mu = \tan\beta$. For a solid, resting on a horizontal plate, β is defined as the inclination angle, which must be achieved to set the solid in motion. Thus, the Shields criterion discussed above can be rewritten as $\theta^c = \phi\tan\beta$. This equation is formally very close to the one calculated for the three spheres in the previous section.

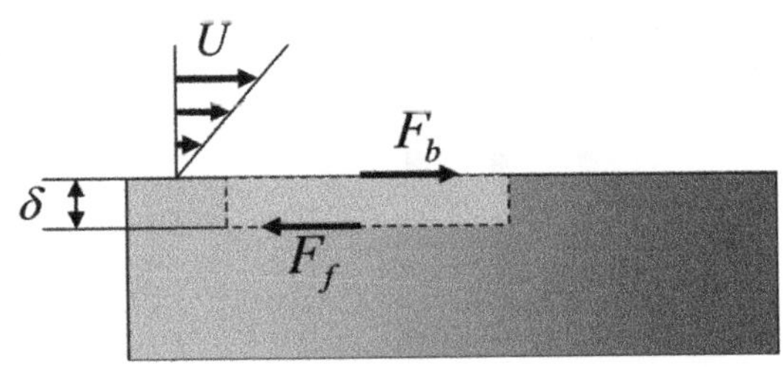

Figure 8.3. *Force balance on a bed of particles;* $F_b = \tau_b dS$, *drag force induced by the fluid flow; and* $F_f = dS\delta(\rho_p - \rho_f)g\phi\mu$ *friction force opposing the motion*

In the case of the spheres described above, considering that their arrangement can be seen as a friction, we then see that both approaches are similar and lead to linking the critical threshold to the packing of the bed and to an effective friction of the medium. This effective friction is divided into two parts: one linked to the grain material and another linked to the grain shape.

8.2.4. *Shields curve*

The experimental measurement of the critical threshold for incipient motion θ^c has triggered interest among researchers since the work of Shields in 1936. Most experiments carried out between the 1940s and 1980s were conducted in open channel mostly with water. On the other hand, a variety of different grains were used: sand, mixture of sands, gravel, river sediment, various minerals, glass beads, etc. A large variety of geometries have been implemented in more recent experiments: Hele-Shaw [LOI 05], Couette geometry [CHA 04], flow in a pipe of cylindrical section [OUR 07], or square section [LOB 08].

A major effort has been made (particularly by Buffington and Montgomery [BUF 97] on the work from the 1940s to 1980s) to synthesize experimental data and represent them on a master curve called the *Shields curve* enabling better comparisons. This curve represents the Shields threshold measured for various grains and fluids (Figure 8.4) versus the particle Reynolds number defined as follows:

$$Re^* = \frac{\rho_f u^* d}{\eta} \qquad [8.6]$$

where u^* is the velocity of the flow around the grain.

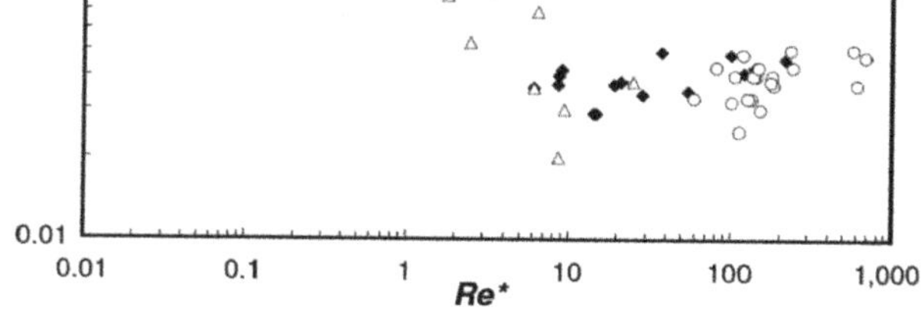

Figure 8.4. *Shields curve: threshold for incipient motion* θ^c *versus Reynolds number* Re^*

The Shields curve shows two main sets of data, those for $Re^* > 10$ and those for $Re^* < 10$. This difference has been attributed to the influence of the flow regime: laminar or turbulent. Thus, the critical threshold is $\theta^c \approx 0.04$ in turbulent regime and $\theta^c \approx 0.2$ in laminar regime. What is also striking when we look at the Shields curve is the scatter of data. This dispersion has several origins, which we now discuss.

The measurement of the threshold for incipient motion is difficult because several problems arise. First, the observation of the threshold itself is difficult. Indeed, visual observation of the grains of the bed shows that the first motions of the grain occur at a threshold of fluid flow rate, but do not present a global motion. It is rather a localized reorganization of the bed of particles. For a slightly higher fluid flow rate, a motion is observed. As discussed by [CHA 04], it is this critical flow rate that must define the threshold. However, scatter in the measurements is also due to the variety for the chosen definition for threshold.

An additional problem comes from the polydispersity of the grains. Even if a global motion is detected, this motion only concerns the smallest grains and by increasing the fluid flow, grains of larger sizes are also set into motion. Thus, it is rather a range of flow rate that is associated with global motion of the grains leading to another possible dispersion of measurements. The initial preparation of the bed can also modify the observed threshold. If the bed is packed initially or prepared by sedimentation without compaction, differences in the critical fluid discharge appear.

Direct observation is, however, not the most common method to measure the threshold. The most widespread method when using an open channel is to extrapolate the measurement of the solid flux. In these studies, the bedload flux is measured for different fluid flux and by linearly extrapolating the evolution of the solid flux versus the Shields number, we obtain a zero bedload flux for a finite Shields number: this defines the threshold for incipient motion θ^c. This extrapolation method also leads to errors in the evaluation of the threshold θ^c.

A final crucial point concerns the definition itself of the Shields number. In fact, while it is historically defined as the stress on the granular bed normalized by the effective hydrostatic pressure across a grain, this definition is problematic because the stress is not directly known but indirectly defined through other measurements. Thus, most of the time, in different works, the stress is either:

– defined by $\tau_b = \rho_f u^{*2}$ with u^* the flow velocity around the grain; or

– defined by $\tau_b = \eta\dot{\gamma}$ with $\dot{\gamma}$ the fluid shear rate at the top of the bed.

In the first case, studies typically involve flows in the turbulent regime. For the second case, the regime is laminar. According to these studies, the variation of definitions of the hydrodynamic stress leads to some confusion. Indeed, the critical Shields number varies according to definitions of this stress and may depend or not on the particle Reynolds number Re^*. We therefore chose to examine in the following two sections the results concerning the threshold measurements obtained in the laminar (section 8.2.5) and the turbulent regime (section 8.2.6).

8.2.5. *Laminar regime*

In laminar regime, the hydrodynamic stress which applies at the interface between the fluid and the bed of particles is viscous: $\tau_b = \eta\dot{\gamma}$ where η is the viscosity of the fluid and $\dot{\gamma}$ the fluid shear rate at the interface.

Compared to the turbulent case, much less data are available. These are summarized in Figure 8.5(a) where, inspired by the historical Shields curve (Figure 8.4), the critical Shields number θ^c is plotted against the particle Reynolds number. Note that for a laminar flow, this particle Reynolds number $\mathrm{Re}_p = \dot{\gamma}d^2\rho_f/\eta$ is constructed by using a characteristic velocity that is obtained by considering the shear rate above the bed at threshold $\dot{\gamma}_c$ applied on a grain of size d.

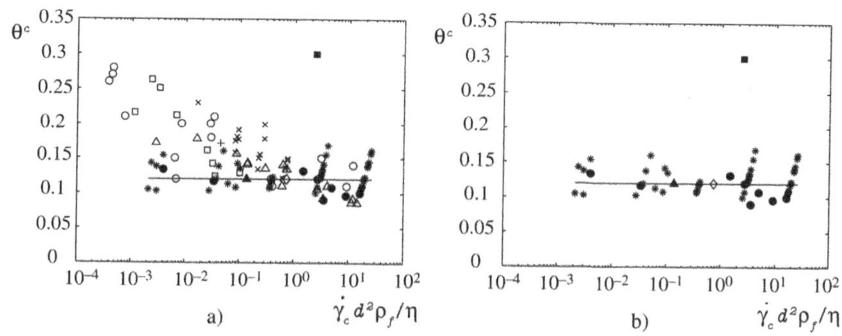

Figure 8.5. *(a) Critical Shields number θ^c versus the particle Reynolds number $\dot{\gamma}d^2\rho_f/\eta$: experimental data of White [WHI 40] quoted by Loiseleux et al. [LOI 05] (+), of White 1970 quoted by Mantz [MAN 77] (□), of Mantz [MAN 77] (×), of Yalin and Karahan [YAL 79] (Δ), of Pilotti and Menduni [PIL 01] (○), of Charru et al. [CHA 04] (▲), of Loiseleux et al. [LOI 05] (●), of Ouriemi et al. [OUR 07] (*), of Lobkovsky et al. [LOB 08] (■) and of Malverti et al. [MAL 08] (◊). The horizontal line represents value $\theta^c = 0.12$; (b) same figure but only representing the most recent data where the threshold is clearly defined*

These data are relatively scattered. Determining the threshold for incipient motion in the laminar regime suffers from the same problems as described above. First, it is difficult to calculate the Shields number because there is no direct measurement of the hydrodynamic stress that applies at the interface between the fluid and the bed of particles. It is evaluated by various methods depending on the experiments and the authors. The second difficulty lies in the definition and the experimental determination of the threshold for incipient motion. This problem was recently highlighted by Charru *et al.* [CHA 04] who examined the individual motion of particles in an annular Couette flow. They showed the existence of two distinct thresholds, the first corresponding to the rearrangement of some particles and the second to the existence of a stationary flux of particles. This second threshold is independent of the initial preparation of the bed, unlike the first, but is not easy to measure because you have to measure the particle flux and ensure that it is saturated.

The oldest data (+, □, ×, Δ) and those of Pilotti and Menduni [PIL 01] (○) were obtained in open channels, from 1 to 10 m long. The particles used come mainly from natural sands. For these experiments in open channel, the stress is evaluated by a momentum balance on the fluid: in uniform and permanent regime, it is written as $\tau_b = \rho_f g h_f \Delta y / \Delta x$ where $\Delta y / \Delta x$ is the slope of the free surface and h_f the water level. This measure proves to be rather difficult because the gradient of the fluid surface at the threshold is generally very low. The threshold is determined by visual observations of the initial motion of grains, and is generally considered to be reached when a significant percentage of the particles is mobile. This definition being very qualitative, the data are relatively noisy. However, the critical Shields number measured lies between 0.1 and 0.3 over a wide range of particle Reynolds number, $4 \times 10^{-4} < Re_p < 14$.

Most of the recent data (▲, •, *, ■) are obtained for confined flows. These flows have the advantage of being well controlled, enabling the evaluation of the shear rate at the top of the bed, $\dot{\gamma}_c$, and the proper calculation of the hydrodynamic stress that applies at the interface between the fluid and the bed of particles $\tau_b = \eta \dot{\gamma}$. Moreover, the definition of the threshold having been clarified in advance, all these measures correspond to the threshold for a stationary flux of particles. The datum of Charru *et al.* [CHA 04] (▲), for example, is obtained in a Couette geometry by measuring the saturated bedload flux. A similar technique is used by Loiseleux *et al.* [LOI 05] (•) who visually observe the grain motion in a Hele-Shaw cell and define the erosion threshold as the lowest flux for which the grains are always eroded after 15 min. Another approach, for experiments not supplied with particles in pipes of circular sections [OUR 07] or rectangular sections [LOB 08] consists of measuring the threshold for cessation of motion. The principle is to impose the liquid flow rate and let the sediment be eroded until this stops. This threshold is effectively the same as that for the onset of motion because a small increase in the fluid flow rate can set the particles into motion again. Also in both experimental procedures, the compactness of the bed has reached a stationary regime.

The recent data of Malverti *et al.* [MAL 08] (◊) are obtained for an open channel. They also assess the Shields number by measuring the slope of the free surface. On the other hand, the threshold is defined by adjusting the variation of the particle flux as a function of the Shields number by the empirical relationship proposed by Meyer-Peter and Muller ([MEY 48]; Table 8.1), and by extrapolating the Shields value to zero flux.

The experiences of [CHA 04], [LOI 05], [OUR 07], and [MAL 08] show that the critical Shields is constant $\theta^c = 0.12 \pm 0.03$, over a large range of small particle Reynolds number $2 \times 10^{-3} < Re_p < 26$ (Figure 8.5). The overestimated value of the critical Shields ≈ 0.3 obtained by Lobkovsky *et al.* [LOB 08] (■) probably comes from the significant side effects not taken into account in the evaluation of the shear rate at the interface between the fluid and the bed of particles.

All these experiments have shown that in the laminar regime, provided that the hydrodynamic stress is properly assessed and the threshold for incipient motion corresponds to the threshold for a steady particle flux, the critical Shields number is then constant $\theta^c = 0.12 \pm 0.03$ over a wide range of small particle Reynolds number $2 \times 10^{-3} < Re_p < 26$. This confirms the choice of a viscous definition of the stress, which is valid in this range. Those recent results showing that θ^c is constant enable us to close a debate in the literature on the variation of θ^c with Re_p and therefore on the Shields curve representation. The latter representation is in fact circular because the particle Reynolds number is directly linked to the critical Shields number by the equation $Re_p = \theta^c Ga$, where $Ga = (\rho_p - \rho_f)\rho_f g d^3/\eta^2$ is the Galileo number and only depends on the used particle–fluid combination. The latest experimental results clearly show that the critical Shields number does not depend on the Reynolds number and, therefore, does not depend on the Galileo number.

8.2.6. *Turbulent regime*

In turbulent regime, the definition of the hydrodynamic stress differs from that used in laminar regime discussed in the previous section. In fact, in most studies concerning measurements of the threshold in turbulent regime the following definition of the Shields number is used:

$$\theta = \frac{\rho u^{*2}}{(\rho_p - \rho_f)gd} \qquad [8.7]$$

with u^* the friction velocity also used in defining the particle Reynolds number of equation [8.6]. The friction velocity can be estimated directly from measurements of flow velocities [MAR 09] or momentum balance. For example, it is the case of open channel where the stress can be written as $\tau_b = \rho_f g h_f \Delta y / \Delta x$. In these conditions the friction velocity is defined as $u^* = (\tau_b / \rho_f)^{1/2}$.

The critical threshold observed for the turbulent regime has a value $\theta^c \approx 0.04$ that remains relatively stable in the Shields curve (Figure 8.4). It raises the question of the difference in value between the critical threshold in laminar and in turbulent regime, which is generally a factor of 3. The reasons for this discrepancy are not yet fully clarified, but two possibilities can be discussed. On the one hand, the formal estimation of the hydrodynamic stress at the surface of the bed may be more difficult in the turbulent regime and it is possible that a gap exists between the stress estimated and the real stress. On the other hand, it is likely that the velocity fluctuations that occur in fully turbulent regime locally trigger an additional hydrodynamic drag force that can set the grains into motion. Similarly, transient lift forces may occur and facilitate the onset of grain motion. It is therefore likely that these phenomena lower the threshold for onset of motion. Recent studies show, in particular, the importance of turbulent fluctuations (pulses) in the initiation of motion [DIP 08, CEL 10].

If the regime is weakly turbulent (characterized by weak fluctuations of the flow parameters and what we call inertial regime), the value of the critical Shields remains identical to that obtained in a laminar regime. But in these conditions, the hydrodynamic stress at the top of the bed must take into account the hydrodynamic inertial effects. Two studies show this result. The experimental results of Loiseleux *et al.* [LOI 05], obtained in a Hele-Shaw cell, lead to $\theta^c = 0.14$ a constant value in the laminar regime (for $Re^* < 1$). Then θ^c decreases when Re^* increases in the range $Re^* \in [1,100]$. The authors show that this decrease comes from inertial effects that appear in the flow and increase the hydrodynamic stress at the top of the bed of particles. They introduce a dimensionless function $\Phi(Re^*) = 1 + 0.15Re^{*0.687}$ that integrates the inertial effect on the stress.

The definition of the Shields number used by the authors is as follows:

$$\theta = \frac{\eta\dot{\gamma}}{(\rho_p - \rho_f)gd}$$

It is interesting to observe that if we define the Shields number by $\theta = \Phi(\mathrm{Re}^*)\eta\dot{\gamma}/[(\rho_p - \rho_f)gd]$ then the critical Shields measured is constant and equal to 0.14 over the entire range of Re^* of these experiments $Re^* \in [10^{-3},100]$.

Similarly, Peysson *et al.* [PEY 09] were able to estimate the threshold for incipient motion for flows in pipes both in laminar and inertial regimes. They use in their approach a definition of the hydrodynamic stress valid in both laminar and inertial regime $\tau_b = 0.5\rho_f(Q/S)^2 f(Re_e)$ where Q is the fluid flow rate in the pipe, S the fluid section of the pipe, and f a dimensionless function that depends on the Reynolds number $Re_e = \rho_f D_e Q/\eta S$ (built with an equivalent diameter D_e).

Quantities f and D_e can be explained in the case of a tube filled with beads (for more details, we can refer to [PEY 09]). By injecting this hydrodynamic stress in the Shields number, the threshold value measured in laminar and inertial regimes remains constant and equal to 0.12 ± 0.03.

8.3. Bedload transport

8.3.1. *Particle flux*

For more than a century, a large number of studies has concerned bedload transport. Beyond the critical Shields θ^c, the granular medium is set into motion. This motion is due to the action of the fluid. It is essential to understand the existing coupling between the fluid and the granular medium. This coupling is usually expressed through the relationship between particle flux q_p and the Shields number.

In the literature, we find many formulas to express this particle flux. Most of these expressions are empirical or semiempirical and have been proposed for turbulent flows. Regardless of the nature of the flow, q_p is generally expressed as $q_p/Q = m\theta^n(\theta - \theta^c)^p$, where m, n, and p are constants that vary according to the authors and Q is a characteristic flow rate obtained mainly by dimensional analysis. For turbulent flows, the scaling is inertial $Q = (\Delta\rho g d^3/\rho_f)^{1/2}$ while for a laminar flow, the scaling is viscous $Q = \Delta\rho g d^3/\eta$.

Table 8.1 summarizes some expressions proposed for turbulent flows. Determining the particle flux requires understanding the mechanism causing grain motion. On the basis of the random nature of the setting in motion, Einstein [EIN 42, EIN 50] presents a model of transport flux based on a statistical approach, the transport flux being connected to the probability that a particle located at a given surface begins to set into motion at a given instant. He assumes that the motion of the particles is mainly due to turbulent velocity fluctuations and that there is no critical stress for the onset of motion. Yalin [YAL 63] addresses the problem of sediment transport in open channel under a different angle. On the basis of the fact that there may be motion of particles without turbulent velocity fluctuations (laminar flow), he proposes that the non-uniform setting in motion of the particles on a bed is due to the random arrangement of particles. He considers that the grains move mainly by saltation and are placed in a configuration in which the height of grains in motion is negligible compared to the height of the fluid. The use of the motion equation for the saltons allows him to obtain the particle flux.

Bagnold [BAG 56] decomposes the total stress in the granular medium in a stress carried by the fluid and a stress carried by the particles to determine the rate of transport of particles. To circumvent problems associated with determining the effects of turbulent velocity fluctuations and those related to the roughness of the solid bed, he considers the average velocity of flow and the drag force as independent data.

Author	Method	$q_p\sqrt{\frac{\rho_f}{\Delta\rho g d^3}}$
[EIN 42]	Semi-empirical	$\frac{e^{(0.391/\theta)}}{0.465}\left(\sqrt{\frac{2}{3}+\frac{36\eta^2}{gd^3\rho_f\Delta\rho}}-\sqrt{\frac{36\eta^2}{gd^3\rho_f\Delta\rho}}\right)$
[MEY 48]	Empirical	$8(c_2c_3\theta-\theta^c)^{1.5}$
[EIN 50]	Semi-empirical	$\frac{1-x}{27f_c(2-x)},\ x=\frac{1}{\sqrt{\pi}}\int_{-2-0.156f/\theta}^{2+0.156f/\theta}e^{-t^2}dt$
[BAG 56]	Semi-empirical	$8.5\sqrt{2\tan\alpha/(3\psi)}(\theta-\theta^c)\theta^{0.5}$
[YAL 63]	Semi-empirical	$0.635\left(\frac{\theta}{\theta^c}-1\right)\left[1-\frac{\log(1+x)}{x}\right],\ x=2.45\left(\frac{\rho_f}{\rho_p}\right)^{0.4}(\theta^c)^{0.5}\left(\frac{\theta}{\theta^c}-1\right)$
[RIB 98]	Empirical	$10.4(\theta-\theta^c)^{1.67}$
[CAM 05]	Empirical	$12\theta^{1.5}e^{-4.5\theta^c/\theta}$
[WON 06]	Empirical	$4.93(\theta-\theta^c)^{1.60}$

Table 8.1. *Some particle flux equations proposed for turbulent flows, where* c_2 *is a correction due to presence of walls,* c_3 *is a correction linked to the shape of the bed,* f_c *the percentage of grains of a given size set into motion,* f *a correction function experimentally obtained for the dispersion of the size of particles,* ψ *a drag coefficient,* ρ_p *the density of particles,* ρ_f *the density of the fluid (*$\Delta\rho=\rho_p-\rho_f$*),* η *the viscosity of the fluid,* d *the diameter of particles, and* $\tan\alpha$ *the dynamic friction coefficient*

Most of the other formulas proposed in the literature are obtained empirically. The most commonly used are the formulas proposed by Ribberink [RIB 98], and Meyer-Peter and Muller [MEY 48], where a correction due to the presence of walls and a correction linked to the shape of the bed are implemented.

Fewer studies have been carried out in the case of laminar flows. We present a non-exhaustive list of some of the formulas proposed in Table 8.2. Charru and Mouilleron-Arnould [CHA 02] propose a semiempirical transport flux formula, with a continuum modeling of the sediment, based on the theory of viscous re-suspension by Leighton and Acrivos [LEI 86]. Cheng [CHE 04] uses a probabilistic approach based on the random distribution of particles at the surface. Charru and Hinch [CHA 06] propose a particle flux based on an erosion/deposition model, where N represents the number of particles in motion per unit area.

Finally, a newer modeling was proposed by Ouriemi *et al.* [OUR 09]. This modeling is based on a continuum approach considering two phases, a particle phase and a fluid phase, and uses a granular rheology stemming from recent research on the subject [GDR 04, FOR 08]. As mentioned in the introduction, we focus on this approach in the following sections.

Author	Method	$q_p\sqrt{\frac{\rho_f}{\Delta\rho g d^3}}$
[CHA 02]	Semi-empirical	$0.42(\theta-\theta^c)^3$
[CHE 04]	Semi-empirical	$\frac{\eta}{\sqrt{\Delta\rho g d^3 \rho_f}} 41\theta^{0.5} Re_*[\sinh(0.139\theta^{1.181} Re_*^{0.39})]^2$
[CHA 04]	Empirical	$0.025\theta(\theta-\theta^c)$
[CHA 06]	Semi-empirical	$0.096N\frac{\theta}{\theta^c}$
[OUR 09]	Semi-empirical	$\frac{\eta_e}{\eta}\phi_0\frac{\theta^c}{24}\left(\frac{\theta}{\theta^c}\right)^3$

Table 8.2. *Some particle flux equations proposed for laminar flows, where* $Re_* = u_* d\rho_f/\eta$ *is a particle Reynolds number based on the shear velocity* u_*, N *is the number of particles in motion by unit area,* η_e *is the effective viscosity of the mixture, and* ϕ_0 *the volume fraction in particles*

8.3.2. *Two-phase modeling*

In principle, a complete modeling of the motion of solid particles in a liquid is given by solving Newton's equations for each particle and the Navier–Stokes equations for the fluid with a no-slip condition on the surface of each particle and at the outer boundaries. This type of direct calculation can now be carried out numerically but only for a limited number of particles compared to reality, and this even with current computing resources. This is why two-phase modeling has been widely used to describe these systems on average.

There are different ways to achieve these averages, see for example [JAC 00] and [DRE 92]: by carrying out ensemble averages at each point in space over different realizations of the system or by carrying out spatial averages over regions smaller than the macroscopic scale of the system, but larger than the size of particles. Each average type is a purely formal process and should essentially give the same results if the calculation is carried out correctly. However the average quantities outnumber the equations available and therefore there is a closure problem. This is the central problem in this type of modeling. In addition, there is no guarantee that the closure of equations is possible.

We present a classic two-phase modeling due to Jackson [JAC 00] or [JAC 97]. From Newton's equations for the solid particles and Navier–Stokes equation for the fluid, Jackson obtained a set of equations for the motion of the fluid and particles in terms of average local variables. This modeling, originally written for a system of fluidized particles [JAC 63] and [AND 68], in fact applies generally to a system of mobile particles in fluid flow [AND 67].

The continuity equations for the fluid phase and particle phase are written respectively as:

$$\frac{\partial \varepsilon}{\partial t} + \frac{\partial (\varepsilon u_i^f)}{\partial x_i} = 0 \qquad [8.8]$$

$$\frac{\partial \phi}{\partial t} + \frac{\partial (\phi u_i^p)}{\partial x_i} = 0 \qquad [8.9]$$

where u_i^f is the average local velocity of the fluid, u_i^p the average local velocity of the particles, ϕ the volume fraction of the particles, and $\varepsilon = 1 - \phi$ the void fraction.

Equations of conservation of the momentum for each phase are:

$$\rho_f \frac{D_f(\varepsilon u_i^f)}{Dt} = \frac{\partial \sigma_{ij}^f}{\partial x_j} - nf_i + \varepsilon \rho_f g_i \qquad [8.10]$$

$$\rho_p \frac{D_p(\phi u_i^p)}{Dt} = \frac{\partial \sigma_{ij}^p}{\partial x_j} + nf_i + \phi \rho_p g_i \qquad [8.11]$$

where g_i is the acceleration of gravity, the density of the fluid, ρ_p that of the particles, and n the number of particles per unit volume. The force f_i is the average value of the resulting force exerted by the fluid on the particles. The stress tensors σ_{ij}^f and σ_{ij}^p can be considered as effective stress tensors associated with the fluid phase and the particle phase, respectively. These latter effective tensors associated with the two phases arise from the averaging process and are connected to the details of the interactions, on the one hand, between the fluid and the particles and, on the other hand, between the particles. The stress tensor of the fluid phase consists of not only the average stress tensor in the fluid, but also a contribution (stresslet) arising from interactions between the fluid and particles and also the Reynolds stress of the fluid (in the general case but which becomes negligible in the viscous case). The stress tensor of the particle phase includes terms arising from interactions between particles (contacts and shocks in the inertial and turbulent cases) and particle Reynolds stresses (again here negligible in the viscous case). For more details on the averaging process and the determination of the interaction force between phases and stresses of the two phases, the reader is referred to the works of Jackson [JAC 97, JAC 00].

In the following section, we present the closures that have been proposed by [OUR 09] for bedload in laminar flow condition. These closures in the laminar case are not definitive and still remain to be reviewed with possibly more details. Obtaining closures in the turbulent case remains a difficult issue and is much more open.

8.3.3. *Closure equations in laminar regime*

Ouriemi *et al.* [OUR 09] have considered both simple and adequate closures of bedload transport. The interphase force was separated into a generalized buoyancy force and a force that brings together all the remaining contributions, denoted by nf_i^1 [JAC 00]:

$$nf_i = \phi \frac{\partial \sigma_{ij}^f}{\partial x_j} + nf_i^1 \qquad [8.12]$$

The generalized buoyancy force is discussed in detail by Jackson [JAC 00]. A term such as the Archimedes force $-\rho_f \phi g$ does not fully take into account the effects of buoyancy as a particle cannot distinguish gravity from other contributions to the fluid pressure gradients (e.g. acceleration). Instead of writing only the Archimedes term, Jackson offers the following decomposition:

$$nf_i = -\rho_f \phi (g_i - \frac{D_f u_i^f}{Dt}) + nf_i^2$$

which can be decomposed again in a similar manner as equation [8.12] with $f_i^1 = \varepsilon f_i^2$.

In the bedload problem, the remaining contribution is reduced to the viscous drag, which is proportional to the velocity difference between the two phases $u_i^p - u_i^f$ but can be re-expressed in terms of $u_i^p - U$, where $U_i = \varepsilon u_i^f + \phi u_i^p$ is the velocity of the mixture. This viscous drag can be expressed in the form of Darcy's law in the case of a dense granular medium in motion:

$$nf_i^1 = \eta \frac{\varepsilon}{K}(U_i - u_i^p) \qquad [8.13]$$

where η the viscosity of the pure fluid and K the permeability of the bed of particles calculated by using the Kozeny–Carman equation $K = \varepsilon^3 d^2 / [180(1-\varepsilon)^2]$.

As explained above, the effective stress tensor of the fluid phase contains the contribution arising from the pure fluid augmented by the contribution arising from interactions between fluid and particles. It represents the hydrodynamic part of the stresses of the whole suspension. To simplify, the effective tensor of the fluid phase is assumed Newtonian with an effective viscosity η_e:

$$\sigma_{ij}^f = -p^f \delta_{ij} + \tau_{ij}^f, \quad \tau_{ij}^f = \eta_e \left(\frac{\partial U_i}{\partial x_j} + \frac{\partial U_j}{\partial x_i} \right) \qquad [8.14]$$

where p^f is the pressure of the fluid. Following Brinkman [BRI 47a, BRI 47b] and Bagnold [BAG 56], we chose to represent this effective viscosity, which is different from the pure fluid, by the viscosity of Einstein $\eta_e = \eta(1+5\phi/2)$. For a more detailed calculation of the effective viscosity, the reader can refer to the article by Kim and Russel [KIM 83] as well as the references cited in this article.

Finally, the stress tensor of the particle phase comes only from direct interactions between particles and is described at first approximation by a Coulomb friction model where:

– the tangential stress is proportional to the load (the pressure of particles p^p) when the granular shear stress is positive, that is μp^p with a friction coefficient μ, which in most cases does not depend on the geometry of particles and which is given by the tangent of the angle of repose;

– the tangential stress is undetermined when the granular shear rate is zero.

This description of the friction between the grains can be extended to a more sophisticated granular rheology [GDR 04]. In this rheology, the stress is decomposed into an isotropic pressure term and a deviatory term, with:

$$\sigma_{ij}^p = -p^p\delta_{ij} + \tau_{ij}^p \,,\; \tau_{ij}^p = \mu(I)\,p^p\frac{\dot{\gamma}_{ij}^p}{\dot{\gamma}^p} \qquad [8.15]$$

$$\mu(I) = (\mu_s + \frac{\mu_2 - \mu_s}{I_0/I + 1}),\; I = \frac{\eta\dot{\gamma}^p}{\alpha p^p},\; \dot{\gamma}^p = \sqrt{\frac{1}{2}\dot{\gamma}_{ij}^p\dot{\gamma}_{ij}^p}$$

where μ_s, μ_2, and I_0 are constants that characterize the granular medium, and I is a dimensionless parameter for submerged granular flows with $\alpha \sim 0.01$. Figure 8.6 shows the typical evolution of the friction coefficient $\mu(I)$. It is observed that for $I >> I_0$, the friction coefficient tends to a constant, which is equivalent to using a simple Coulomb-type rheology. In fact, in the transport of particles, the upper layers of the particles are subjected to low confinement pressure (tending to zero at the interface), which corresponds to large I. Moreover, these layers are the fastest and therefore are those that contribute most to the particle flux. Using a Coulomb-type rheology for the particle phase seems sensible at first approximation. For a more complete calculation, the granular rheology is nonlinear, and thus numerical resolutions have to be undertaken.

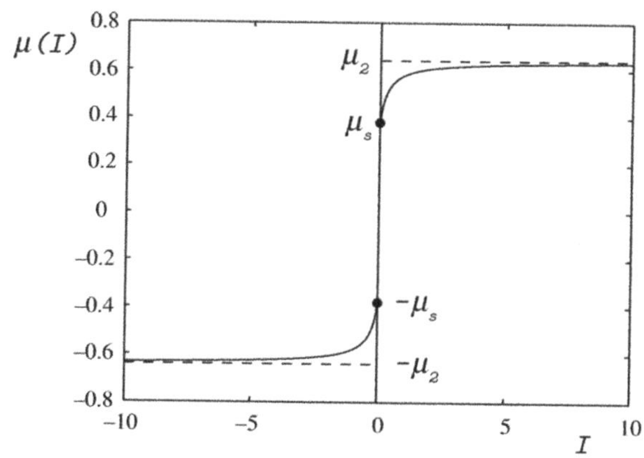

Figure 8.6. *Evolution of the friction coefficient $\mu(I)$ for glass beads with $\mu_s \sim 0.38$, $\mu_2 \sim 0.64$, and $I_0 \sim 0.3$. The dotted lines represent the asymptotic value*

8.3.4. *Resolution for shear flows*

These equations were numerically and analytically [OUR 09] solved for shear flows not only in the two-dimensional case (Couette and Poiseuille), but also numerically in the three-dimensional case [CHA 10]. An example of the numerical solution in cylindrical and rectangular section pipes carried out by Chauchat and

Médale is given in Figure 8.7. The interest of the numerical calculation in full three dimensions is that we can use the granular rheology discussed previously and can also adapt to real geometries. We show in this section a much simpler calculation in the case of a Poiseuille plane flow, which gives an idea of the physical mechanisms involved, but the reader can refer to the original articles for more details and complete calculations [OUR 09, CHA 10].

In this calculation, we have considered a flat bed of thickness h_p, which consists of particles having a diameter d. This bed is subjected to a stationary and uniform Poiseuille flow in a two-dimensional channel of thickness D (see Figure 8.8). The two-phase equations shown above are reduced to the equation of conservation of momentum for the mixture (particles + fluid) and the Brinkman equation for the fluid; see the detail of calculations in [OUR 0] and Figure 8.9(a).

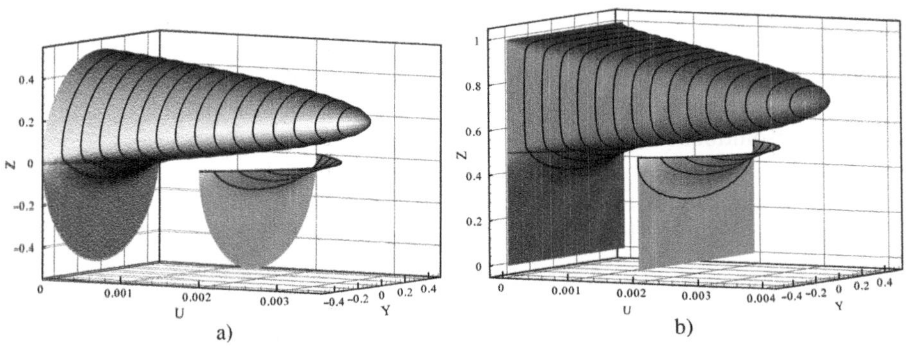

Figure 8.7. *Example of numerical solution in cylindrical and rectangular section pipes carried out by Chauchat and Médale [CHA 10]. The pipes are half filled with particles. The velocity profile of particles is artificially shifted to the right for more clarity*

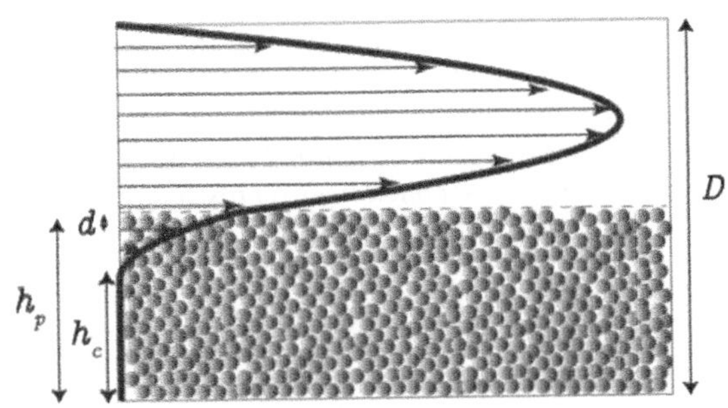

Figure 8.8. *Sketch of the Poiseuille plane flow*

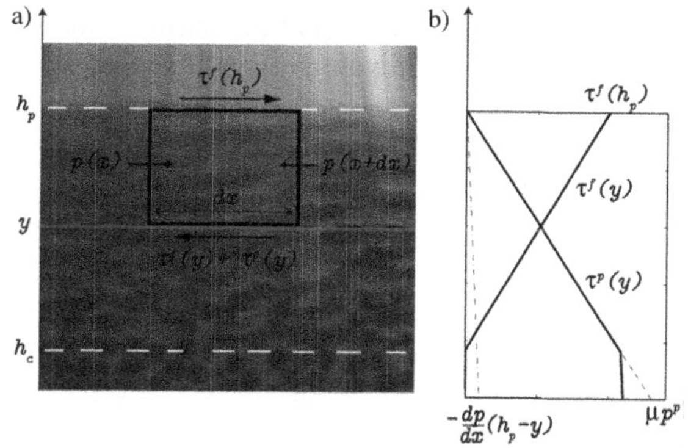

Figure 8.9. *(a) Conservation of momentum for the mixture (particles + fluid) for 2 mm PMMA particles subjected to a Poiseuille flow in a pipe of rectangular section. The photos are superimposed which permits us to show the bed in motion; (b) diagram of the stress distribution in the mixture*

The equation of conservation of momentum of the mixture is written as:

$$\tau^p(y)+\tau^f(y)=\tau^f(h_p)-(h_p-y)\frac{\partial p^f}{\partial x} \qquad [8.16]$$

where $\partial p^f/\partial x$ is the horizontal pressure gradient generating the Poiseuille flow. This equation shows that the shear stress of the mixture $\tau^p(y)+\tau^f(y)$ linearly increases with depth from its value at the bed surface $\tau^f(h_p)$ due to the horizontal pressure gradient. It also describes the exchanges between the shear stress of the fluid phase τ^f and that of the solid phase τ^p (Figure 8.9(b)). At the surface of the bed $y=h_p$, the shear stress of the particles is equal to zero. It builds up inside the bed until it reaches μp^p. The granular medium therefore starts to be sheared. The pressure of the particles is proportional to the apparent weight of the solid phase $\Delta\rho=\rho_p-\rho_f$ and grows inside the bed:

$$p^p=\phi_0\Delta\rho g(h_p-y) \qquad [8.17]$$

The shear stress of the particles keeps this value μp^p up to $y=h_c$ inside the bed. On the other hand, at the surface of the bed, the shear stress of the fluid phase is equal to $\tau^f(h_p)$ and tends to zero at $y=h_c$. Below, the granular medium is immobile and behaves as a porous medium. In addition, the Darcy drag term is dominant in the Brinkman equation. Consequently, within the bed, there is very little slip between the two phases, and the fluid and the particles move with the mixture velocity. Under these conditions, the velocity profile is obtained by solving the equation:

$$\mu\phi_0\Delta\rho g(h_p - y) + \eta_e \frac{\partial U}{\partial y} = \tau^f(h_p) - \frac{\partial p^f}{\partial x}(h_p - y) \qquad [8.18]$$

with zero shear condition at $y = h_c$. This condition means that the thickness of the mobile bed linearly varies with the Shields number:

$$h_p - h_c \approx \theta d\left(\mu\phi_0 + \frac{1}{\Delta\rho g}\frac{\partial p^f}{\partial x}\right)^{-1} \qquad [8.19]$$

The velocity follows the parabolic profile:

$$u^p \approx u^f \approx U \approx \frac{1}{2\eta_e}\left(\mu\phi_0\Delta\rho g + \frac{\partial p^f}{\partial x}\right)(y - h_c)^2 \qquad [8.20]$$

Assuming that the critical Shields number of the threshold of motion corresponds to a thickness of the mobile layer of a particle size, we obtain:

$$\theta^c \approx \mu\phi_0 \qquad [8.21]$$

and we find the predictions previously mentioned in section 8.3. Taking typical values for the constants ($\mu \sim 0.4$ and $\phi_0 \sim 0.55$) we obtain $\theta^c \sim 0.22$. This value is of the correct order of magnitude but is still two times larger compared to the experimental results shown in section 2.5. However, we can refine this calculation [OUR 09] by taking into account the fact that the first layer has a volume fraction two times less than the sediment, which gives a value consistent with the experimental result.

The particle flux $q_p = \int_0^h \phi_0 u^p dy$ is also expressed by a simple law, a cubic variation in the Shields number (given in Table 8.2):

$$\frac{q_p}{\Delta\rho g d^3 \eta^{-1}} = \frac{\eta_e}{\eta}\phi_0 \frac{\theta^c}{24}\left(\frac{\theta}{\theta^c}\right)^3 \qquad [8.22]$$

where ϕ_0 is the volume fraction considered constant within the bed.

8.3.5. *Comparison with experiments in laminar regime*

These predictions were experimentally tested by Ouriemi *et al.* [OUR 09] in a circular pipe. The principle of the experiment is first to fill the tube with the fluid and particles and form a flat bed, then to measure the variation in bed height as a function of time for a given flow. By choosing to be in bedload conditions, that is, above the critical Shields, we observe that the height of the bed always decreases as the measuring section is not continuously fed with particles. The indirect method to measure the particle flux is to use the equation of conservation of particles:

$$\phi_0 \frac{\partial h_p}{\partial t} + \frac{dq_p}{dh_p} \frac{\partial h_p}{\partial x} = 0 \qquad [8.23]$$

which can be written as a kinematic wave equation in dimensionless form in relation to the Shields number. The cubic law in Shields number for the particle flux seems quite satisfactory to describe the time evolution of the height of the bed.

The experimental work of Ouriemi *et al.* [OUR 09] enabled us to measure height variations, but does not allow direct access to the discharge or velocity profile within the sediment motion. It is possible to look inside the bed using an index matched technique (see for instance, Lobkovsky *et al.* [LOB 08] in a rectangular pipe and in laminar flow conditions). They were able to obtain the vertical profile of horizontal velocities as well as the volume fraction inside the bed, and were able to deduce the particle flux from it. Ouriemi *et al.* [OUR 09] have reinterpreted their experimental data and have shown that these results are compatible with a cubic variation in θ. The prefactor in equation [8.22] is overestimated. This discrepancy is probably due to wall effects not taken into account. In addition, the spatial resolution is too low to carry out accurate comparison with the parabolic profile predicted by equation [8.20] or with the linear variation of the thickness of the mobile bed with the Shields number given by equation [8.19]. The same technique was used by [MOU 09] in an annular Couette flow in laminar flow conditions. Although the spatial resolution and the number of mobile layers are relatively small, the velocity profiles are consistent with the parabolic profiles predicted by equation [8.20].

These two experiments do not, however, enable us to accurately test the scaling laws obtained in the two-phase model as the size and density of particles as well as the viscosity of the fluid were not varied and the flow rate range is relatively limited. New and more complete experiments and numerical simulations are now expected to see in what range the analytical model and numerical resolution reproduce the observations. In particular, when a large number of the particle layer is flowing, the flow should be more complex than that predicted by the simple model shown using a Coulomb friction and it is probably essential to introduce the granular friction to predict the flows in detail.

8.4. Conclusion and outlook

In this chapter, we have discussed the two key points of bedload transport which are the threshold for incipient motion and the sediment flux beyond this threshold.

The threshold for incipient motion is described by a critical Shields number defined as the critical hydrodynamic stress that must be applied to the bed of particles to be set into motion. This stress is normalized by the apparent hydrostatic pressure across a grain. Simple models that provide a force balance either by a discrete approach on a grain or by a continuous approach on a sediment layer show that this threshold depends primarily on a friction coefficient of granular material and on the volume fraction. We find a critical threshold consistent with these models in the laminar case. This is also found in the turbulent case, but the difficulty remains in the estimation of the hydrodynamic stress at the top of the bed.

Different approaches, either empirical or semiempirical, have been proposed to modelize the particle flux. In this chapter, we have focused on a continuum approach, which considers a mixture of two phases, a granular phase and a fluid phase. The great difficulty with this type of approach is to determine the closures, which govern the interaction force between the two phases as well as the stresses of the solid and the fluid phase. We have proposed some closures in the laminar case: the interphase force is separated into buoyancy and a viscous drag of Darcy type, the stresses associated with the liquid phase are Newtonian with an effective viscosity whereas the stresses of the particle phase are described with a friction model stemming from recent studies on the rheology of granular media. Recent experiments, in particular, which measure the flow of particles and the fluid within the mobile sediment, are consistent with the predictions of the model. However, these constitutive laws need to be tested in more detail. In the turbulent case, the situation is much more open. The closures of these equations (concerning not only the stress of the fluid phase, but also the interaction force between phases and the stress of the granular phase) must integrate the inertial and turbulent character of the flow. A number of recent studies have proposed closures adequate to this turbulent regime [JEN 98, AMO 08, HSU 09]. In particular, in the case where the stress on the bed is very high and where collisions between particles are important, closures based on the kinetic theory of granular media have been proposed. The challenge is to achieve an understanding of the different physical mechanisms at work: turbulent interactions, particle collisions, friction between grains, etc.

8.5. Acknowledgments

Part of the studies that led to this chapter was funded by IFP Energies nouvelles and the National Research Agency (Dunes Project ANR- 07-3.18-3892). This chapter stems from work undertaken with J. Chauchat, M. Ouriemi, M. Médale and M. Pailha.

8.6. Bibliography

[AMO 08] AMOUDRY L., LIU T.-J., LIU P.-F., "Two-phase model for sand transport in sheet flow regime", *Journal of Geophysical Research*, vol. 113, C03011, 2008.

[AND 67] ANDERSON T.B., JACKSON R., "A fluid mechanical description of fluidized beds: equation of motion", *Industrial and Engineering Chemistry Fundamentals*, vol. 6, pp. 527–539, 1967.

[AND 68] ANDERSON T.B., JACKSON R., "A fluid mechanical description of fluidized beds: stability of the state of uniform fluidization", *Industrial and Engineering Chemistry Fundamentals*, vol. 7, pp. 12–21, 1968.

[AND 04] ANDREOTTI B., "A two species model of aeolian sand transport", *Journal of Fluid Mechanics*, vol. 510, pp. 47–70, 2004.

[BAG 56] BAGNOLD R.A., "The flow of cohesionless grains in fluids", *Proceedings of the Royal Society of London: Series A*, vol. 249, pp. 235–297, 1956.

[BAG 66] BAGNOLD R., "An approach to the sediment transport problem from general physics", *Geological Survey Professional Paper 422–I*, pp. 1–37, 1966.

[BRI 47a] BRINKMAN H., "A calculation of the viscous force exerted by a flowing fluid on a dense swarm of particles", *Applied Scientific Research*, vol. A1, pp. 27–34, 1947.

[BRI 47b] BRINKMAN H., "On the permeability of media consisting of closely packed porous particles", *Applied Scientific Research*, vol. A1, pp. 81–86, 1947.

[BUF 97] BUFFINGTON J., MONTGOMERY D., "A systematic analysis of eight decades of incipient motion studies, with special reference to gravel-bedded rivers", *Water Resources Research*, vol. 33, no. 8, pp. 1993–2029, 1997.

[BUF 99] BUFFINGTON J., "The legend of A.F. Shields", *Journal of Hydraulic Engineering*, vol. 125, pp. 376–387, 1999.

[CAM 05] CAMEMEN B., LARSON M., "A general formula for non-cohesive bed load sediment transport", *Estuarine Coastal*, vol. 63, pp. 249–260, 2005.

[CEL 10] CELIK A., DIPLAS P., DANCEY C., VALYRAKIS M., "Impulse and particle dislodgement under turbulent flow conditions", *Physics of Fluids*, vol. 22, no. 4, 046601, 2010.

[CHA 02] CHARRU F., MOUILLERON-ARNOULD H., "Instability of a bed of particles sheared by a viscous flow", *Journal of Fluid Mechanics*, vol. 452, pp. 303–323, 2002.

[CHA 04] CHARRU F., MOUILLERON-ARNOULD H., EIFF O., "Erosion and deposition of particles on a bed sheared by a viscous flow", *Journal of Fluid Mechanics*, vol. 519, pp. 55–80, 2004.

[CHA 06] CHARRU F., HINCH J., "Ripple formation on a particle bed sheared by viscous liquid. Part One: steady flow", *Journal of Fluid Mechanics*, vol. 550, pp. 111–121, 2006.

[CHA 10] CHAUCHAT J., MÉDALE M., "A three-dimensional numerical model for incompressible two-phase flow of a granular bed submitted to a laminar shearing flow", Computer *Methods in Applied Mechanics and Engineering*, vol. 199, pp. 439–449, 2010.

[CHE 04] CHENG N.S., "Analysis of bed load transport in laminar flows", *Advances in Water Resources*, vol. 27, pp. 937–942, 2004.

[DIP 08] DIPLAS C., DANCEY A., CELIK M., VALYRAKIS M., GREER K., AKAR T., "The role of impulse on the initiation of particle movement under turbulent flow conditions", *Science*, vol. 322, pp. 717–720, 2008.

[DRE 92] DREW D., LAHEY R., "Analytical modeling of multiphase flow", in ROCCO M.C. (ed.), *Particulate Two-Phase Flow*, Butterworth-Heinemann, 1992.

[EIN 42] EINSTEIN H.A., Formulas for the transportation of bed load, Report no. 2140, *Transactions of the American Society of Civil Engineers*, 1942.

[EIN 50] EINSTEIN H.A., The bed load function for sediment transportation in open channel flow, Report no. 1026, U.S. Department of Agriculture, 1950.

[FOR 08] FORTERRE Y., POULIQUEN O., "Flows of dense granular media", *Annual Review of Fluid Mechanics*, vol. 40, pp. 1–24, 2008.

[FRA 73] FRANCIS J., "Experiments on the motion of solitary grains along the bed of a waterstream", *Proceedings of the Royal Society London: Series A*, vol. 332, pp. 443–471, 1973.

[GDR 04] GDR Midi, "On dense granular flows", *European Physical Journal*, vol. 14, pp. 341–365, 2004.

[GON 02] GONDRET P., LANCE M., PETIT L., "Bouncing motion of spherical particles in fluids", *Physics of Fluids*, vol. 14, pp. 643–652, 2002.

[HSU 09] HSU T., JENKINS J., LIU P.-F., "On two-phase sediment transport: sheet flow of massive particles", *Proceedings of the Royal Society London: Series A*, vol. 460, pp. 2223–2250, 2009.

[JAC 63] JACKSON R., "The mechanics of fluidized beds. Part I: the stability of the state of uniform fluidization", *Transactions of the Institution of Chemical Engineers*, vol. 41, pp. 13–21, 1963.

[JAC 97] JACKSON R., "Locally averaged equations of motion for a mixture of identical spherical particles and a Newtonian fluid", *Chemical Engineering Science*, vol. 52, pp. 2457–2469, 1997.

[JAC 00] JACKSON R., *The Dynamics of Fluidized Particles*, Cambridge University Press, Cambridge, 2000.

[JEN 98] JENKINS J., HANES D., "Collisional sheet flow of sediment driven by a turbulent fluid", *Journal of Fluid Mechanics*, vol. 370, pp. 29–52, 1998.

[KIM 83] KIM S., RUSSEL W., "Modelling of porous media by renormalization of the Stokes equations", *Journal of Fluid Mechanics*, vol. 154, pp. 269–286, 1983.

[KIN 97] KING M., LEIGHTON D., "Measurement of the inertial lift on a moving sphere in contact with a plane wall in a shear flow", *Physics of Fluids*, vol. 9, pp. 1248–1255, 1997.

[LEI 86] LEIGHTON D., ACRIVOS A., "Viscous resuspension", *Chemical Engineering Science*, vol. 41, pp. 1377–1384, 1986.

[LOB 08] LOBKOVSKY A.E., ORPE A.V., MOLLOY R., KUDROLLI A., ROTHMAN D., "Erosion of granular bed driven by laminar fluid flow", *Journal of Fluid Mechanics*, vol. 605, pp. 47–58, 2008.

[LOI 05] LOISELEUX T., GONDRET P., RABAUD M., DOPPLER D., "Onset of erosion and avalanche for an inclined granular bed sheared by a continuous laminar flow", *Physics of Fluids*, vol. 17, 103304–103309, 2005.

[MAL 08] MALVERTI L., LAJEUNESSE E., MÉTIVIER F., "Small is beautiful: upscaling from microscale laminar to natural turbulent rivers", *Journal of Geophysical Research*, vol. 113, F04004, 2008.

[MAN 77] MANTZ P., "Incipient transport of fine grains and flanks by fluids extended Shields diagram", *Journal of the Hydraulics Division*, vol. 103, pp. 601–615, 1977.

[MAR 09] MARTINO R., PATERSON A., PIVA M., "Onset of motion of a partly hidden cylinder in a laminar shear flow", *Physical Review E*, vol. 79, 036315, 2009.

[MAT 05] MATOUŠEK V., "Research developments in pipeline transport of settling slurries", *Powder Technology*, vol. 156, pp. 43–51, 2005.

[MEY 48] MEYER-PETER E., MULLER R., "Formulas for bed-load transport", *Proceedings of the 2nd Meeting of the International Association for Hydraulic Structures Research*, International Association for Hydraulic Research, Delft, Netherlands, pp. 39–64, 1948.

[MOL 96] MOLLINGER A., NIEUWSTADT F., "Measurement of the lift force on a particle fixed to the wall in the viscous sublayer of a fully developed turbulent boundary", *Journal of Fluid Mechanics*, vol. 316, pp. 285–306, 1996.

[MOU 09] MOUILLERON H., CHARRU F., EIFF O., "Inside the moving layer of a sheared granular bed", *Journal of Fluid Mechanics*, vol. 628, pp. 229–239, 2009.

[OUR 07] OURIEMI M., AUSSILLOUS P., MEDALE M., PEYSSON Y., GUAZZELLI E., "Determination of the critical Shields number for particle erosion in laminar flow", *Physics of Fluids*, vol. 19, 061706, 2007.

[OUR 09] OURIEMI M., AUSSILLOUS P., GUAZZELLI E., "Sediment dynamics. Part 1: Bed load transport by shearing flows", *Journal of Fluid Mechanics*, vol. 636, pp. 295–319, 2009.

[PEY 09] PEYSSON Y., OURIEMI M., MEDALE M., AUSSILLOUS P., GUAZZELLI E., "Threshold for sediment erosion in pipe flow", *International Journal of Multiphase Flow*, vol. 35, pp. 597–600, 2009.

[PIL 01] PILOTTI M., MENDUNI G., "Beginning of sediment transport of incoherent grains in shallow shear flows", *Journal of Hydraulic Research*, vol. 39, pp. 115–124, 2001.

[RIB 98] RIBBERINK J.S., "Bed-load transport for steady flows and unsteady oscillatory flows", *Coastal Engineering*, vol. 34, pp. 58–82, 1998.

[SHO 91] SHOOK C., ROCO M.C., *Slurry Flow: Principles and Practice*, Butterworth-Heinemann Series in Chemical Engineering, Stoneham, 1991.

[VAN 66] VANONI V., "Sediment transportation mechanics: initiation of motion", *Journal of the Hydraulics Division*, vol. 92, pp. 291–314, 1966.

[WHI 40] WHITE C., "The equilibrium of grains on the bed of a stream", *Proceedings of the Royal Society of London: Series A*, vol. 174, pp. 322–338, 1940.

[WON 06] WONG M., PARKER G., "Reanalysis and correction of bed-load relation of Meyer-Peter and Muller using their own database", *Journal of Hydraulic Engineering*, vol. 132, pp. 1159–1168, 2006.

[YAL 63] YALIN M.S., "An expression for bed-load transportation", *Journal of the Hydraulics Division*, vol. HY3, pp. 221–250, 1963.

[YAL 79] YALIN M.S., KARAHAN E., "Inception of sediment transport", *Journal of the Hydraulics Division*, vol. 105, pp. 1433–1443, 1979.

Chapter 9

Characterization of Natural Cohesive Sediments and Water Quality of Rivers

9.1. Introduction

Rivers are systems characterized by permanent or temporary fluxes subjected to inputs and hydrological characteristics of their watersheds. Mainly composed of water and various materials, they are areas of biodiversity that shelter a wide range of live microbial organisms, animals, and plants. These complex and fragile ecosystems are subjected to anthropic pressure due to the fact that they are natural lines of communication, water resources, energy resources, fishery resources, recreational areas, and sewage outfalls. Alongside these direct pressures, they are likely to be influenced by most of the activities that impact on their watershed because they are the convergence of most surface and underground water flows. Thus, there is a broad range of direct and diffuse pollution, of industrial and domestic origins, metallic, organic, chemical, radioactive and heat, that rivers are likely to be subjected to.

To limit, control, and assess the impacts of these pollutions, regulatory frameworks and management plans are developed. These steps need us to understand the overall functioning of these ecosystems, including the pollutant transfer mechanisms through their own reactivity and their mobility under the effect of water and matter fluxes. Note that the quality assessment involves the choice of descriptive variables, which, in our case and in accordance with the European

---

Chapter written by Fabien TERNAT, Patrick BOYER, Fabien ANSELMET and Muriel AMIELH.

environmental quality standards, are defined in terms of average values of concentrations of pollutants.

This chapter focuses on the relationship between the water quality of rivers and the behavior of their sediments in view of the assessment of the fate of sedimentary pollutants. It consists of five main sections. Section 9.2 addresses the general behavior of pollutants in a river in order to specify the role of hydro-sedimentary mechanisms. Section 9.3 shows an approach to address these mechanisms by modeling. Section 9.4 describes experimental methods that can be applied to parameterize the calculations. Section 9.5 shows the operational application of this methodology to radioactive contamination of the Techa River in Russia. For this, our model is declined under an operational form to provide access to the spatiotemporal changes of water quality that is considered through the concentrations of radioactive elements, here the ^{90}Sr (strontium isotope), the ^{137}Cs (cesium isotope), and $^{239,240}Pu$ (plutonium isotope). Finally, section 9.6 concludes the chapter. However, any contamination by metal ions could also be modeled as a function of their own affinity with the solid particles.

9.2. Behavior of pollutants in rivers

Rivers arise from the drainage of their watersheds and are, as such, likely to receive a portion of the different pollutants generated by anthropic activities located in these basins. In fact, these pollutants cover a broad spectrum and, without attempting to be exhaustive, they include:

– heavy metals emitted by different industries, domestic waste incinerators, and mines;

– radionuclides from atmospheric fallouts of nuclear testing, nuclear power plants, reprocessing facilities, and hospitals;

– chemical substances from fertilizers (e.g. nitrates and phosphates) and physiosanitary products (e.g. insecticides, pesticides, and fungicides);

– organic materials such as hydrocarbons, PCBs[1], and pathogenic genes (e.g. viruses, bacteria, and parasites) from farms and domestic waste.

In this context, the purpose is not to describe the behavior of each of these forms of pollution but to outline the broad lines of the direction of their transfer along a river. The aim is to specify the role of fine cohesive sediment erosion within this context.

1 PCB: polychlorinated biphenyls.

Schematically, rivers can be described according to two main compartments: the water column (denoted by Ω_w in Figure 9.1), limited by the interfaces with the atmosphere and the bed sediments, and the sediment column (denoted by Ω_s in Figure 9.1).

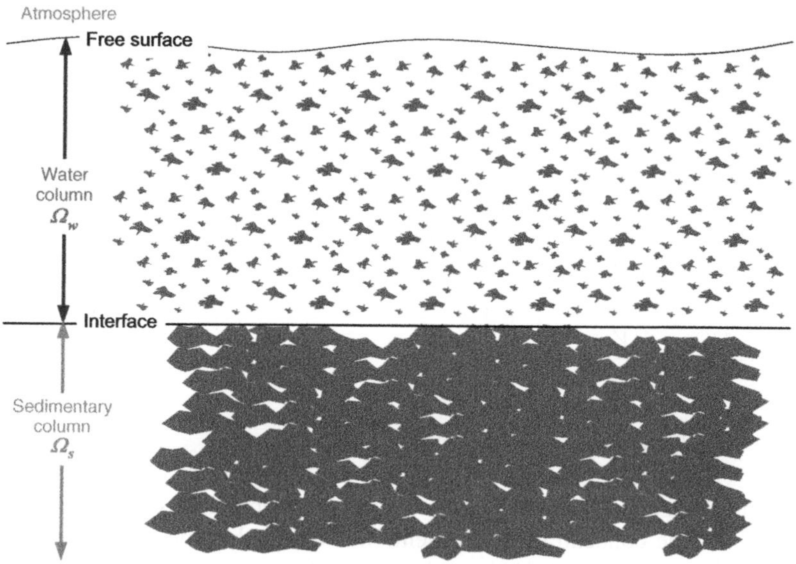

Figure 9.1. *Schematic representation of a river. Water column Ω_w contains water and suspended matter; the sediment column Ω_s contains the sedimentary matrix and pore water (void)*

These two compartments are mainly composed of water and solid particles of various sizes and various natures, organic (e.g. algae, bacteria, and biological debris) or inorganic (e.g. clay minerals, oxides, hydroxides, and carbonates). Thus, whatever pollutant is introduced into this environment, it will likely be found in a dissolved form or in forms linked with these particles. Therefore, the water quality assessment of rivers is based on the determination of the spatiotemporal distribution of these different forms, which are the main routes of contamination of animals and plants in the hydrosystem.

On this basis, the concentrations of pollutants in the environment can be understood through the following equations:

$$R_w = R_{d,w} + C_w R_{s,w} \quad [9.1]$$

$$R_s = nR_{d,s} + \rho_s(1-n)R_{s,s} \quad [9.2]$$

where R_w (mol.m^{-3}) is the total volume concentration of the pollutant in the water column, $R_{d,w}$ (mol.m^{-3}) is the dissolved volume concentration of the pollutant in the water column, $R_{s,w}$ (mol.kg^{-1}) is the mass concentration of the pollutant of matter in the water column, C_w (kg.m^{-3}) is the volume concentration of the suspended matter in the water column, R_s (mol.m^{-3}) is the total volume concentration of the pollutant in the sediment column, $R_{d,s}$ (mol.m^{-3}) is the dissolved volume concentration of the pollutant in the sediment column, $R_{s,s}$ (mol.kg^{-1}) is the mass concentration of the pollutant of sediment matter, ρ_s (kg.m^{-3}) is the density of the sediments, and n is the porosity of the sediments.

Porosity is expressed as the ratio between the void volume and the total sediment volume. Assuming that the sediments are saturated in water, the porosity becomes the ratio between the volume of pore water $V_{w,s}$ (m^3) and the total sediment volume $V_{t,s} = V_{w,s} + V_{s,s}$ (m^3), that is:

$$n = \frac{V_{w,s}}{V_{t,s}} = \frac{V_{w,s}}{V_{w,s} + V_{s,s}} \qquad [9.3]$$

where $V_{s,s}$ (m^3) is the volume of the solid matter.

To characterize the pollutant content of the sediments, it is sometimes useful not to use the volume concentrations as mentioned above, but the mass concentrations reduced to dry weight (kg.kg^{-1}, which is often measured) or even surface concentrations (kg.m^{-2}):

– Surface concentrations are obtained by multiplying the volume concentrations by the thickness of the sediment layer considered.

– Mass concentrations reduced to dry weight are obtained by dividing R_s by $\rho_s(1-n)$.

Transfers of these concentrations along a river depend first on inputs that can be diffuse (e.g. surface deposition, runoff, and infiltration) or localized (nozzles and outfalls). The fate of introduced pollutants then depends on spatial fluxes associated with the movement of water, of fluxes between dissolved and particle forms related to physicochemical reactivities of pollutants to the surface of particles, and then fluxes between the water column and the bottom sediments through deposit mechanisms, erosion and interstitial diffusion induced by hydro-sedimentary dynamics of the environment (Figure 9.1).

Each pollutant shows a specific reactivity (e.g. speciation, chemical reaction, diagenetic degradation, and radioactive decay) likely to alter its physicochemical form, its toxicity, and its ability to remain dissolved or attached to materials. In particular, the distribution between the dissolved and particle forms plays a major role because it determines the primary mechanism for the transfer to the bottom sediments, namely the deposit of adsorbent materials. Subsequently, as long as the pollutants are in the water column or in the sediment column, the kinetics depuration of the hydrosystem can be very different:

– Fast in the water column as they are dominated by the mechanisms of transport and dispersion generated by the water flow.

– Slow in the sediment column where they depend on transfers by interstitial diffusion of dissolved species, bioturbation produced by benthic organisms, and especially the erosion of bottom materials in relation to the friction exerted on the sediment bed by the water column in motion: bedload, saltation and suspension.

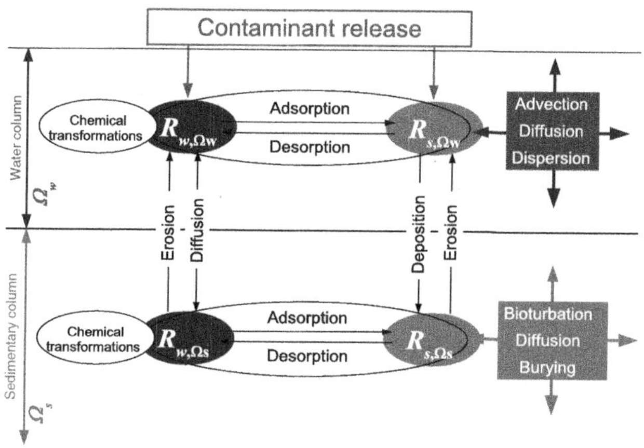

Figure 9.2. *Primary mechanisms of transfer of pollutants*

Thus, depending on the intensity and frequency of these mechanisms, it is not uncommon that the depuration of sediments is slower than local inputs by deposit. Sediment stocks likely to alter the water quality of rivers in the medium and long terms are then formed. A common example is that of polychlorinated biphenyls (PCBs) issued at the end of the 20th Century, which, trapped in the sediments of many rivers, carry on today and for many more years to degrade their quality.

The solid–liquid distributions are specific to each pollutant, to the nature of the particles, and to environmental conditions. However, transfer mechanisms (adsorption and desorption) that sit at the surface of particles show a significant correlation with their size because they become more effective when the specific surface of particles increases [SIG 01]. This behavior emphasizes the key role of the smallest particles on the particle transfer of pollutants and consequently on sediment contamination.

At this level, it is necessary to characterize the particles that may be encountered in a river. For this, there are several classifications in relation to granulometry. Figure 9.3 shows a few examples of them. These different classifications reveal three major families of particles: colloids (<1 μm), cohesive particles (<75 μm), and non-cohesive particles.

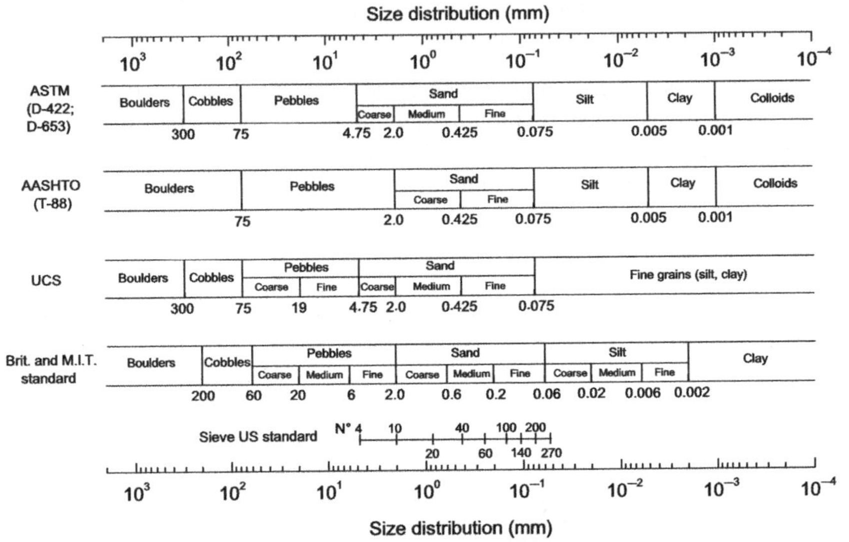

Figure 9.3. *Granulometric classification of matters in a river (according to Al-Hussaini (1977), quoted by Holtz [HOL 81])*

Cohesive particles are mainly composed of silts and clays. They have electrostatic properties that allow them to flocculate, to aggregate with each other in the water column, and are the source of cohesion and consolidation of the sediments. Colloids are the finest cohesive particles. Biological in nature (humic matter) or mineral (e.g. clays and metal oxides), they are small enough that the action of gravity has no effect on their movements, but are big enough to confer surface properties and to distinguish them from a molecule in a solution [CIT 04]. From the perspective of transfer of pollutants, they are the most reactive particles. In practice, it is customary to set the threshold at 1 μm between dissolved and particle phases and to link these particles to the dissolved phase.

Non-cohesive particles are of larger size. They include sand, gravel, stones, and blocks, which play a lesser role in the transfer of pollutants because, on the one hand, they consist of less reactive materials and, on the other hand, they show the lowest specific surfaces (Figure 9.4).

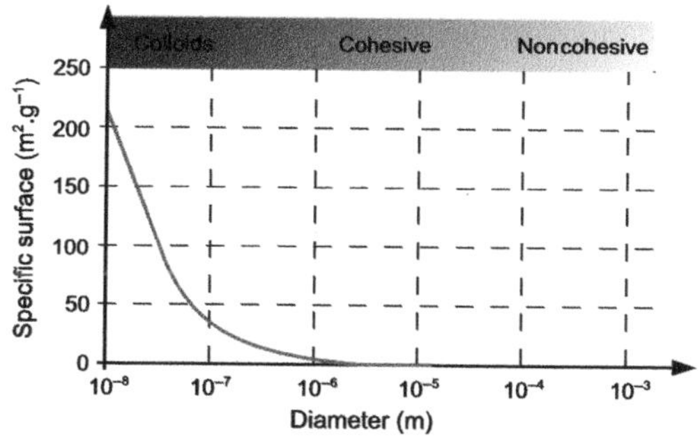

Figure 9.4. *Evolution of the specific surface with the diameter for spherical particles (ρ_s = 2,650 kg.m^{-3})*

These considerations highlight the relevance of distinguishing between the different types of particles present in a river because each has its own kinetics either in terms of affinity with the pollutants or in terms of their own dynamics [ABR 96, BOR 00, ELG 04]. In this context, the following section shows a set of general balance equations suited to take into account several types of particles.

9.2.1. *General balance equations*

To understand the role of material heterogeneity on the transfer of pollutants, the general balance equations for dissolved and particle phases of the water column and the sediment column are presented below in a multiclass context (different classes of particles are identified by the index *i*).

Under these conditions, the terms of the total volume concentrations are as follows:

$$R_w = R_{d,w} + \sum_i C_{i,w} R_{i,w}$$

$$R_s = nR_{d,s} + \sum_i C_{i,s} R_{i,s}$$

These equations derived in respect of time are written as:

$$\frac{\partial R_w}{\partial t} = \frac{\partial R_{d,w}}{\partial t} + \sum_i \frac{\partial (C_{i,w} R_{i,w})}{\partial t} \qquad [9.4]$$

$$\frac{\partial R_s}{\partial t} = \frac{\partial (nR_{d,s})}{\partial t} + \sum_i \frac{\partial (C_{i,s} R_{i,s})}{\partial t} \qquad [9.5]$$

General balance equations of these components can be written in the following forms:

$$\frac{\partial R_{d,w}}{\partial t} + \vec{\nabla} \cdot \left(R_{d,w} \vec{v}_{d,w} - \mathbf{K}_{d,w} \cdot \vec{\nabla} R_{d,w} \right) = J_{d,A,w} + J_{d,S,w} + J_{d,R,w} + J_{d,D,w} + J_{d,E,w} \quad [9.6]$$

$$\frac{\partial (C_{i,w} R_{i,w})}{\partial t} + \vec{\nabla} \cdot \left(C_{i,w} R_{i,w} \vec{v}_{i,w} - \mathbf{K}_{i,w} \cdot \vec{\nabla} (C_{i,w} R_{i,w}) \right) = J_{i,A,w} + J_{i,S,w} + J_{i,R,w} + J_{i,D,w} + J_{i,E,w} + \sum_{i \neq j} J_{i,j,F,w} \quad [9.7]$$

$$\frac{\partial (n R_{d,s})}{\partial t} + \vec{\nabla} \cdot \left(n R_{d,s} \vec{v}_{d,s} - \mathbf{K}_{d,s} \cdot \vec{\nabla} (n R_{d,s}) \right) = J_{d,A,s} + J_{d,S,s} + J_{d,R,s} + J_{d,D,s} + J_{d,E,s} \quad [9.8]$$

$$\frac{\partial (C_{i,s} R_{i,s})}{\partial t} + \vec{\nabla} \cdot \left(C_{i,s} R_{i,s} \vec{v}_{i,s} - \mathbf{K}_{i,s} \cdot \vec{\nabla} (C_{i,s} R_{i,s}) \right) = J_{i,A,s} + J_{i,S,s} + J_{i,R,s} + J_{i,D,s} + J_{i,E,s} \quad [9.9]$$

Terms that make up these equations represent the different mechanisms likely to influence the transfer of pollutants. They are explained below.

9.2.1.1. *Sources*

Terms $J_{d,A,w}$, $J_{i,A,w}$, $J_{d,A,s}$ and $J_{i,A,s}$ (mol.m^{-3}.s^{-1}) integrate all forms of pollutant inputs of external origins into the river, for example, direct discharges, transfers with the watershed, or even atmospheric deposit.

9.2.1.2. *Diffusive mechanisms*

In the water column, the diffusions of dissolved and particle phases, respectively, express themselves through terms $\mathbf{K}_{d,w} \cdot \vec{\nabla} R_{d,w}$ and $\mathbf{K}_{i,w} \cdot \vec{\nabla}(C_{i,w} R_{i,w})$. For all three spatial directions, the diffusion coefficients denoted by $\mathbf{K}_{d,w}$ (m^2.s^{-1}) and $\mathbf{K}_{i,w}$ (m^2.s^{-1}) can be divided into two contributions: the molecular diffusion (isotropic) and turbulent diffusion (non-isotropic) generated by the flow turbulence [GRA 84]. As the latter is largely prevailing in the water column, molecular diffusion is usually neglected. Although, strictly speaking, turbulent diffusion of particle phase is different from that of the dissolved phase, it is usual to apply the same coefficient of turbulent diffusion to the dissolved phase and to all the fine particles.

In the sediments, diffusions of dissolved and particle phases are linked with terms $\mathbf{K}_{d,s} \cdot \vec{\nabla}(nR_{d,s})$ and $\mathbf{K}_{i,s} \cdot \vec{\nabla}(C_{i,s} R_{i,s})$. Diffusion coefficients $\mathbf{K}_{d,s}$ ($m^2.s^{-1}$) and $\mathbf{K}_{i,s}$ ($m^2.s^{-1}$) that appear in these equations may contain two types of diffusion: interstitial diffusion within the porous environment and biodiffusion generated by some benthic organisms. The interstitial diffusion is mainly of molecular nature and its effectiveness is reduced due to the low porosity of the environment. Thus, it mainly concerns the dissolved and colloidal phases and alleviates itself with the depth of the sediments along with the porosity. Biodiffusion depends on the activity of organisms living in the sediments and can involve all components of dissolved and sediment particles [BOU 97, CAR 03].

At the interface of the water column and sediment column, it should be noted that these diffusive mechanisms contribute to transfers between these two environments.

9.2.1.3. *Solid–liquid transfers*

Terms $J_{d,S,w}$, $J_{i,S,w}$, $J_{d,S,s}$, and $J_{i,S,s}$ ($mol.m^{-3}.s^{-1}$) represent the solid–liquid transfers that determine the distribution of a pollutant between its dissolved and particle forms. These transfers can be treated with approaches involving the combination of several kinetics. However, parameterizing these approaches can be quite complex and, in practice, it is the assumption of equilibrium and reversible transfer that is most frequently used. This assumption results in the use of a solid–liquid partition coefficient at equilibrium denoted by $K_{d,i}$ ($m^3.kg^{-1}$):

$$K_{d,i} = \frac{R_i}{R_d} \qquad [9.10]$$

9.2.1.4. *Chemical reactions*

Terms $J_{d,R,w}$, $J_{i,R,w}$, $J_{d,R,s}$ and $J_{i,R,s}$ ($mol.m^{-3}.s^{-1}$) quantify the chemical changes that alter the nature of pollutants or their solid supports. In the water column, they cover various chemical reactions (e.g. oxidation, reduction, and precipitation) specific to each pollutant. In the sediments, these reactions are dominated by diagenetic changes linked to the degradation of organic matter by bacterial populations.

9.2.1.5. *Flocculation – aggregation disaggregation*

In the context of multiclass, it may be worthwhile to consider transfers between different classes of particles. These transfers from a class i to a class j, quantified by the term $J_{i,j,F,w}$ in equation [9.7], are made mostly from finer to larger particles via flocculation or aggregation mechanisms. These mechanisms become significant when we observe physicochemical changes in the environment (e.g. pH, temperature and salinity). They are largely responsible for transfers toward sediments of estuarine areas because the increase in salinity present there causes the flocculation of colloidal phases (most contaminated) and causes them to precipitate to the bottom sediments [VER 09].

9.2.1.6. *Advective mechanisms*

In the water column, the advection of the dissolved phase is represented by the term $\vec{\nabla}\cdot(R_{d,w}\vec{v}_{d,w})$ and the advection of the particle phase is represented by $\vec{\nabla}\cdot(R_{i,w}\vec{v}_{i,w})$. $\vec{v}_{d,w}$ ($\mathrm{m.s^{-1}}$) is the average velocity field of water and $\vec{v}_{i,w}$ ($\mathrm{m.s^{-1}}$) is the velocity field of class i particles under the effect of their own weight and the movement of water, that is, $\vec{v}_{i,w} = \vec{v}_{d,w} + \vec{W}_{i,c}$.

In the sediment column, the advection of the dissolved phase is expressed by $\vec{\nabla}\cdot(R_{i,s}\vec{v}_{d,s})$, where $\vec{v}_{d,s}$ ($\mathrm{m.s^{-1}}$) is the velocity field of pore water under the effects of the pressure gradients induced by the slope of the bed, the transfers with the groundwater, and the compaction of sediment layers. Term $\vec{\nabla}\cdot(R_{i,s}\vec{v}_{i,s})$ corresponds to the advection of phases fixed on suspended matter with the velocity of particles $\vec{v}_{i,s}$. This velocity depends mainly on the burial of successive layers of deposit and compaction generated by the weight of accumulated material. Some forms of bioturbation may also be the source of advective-type particle displacements [CAR 03].

9.2.2. *Term of deposit flow and erosion*

Transfers between the water and sediment columns by erosion and deposition occur at the interface of these media and do not appear explicitly in the balance equations. Insofar as this is the behavior of fine cohesive particles that is relevant to the transfer of pollutants, it can be assumed that erosion occurs by particle resuspension. Therefore, the deposit and erosion constitute the boundary conditions for the advection of the particle phases at the interface of the sediments and the water column.

Pollutant fluxes by erosion $J_{,E,}$ ($\mathrm{mol.m^{-3}.s^{-1}}$) or by deposit $J_{,D,}$ ($\mathrm{mol.m^{-3}.s^{-1}}$) that appear in the balance equations can be reduced to mass fluxes of erosion E or deposit D ($\mathrm{kg.m^{-2}.s^{-1}}$). Their respective equations are proposed in the following sections.

9.2.2.1. *Deposit fluxes*

The deposit follows a Krone-type law [KRO 62]. The latter introduces the concept of critical stress deposit $\tau_{cd,i}$ ($N.m^{-2}$), limited value of fluid shear stress τ_f ($N.m^{-2}$) beyond which the particles of class i stop decanting:

$$D_{i,w} = \begin{cases} W_{i,c} C_{i,w}\left(1 - \dfrac{\tau_f}{\tau_{cd,i}}\right) \text{ if } \tau_f < \tau_{cd,i} \\ 0 \text{ else} \end{cases} \qquad [9.11]$$

where $C_{i,w}$ ($kg.m^{-3}$) is the concentration of suspended matter. The velocity $W_{i,c}$ ($m.s^{-1}$) is that of the particles freely falling into the water at rest. In moving water, this kinetic is mainly related to the dynamics of water, gravity, and the hydrodynamic resistance[2]. Finer modeling allows for integrating other factors such as the acceleration experienced by the particle, the pressure forces, the inertial term, the pressure due to the relative acceleration between the particle and the fluid, and the Basset term [GRA 84].

The uncertainty of the critical stress of deposit $\tau_{cd,i}$ can be overcome through the balance of considerations, which will be discussed later.

9.2.2.2. *Erosion fluxes*

Just as for the deposit, the erosion flux is often modeled by threshold laws:

$$E_{i,w} = \begin{cases} E_i \text{ if } \tau_f > \tau_{ce,i} \\ 0 \text{ else} \end{cases} \qquad [9.12]$$

The ratio of the fluid shear stress τ_f and of the erosion threshold $\tau_{ce,i}$ controls mainly the amount of remobilized matter, and characterizes the mode of erosion, the mechanism by which sediment material is resuspended. The corresponding law of erosion must be chosen accordingly [WIN 04]. We can find different equations for the flux E, such as the power or exponential laws [SAN 01]. The linear relationship of Partheniades [PAR 65] appears suitable for cohesive sediments:

$$E_{i,w} = \begin{cases} e_i\left(\dfrac{\tau_f}{\tau_{ce,i}} - 1\right) \text{ if } \tau_f > \tau_{ce,i} \\ 0 \text{ else} \end{cases} \qquad [9.13]$$

2 Which we can linearly write by Stokes' law and which introduces the resistance coefficient at equilibrium.

where τ_f is the shear stress imposed by the moving fluid, $\tau_{ce,i}$ is the critical erosion stress, threshold value beyond which the erosion occurs, and e_i ($kg.m^{-2}.s^{-1}$) is the rate of erosion. If the friction exerted by the fluid can be obtained from a hydraulic model ($\tau_f = \rho_w u^{*2}$, where u^* is the friction velocity), two parameters will remain undetermined: erosion rate and threshold.

The uncertainty of the critical stress of deposit $\tau_{cd,i}$ can be overcome through the consideration of equilibrium state in a theory called multiclass. In this approach, the assumption that the deposit and erosion of matter can take place simultaneously allows us to formulate an overall equation of the matter flux at the interface. The resulting deposit–erosion law therefore becomes

$$E_{i,w} = W_{i,c}(C^*_{i,w} - C_{i,w}) \qquad [9.14]$$

with

$$C^*_{i,w} = \begin{cases} \dfrac{e_i}{W_{i,c}}\left(\dfrac{\tau_f}{\tau_{ce,i}} - 1\right) & \text{if } \tau_f > \tau_{ce,i} \\ 0 & \text{else} \end{cases} \qquad [9.15]$$

9.3. Erosion of fine cohesive sediments

Modeling of the erosion of fine cohesive sediments must take into account the cohesion state of the particles. Thus, wc distinguish the erosion of fine non-consolidated sediments and the erosion of fine consolidated sediments.

9.3.1. *Erosion threshold of a non-consolidated layer*

The erosion flux is described by a Partheniades-type erosion law [PAR 65]. In the absence of stabilizing effects (biofilm), the non-consolidated layer does not generally show cohesive character [ELG 04]. Its threshold of erosion is primarily a function of grain characteristics (size and form factor) whose value is tabulated, for example, in the Shields diagram (Figure 9.5).

In this diagram, the knowledge of the size of the particle d(m) provides the critical stress for erosion. A summary of experimental data was collected from the literature, notably by Soulsby (1997) (see [ZIE 03]). Many models theoretically estimate this value [GRA 84, WIB 87, DEY 99, PIL 01]. An assessment of the actions on a particle (Figure 9.6), without considering the cohesive force, leads to a theoretical expression that connects the dimensionless erosion threshold and the particle Reynolds number:

$$\tau^* = \frac{\tau_{ce}}{(\rho_s - \rho_w)gd}, \ \mathrm{Re}^* = \frac{u^* d}{\nu} \qquad [9.16]$$

by the following equation:

$$\tau^* = \tau^*_{nc}(\mathrm{Re}^*) \qquad [9.17]$$

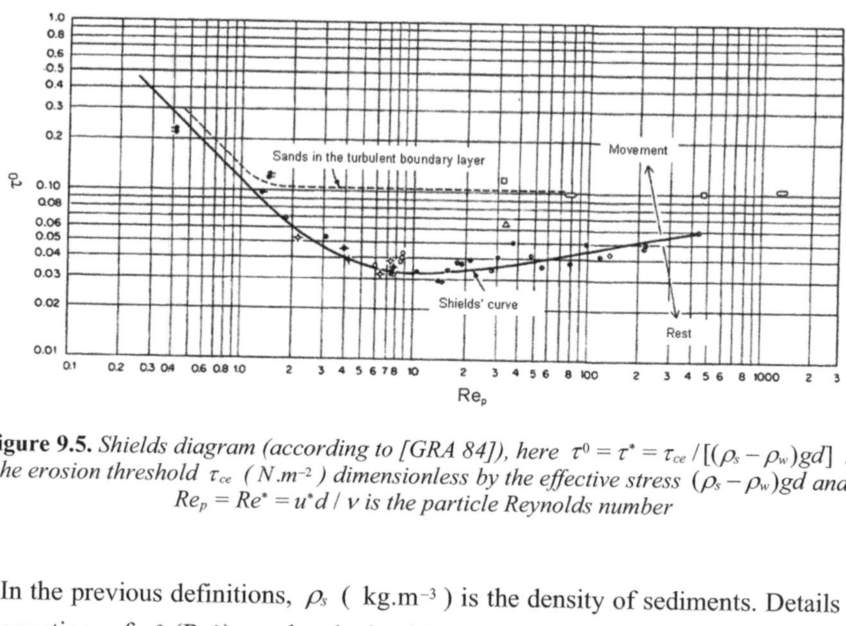

Figure 9.5. *Shields diagram (according to [GRA 84]), here* $\tau^0 = \tau^* = \tau_{ce} / [(\rho_s - \rho_w)gd]$ *is the erosion threshold* τ_{ce} *(* $N.m^{-2}$ *) dimensionless by the effective stress* $(\rho_s - \rho_w)gd$ *and* $Re_p = Re^* = u^*d / \nu$ *is the particle Reynolds number*

In the previous definitions, ρ_s ($\mathrm{kg.m^{-3}}$) is the density of sediments. Details on the equations of $\tau^*_{nc}(\mathrm{Re}^*)$ can be obtained from [TER 07].

9.3.2. *Erosion threshold of a consolidated and saturated layer*

When a deposit is sufficiently compact and contains cohesive particles, it has a bulk behavior called cohesion in which particles interact. The sources of interactions, which may be related to the presence of clays [GRA 84] or to other factors such as the presence of microorganisms [BLA 02], are very sensitive to the state of compaction and consolidation[3]. Variability of erosion threshold in relation to the compaction has already been mentioned [SUN 56, POS 67, KEL 81, MEH 82,

3 By these two mechanisms, the time of consolidation [MIG 68] and the burial of the layer then become variables of the erosion threshold.

KAM 83, PAR 85, LEE 04]. The introduction of a cohesive force between the particle and the sediment bed alters the expression of the erosion threshold (equation [9.17]), which can be rewritten as:

$$\tau^* = \tau_c^* = \tau_{nc}^*(\mathrm{Re}^*)\left(1+\frac{F_c}{F_g'}\right) \qquad [9.18]$$

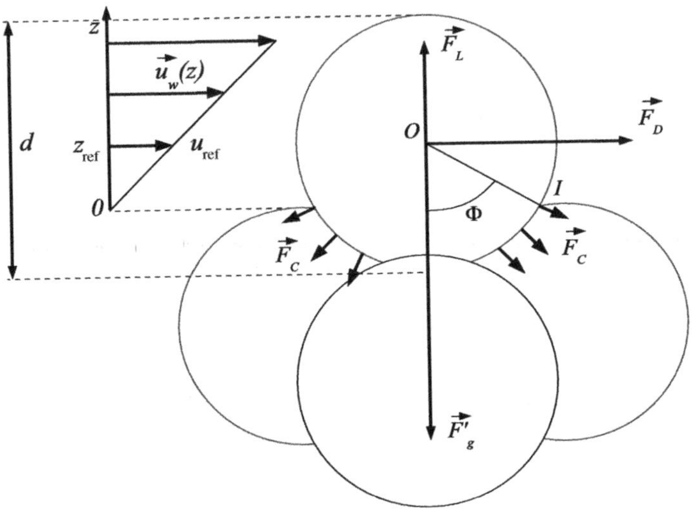

Figure 9.6. *Forces acting on a particle; $\vec{F}_L$ denotes the hydrodynamic lift, $\vec{F}_D$ denotes the hydrodynamic drive, $\vec{F}_g'$ denotes the submerged weight, and $\vec{F}_c$ denotes the cohesive strength*

The term that appears in parentheses denotes itself as the cohesion function. It shows the relationship between the cohesive strength F_c and the apparent weight of the particle F_g'. When interparticle effects are comparable in magnitude to the weight of the particle, this factor induces a significant increase in the erosion threshold[4].

9.3.2.1. *Consolidation model*

The clay particles in a saturated environment are surrounded by a double layer of adsorbed water whose thickness can reach several times the size of a grain and give it a look closer to a sphere than to a sheet [LAD 67].

---

4 In each of the definitions of the Reynolds number and the critical stress of dimensionless erosion (equation [9.16]), the variable u^* appears, which makes the direct solving of equation [9.18] impossible. A method is proposed in [TER 07].

A study proposes to model the cohesion function [TER 08] by counting the long-range Van der Waals interactions between the double ionic layers surrounding the particles. The Van der Waals interaction varies primarily with the distance between the particles d_i (m) – and thus the local density of the medium – and with the size of particles. It depends on the state of compaction of the sediments. Its equation in the case of spherical particles of respective sizes d_1 (m) and d_2 (m) is given by [ISR 85]:

$$F_H(n,d_1,d_2) = \frac{A_H d_1 d_2}{12(d_1 + d_2) d_i(n)^2} \qquad [9.19]$$

where $A_H \simeq 10^{-20}$ J is the Hamaker constant, whose values vary according to the clay mineral: 2.2×10^{-20} J for the montmorillonite and 3.1×10^{-20} J for the kaolinite (according to Bergström 1997, quoted by [GEL 05]).

The interparticle distance is assumed to be a function of the state of compaction of the porous medium, itself described by the porosity. When we assume a cubic arrangement of particles, the proposed equation is as follows:

$$d_i = K(n)d \ , \ K(n) = \left(\frac{n_{\max} - n_c}{n_{\max} - n}\right)^{1/3} - 1 \qquad [9.20]$$

where $K(n)$ is the compaction function. Constants $n_{\max}$ and n_c are descriptive parameters of the arrangement of grains between one another in the porous medium. Their values are taken as constant: $n_{\max} = 1$ and $n_c = 1 - \pi/6$.

In this approach, the state of consolidation is parameterized by the number of interactions and the intensity of interparticle forces. We have already seen that the values of the force intensity depend on the type of clay and the electrochemical conditions of the medium, through the Hamaker constant. Therefore, it is necessary to count the number of interactions, often called the number of bonds, acting on a particle.

For this, the concept of coordination is borrowed from the chemistry of crystals. Let us consider two networks of particles of sizes d_b and d_s, respectively, close to each other such that $d_b > d_s$ (Figure 9.7). The partial coordination between these networks is defined by the number of particles of size d_s that can fit into a shell surrounding a particle of size d_b. Its estimate is derived by volume considerations, which, without going into details, lead to the following equation:

$$C_I(n,d_b,d_s)=\left(\frac{1-\cos\Phi}{2}\right)\left(\frac{d_b}{(1+2K(n))d_s}\right)^3 \times\left[\left(1+2(1+2K(n))\frac{d_s}{d_b}\right)^3-1\right] \qquad [9.21]$$

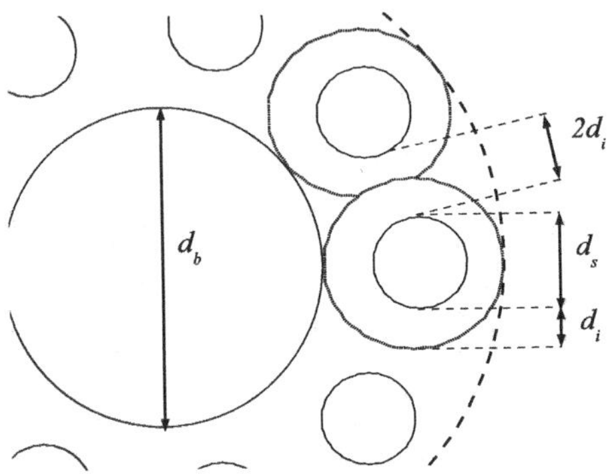

Figure 9.7. *Coordination between two networks of particles*

Angle Φ (Figure 9.5) is a parameter indicating the burying of the particle in the sediment bed. Its value is set at 52.5° [TER 07]. All forces of cohesive nature that apply on a particle are obtained by the integration of the unit force (equation [9.19]) on the size of small particles weighted by their volume proportion $s(d)$ [5]:

$$F_C(n,d_b)=\int_0^{d_b} C_I(n,d_b,x)F_H(n,d_b,x)s(x)dx \qquad [9.22]$$

Thus, an expression of the cohesive force was expressed in relation to not only the size of the particles, but also the state of compaction. It is also a function of the type of particle and electrochemical characteristics of the medium. Through it, it is theoretically possible to involve diagenetical reactions such as mechanisms of change of the cohesion. In the first approach, we will not take into account these final parameters by considering them as constant. Moreover, the number of bonds between particles plays a significant role. It is through a model of coordination borrowed from the chemistry of crystals that an estimate of this number was suggested. Coordination becomes not only a function of the relative size of particles, but also a function of the medium local density.

5 This quantity is measured by granulometry.

The essential role of particles in the granulometric class of clays as origins of intermolecular forces was confirmed by the calculation (Figure 9.8). In fact, the cohesion function takes increasingly large values as the size decreases, and significantly, when the size reaches a few micrometers. The importance of estimating the coordination between particles (Figure 9.9) is also highlighted. Although the shape of the curves is close to the previous case, cohesion takes considerable values from a few tens of micrometers. At this size, cohesion is not from the particles themselves but rather from the finest particles around them. In both cases, compaction, parameterized by the porosity, in fact acts on the cohesion of the medium: the more compact the medium, the greater the cohesion.

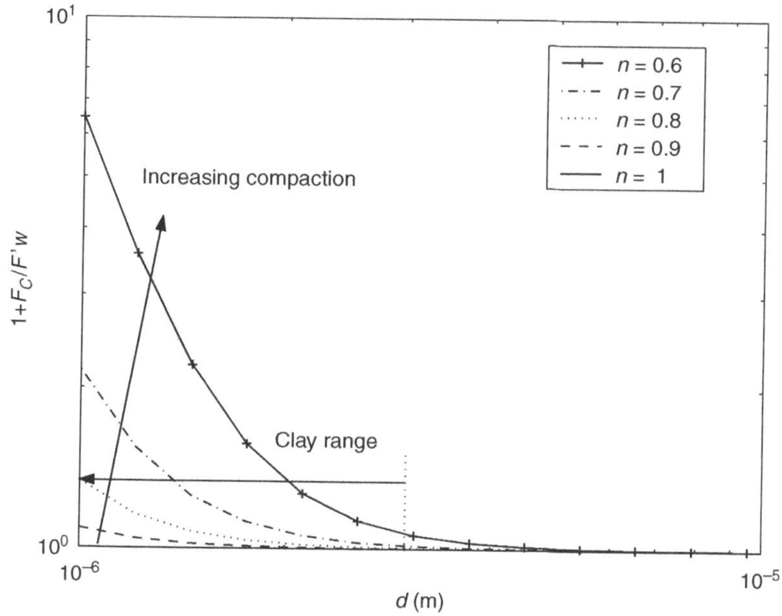

Figure 9.8. *Values of the cohesion function* $1+F_C/F_g'$ *between two particles of equal diameter (equation [9.19] with* $d_1 = d_2$*) in relation to the granulometry for different values of porosity. We will note that this function takes significant values for the granulometric class of clays*

Returning to the erosion threshold (equation [9.18]), the erosion diagram using Shields variables thus becomes parameterized by the porosity and takes into account the continuous granulometric spectrum (Figure 9.10). The porosity can be either measured or inferred from a model of compaction that we show below.

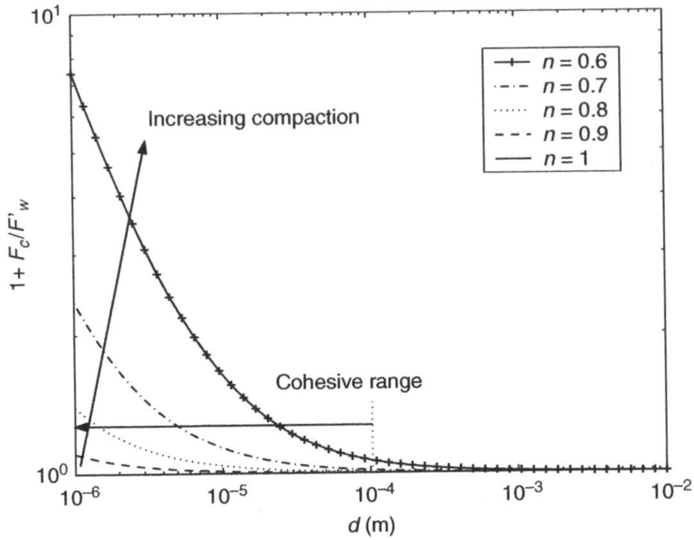

Figure 9.9. *Values of the cohesion function* $1+F_C/F_g'$ *between two particles accounting for coordination (multiple bonds between one particle and the surrounding particles and with s(x) = constant in equation [9.22]) in relation to the granulometry for different values of porosity. We note that all granulometric classes to fine sands are affected by the cohesion*

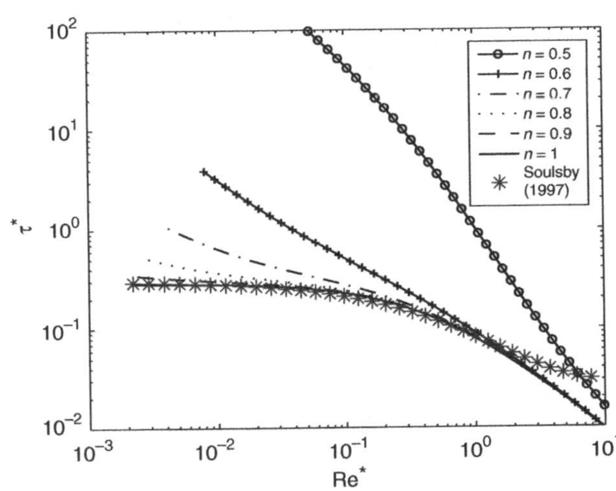

Figure 9.10. *Shields-type diagram for consolidated cohesive sediments, parameterized by the porosity n, in the range of the particle Reynolds number less than 1*

9.3.2.2. *Compaction model*

From the kinematic point of view, the deposited particles define the interface between water and sediments, also called sedimentation or thickening front (Figure 9.11). Many works report experimental and theoretical studies [TIL 81, CON 87, BUR 98, BUS 99, GAR 00], [BUR 05]). Dynamically, as soon as the grains accumulate, the porous matrix becomes denser by the effect of their own weight. This compaction involves two simultaneous processes. Grains rearrange their positions and cause excessive pressure in the pore water, which tends to be released. By combining the deformability of the sediment matrix and the modification of the interstitial flow, it is possible to obtain a modeling of the dynamics of compaction, as in the Terzaghi model (1943) or that of Gibson [GIB 67, LEE 68]. We reduce the framework to the vertical variable z. Sediment matrix is compacted with velocity v^S by means of the release of interstitial fluid with velocity v^F.

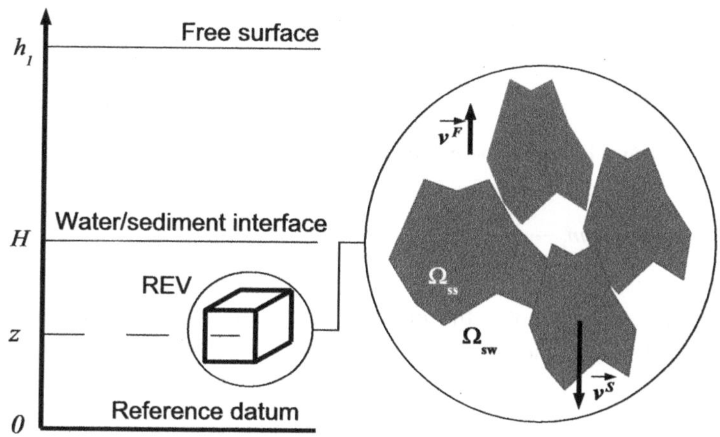

Figure 9.11. *Mapping of the bottom sediments and localization of a representative volume element (REV). Grains (in gray) form the sediment matrix (denoted by Ω_{ss}, of velocity $\vec{v}^S$) and delimit the interstitial fluid (void between the grains, denoted by Ω_{sw}, animated by velocity $\vec{v}^F$)*

Taking into account the relative proportion in fluid, given by the porosity n, and in matter, given by its complement $1-n$, the continuity equation is written as:

$$n\frac{\partial v^F}{\partial z}+(1-n)\frac{\partial v^S}{\partial z}=0 \qquad [9.23]$$

Considering the action of the vertical pressure p in a volume element and the specific weight of water γ_w (N.m^{-3}), the volume density of force is written as:

$$F_z = \frac{\partial p}{\partial z} + \gamma_w \quad [9.24]$$

This force is balanced with that induced by the flow moving in a porous medium of porosity n described by Darcy's law:

$$F_z = -n\frac{\gamma_w}{k}(v^F - v^S) \quad [9.25]$$

where k (m.s^{-1}) is the permeability of the porous medium. The deformation velocity of the porous matrix is, on the one hand, connected to the velocity gradient (geometric relationship):

$$\frac{\partial \varepsilon}{\partial t} = -\frac{\partial v^S}{\partial z} \quad [9.26]$$

And, on the other hand, to the effective stress rate (behavior law of the matrix):

$$\frac{\partial \varepsilon}{\partial t} = m_v \frac{\partial \sigma'}{\partial t}, \quad m_v = \frac{\partial \varepsilon}{\partial \sigma'} \quad [9.27]$$

Here, σ' (N.m^{-2}) denotes the effective stress in the sediment matrix that is deduced by the Terzaghi theory:

$$\sigma = \sigma' + p \quad [9.28]$$

where σ (N.m^{-2}) is the total stress in the sediment bed and p (N.m^{-2}) is the pressure of the interstitial fluid. The total stress changes arise from the flow of matter by deposit J_D and erosion J_E:

$$\frac{\partial \sigma}{\partial t} = g(J_D - J_E) \quad [9.29]$$

The previous equations finally lead to the relation:

$$\frac{\partial v^S}{\partial z} = -m_v\left[g(J_D - J_E) - \frac{\partial p}{\partial t}\right] \quad [9.30]$$

The dynamic equation of the sediment bed is written as:

$$\frac{\partial p}{\partial t}=C_v\left[\frac{\partial^2 p}{\partial z^2}+\left(\frac{1}{k}\frac{\partial k}{\partial z}-\frac{1}{n}\frac{\partial n}{\partial z}\right)\left(\frac{\partial p}{\partial z}+\gamma_w\right)\right]+g(J_D-J_E) \qquad [9.31]$$

where C_v ($m^2.s^{-1}$) denotes the coefficient of compaction[6]. Transfers by deposit or by erosion are represented by the term $g(J_D-J_E)$. The surface flux of pollutants released toward the water column by compaction J_T is then expressed by means of the velocity of the fluid leaving the porous medium $v^F(z=H)$ and the concentration of contaminants dissolved in pore water $R_{d,s}(z=H)$: $J_T=v^F R_{d,s}(z=H)$.

The stack of layers deposited causes their compaction under the effect of their own weight. We have seen that compaction affects the erosion threshold through porosity. But another effect on the rate of erosion is expected. In fact, dimensional analysis of the erosion rate e_i (equation [9.13]) can be rewritten as follows:

$$e=\rho_s(1-n)W_e \qquad [9.32]$$

where W_e is the displacement velocity of the interface under the action of erosion.

Thus, the kinetics of the erosion of sediments W_e is separated from the amount of eroded matter $\rho_s(1-n)$. The denser the sediment is (small porosity), the larger is the amount of matter available to erosion per unit area and, therefore, the greater is the value of the erosion rate. However, the low values of porosity increase the level of cohesion. This suggests that changes in the rate of erosion are due to two antagonistic mechanisms that partially cancel out each other. The form that is proposed opens possible means for a better understanding of the erosion flux changes. However, this parameter remains little modeled and it is usually only obtained experimentally. Thus, further research is required on this point.

9.3.2.3. *Coupling model of compaction to erosion threshold*

Through the constitutive relationship linking pressure to deformability (equation [9.27]), it is possible to obtain a dynamic change of porosity. Coupling between compaction and the erosion law occurs through this variable. Thus, aging (time) and burial of layers (depth) appear as variables in the erosion law, as shown in Figure 9.12 (taken from [TER 07]).

---

6 This coefficient is considered constant due to offsetting changes in permeability and deformability of the medium.

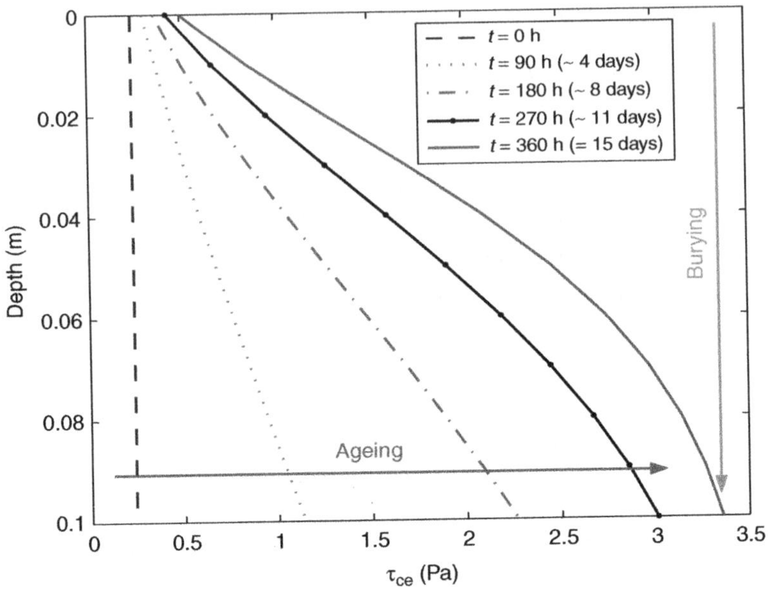

Figure 9.12. *Changes in the erosion critical stress τ_{ce} in relation to the depth for different simulation times (beginning, 1/4, half, 3/4, and final)*

9.4. Experimental characterization

9.4.1. *Characterization of sediments in the water column*

According to the model described in the previous section, the parameterizing of the deposit concerns three parameters: size distribution, concentration, and settling velocity of the suspended matter. For fine sediments, sieving techniques conventionally used for sands are not suitable for the determination of size distribution. Techniques such as the Andreasen pipette method and the Atterberg method are recommended for this purpose. Following the assumption that the fine grains are spherical, the settling velocity W_c is then deduced from Stokes' law (Table 9.1). Another approach, compatible with the multiclass one [ELG 04], consists of considering the suspension as a continuum of heterogeneous particles that can be divided into classes according to their settling velocities [BRA 06]. The settling velocities of each class is derived from monitoring, at different heights in a large-size settling tank, the temporal evolutions of the concentrations of suspended matter.

Laminar regime (Re < 1)	Transitory regime (1 < Re < 10^3)	Turbulent regime (10^3 < Re)
Stokes' law (fine sediments)	Budryk law (sands)	Rittinger's law (coarse sediments)
$W_c = \frac{gR}{18\nu} D^2$	$W_c = \frac{18\nu}{D}\left[\left(1+\frac{gR}{162\nu^2}D^3\right)^{1/2} - 1\right]$	$W_c = 87(RD)^{1/2}$

Table 9.1. *Settling velocity in function of particles and flow regime [SZT 04]. $R = (\rho_s - \rho_w)/\rho_w$ is the specific density of the immersed sediments. $Re = W_c D/\nu$ denotes the Reynolds number of the particle in decantation*

During decantation, the concentration is a decreasing function of time, which can be expressed, according to the multiclass approach, by the sum of decreasing exponential functions:

$$C(z,t) = \sum_i C_i(z,t) = \sum_i C_{0,i}(z) e^{-\lambda_i(z)(t-t_0)} \qquad [9.33]$$

where t_0 (*s*) is the initial time, $C(z,t)$ (kg.m^{-3}) is the total concentration of the suspended matter at depth *z* at time *t*, $C_i(z,t)$ (kg.m^{-3}) is the concentration of the suspended matter of class *i* at depth *z* at time *t*, $C_{0,i}(z) = C_i(z,t_0)$ (kg.m^{-3}) is the initial concentration ($t = t_0$) of the suspended matter of class *i* at depth *z*, and $\lambda_i(z)$ (s^{-1}) is the constant of the concentration decrease of suspended matter of class *i* at depth *z*.

The average concentration of particles of class *i* at height *h* (m) of suspension at time *t* (s), given by

$$C_i(t) = \frac{1}{h}\int_0^h C_i(z,t)dz \qquad [9.34]$$

is linked to the settling velocity $W_{c,i}$, according to equation

$$\frac{dC_i(t)}{dt} = -\frac{W_{c,i}(t)}{h} C_i(t) \qquad [9.35]$$

As $C_i(z,t) = C_{0,i}(z) e^{-\lambda_i(z)(t-t_0)}$, we obtain:

$$W_{c,i}(t) = h\frac{\int_0^h \lambda_i(z) C_i(z,t)dz}{\int_0^h C_i(z,t)dz} = h\frac{\int_0^h \lambda_i(z) C_{0,i}(z) e^{-\lambda_i(z)(t-t_0)}dz}{\int_0^h C_{0,i}(z) e^{-\lambda_i(z)(t-t_0)}dz} \qquad [9.36]$$

Assuming that, for each class, the constant of concentration decrease varies linearly with z and stating that $\lambda_i(z) = a_i z + b_i$, the settling velocity of particles of class i is given by:

$$W_{c,i}(t) = \frac{h}{t - t_0} - a_i h^2 \frac{\exp\left[-a_i h(t - t_0)\right]}{1 - \exp\left[-a_i h(t - t_0)\right]} + b_i h \qquad [9.37]$$

Adjusting the decreasing exponential laws on concentration time signals allows us to determine λ_i, and then the associated a_i and b_i. We then deduce the contributions to the total load and the settling velocities of each class i from the initial time t_0.

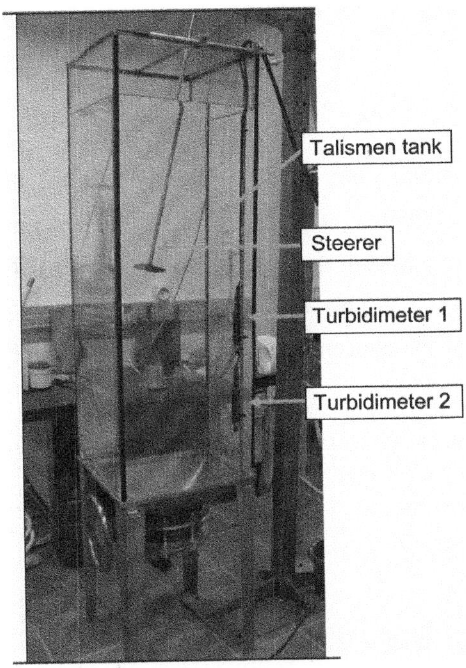

Figure 9.13. *Overview of the TALISMEN device (measurements of the settling velocity)*

This technique for measuring the decantation velocity by class is carried out in the TALISMEN[7] tank (Figure 9.13). This facility consists of a glass tank of 350 L ($1.5\,\text{m} \times 0.46\,\text{m} \times 0.46\,\text{m}$), whose bottom has a collector for the isolation and

7 TALISMEN is the acronym for transfer liquid/solids of metals in natural water.

extraction of deposited materials. A mechanical stirrer (shaft length 1 m and helix with diameter 80 mm) is used for the homogenization and the maintenance of particles in the suspension before letting them decant. The velocity is deduced by monitoring the time changes of the concentration at different heights during the decanting stage. To this end, two turbidimeters (range 0–20 NTU–nephelometric turbidity unit) are installed one above the other in the tank at heights $z = 0.50$ and 1.20 m. The turbidimetric signal issued at the height z is proportional to the local concentration of the suspended matter. The turbidimetric signal is systematically calibrated from water material samples taken from the column. Concentration is obtained by filtration, drying, and weighing of the collected matter. Further, the size distribution of the collected material is measured by a laser sizer (Master Sizer S, MALVERN).

For the purpose of illustration, the adjustment of the multiclass deposit model on the decrease of a sample of suspended matter collected in the Rhône river, located in the south-East of France (Figure 9.14) reveals the presence of two classes whose characteristics are shown in Table 9.2. The settling velocity associated with each class is then estimated from equation [9.37] (Figure 9.15). The average diameters associated with each class, 38 and 201 μm, respectively, are deduced from it by using Stokes' law (Table 9.1).

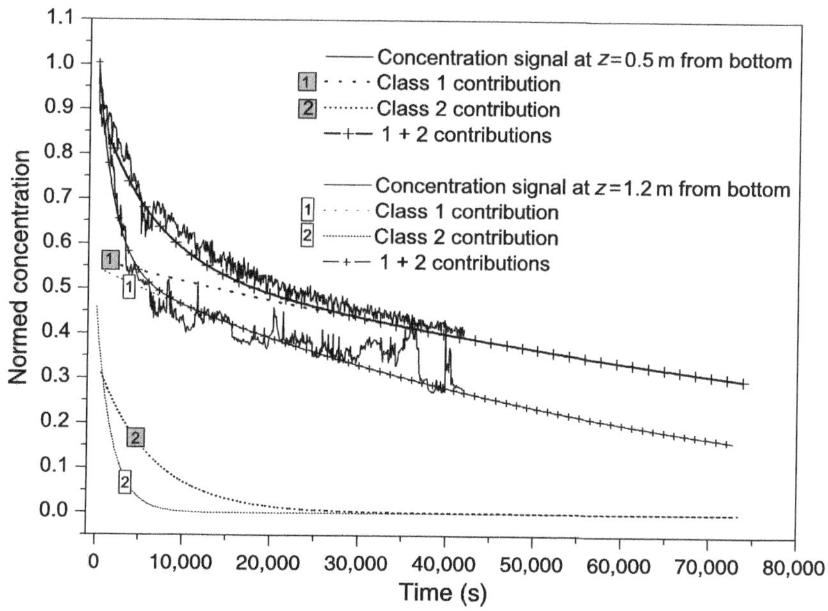

Figure 9.14. *Decrease in concentration during the decantation phase in the TALISMEN facility: contributions of the different classes of suspended matter*

Parameters	Class 1	Class 2
$\lambda_i(z=0.5\text{ m})\ (\text{s}^{-1})$	8.56×10^{-6}	1.60×10^{-4}
$\lambda_i(z=1.2\text{ m})\ (\text{s}^{-1})$	1.67×10^{-5}	5.21×10^{-4}
$C_{0,i}(z=0.5\text{ m})$	0.562	0.342
$C_{0,i}(z=1.2\text{ m})$	0.545	0.458
$a_i\ (\text{m}^{-1}.\text{s}^{-1})$	-1.18×10^{-7}	-5.25×10^{-6}
$b_i\ (\text{s}^{-1})$	2.27×10^{-5}	7.84×10^{-4}

Table 9.2. *Adjusted parameters of a multiclass deposit model for n = 2*

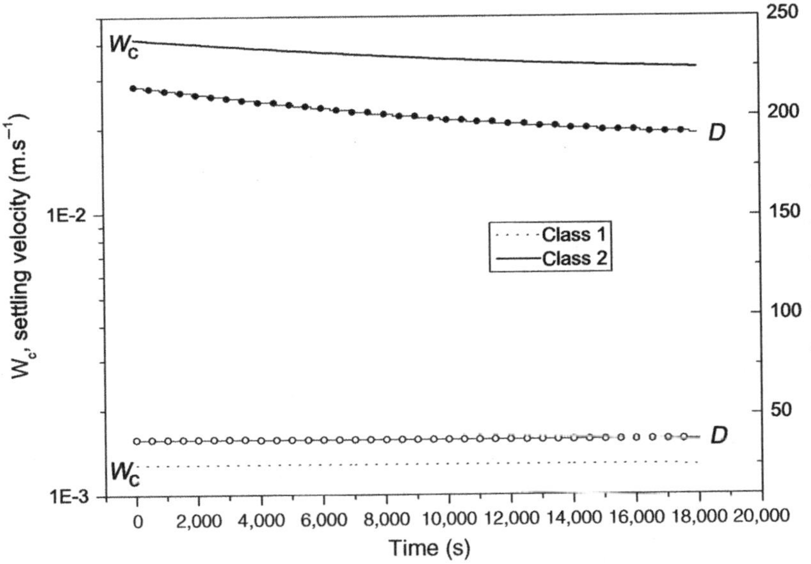

Figure 9.15. *Changes of the settling velocity and the diameter linked to two classes of particles shown in the suspension during the decantation phase in the TALISMEN facility*

9.4.2. *Characterization of the sediment column*

The proposed models are based on four parameters of the sediment bed: water content, size distribution, erosion rate, and critical erosion stress. To determine the critical erosion stress, several protocols are available. For non-consolidated sediments, it can be deduced from the size distribution via the Shields diagram. In the case of fine and cohesive sediments, this equation is no longer unequivocal. The types of erosion [WIN 04] and the associated models are multiple. Consequently, direct measurements on the sediments remain to this day a privileged process. *In situ* measurements are not straightforward. Recent developments show that the critical erosion stresses can be derived from current measurements by ADV [AND 07]. Erodimeters based on the generation of a hydraulic stream are often used. We can name the portable erodimeter developed by Le Hir *et al.* [LEH 06]. The protocol consists of eroding a sediment core set up on the bottom of this small flume. Another approach is to characterize the bottom sediments, collected by *in situ* core sampling, in a large laboratory channel current where a hydrodynamically controlled turbulent boundary layer is generated above the test zone [TER 07]. This type of analysis was conducted by IRPHE[8] and IRSN[9] from a protocol originally developed in collaboration with COM[10] for the experimental characterization of sediments from the Rhône. The various steps involved in this protocol are described below. The HEROD[11] channel (see Figure 9.16) built at IRPHE is dedicated to this measurement.

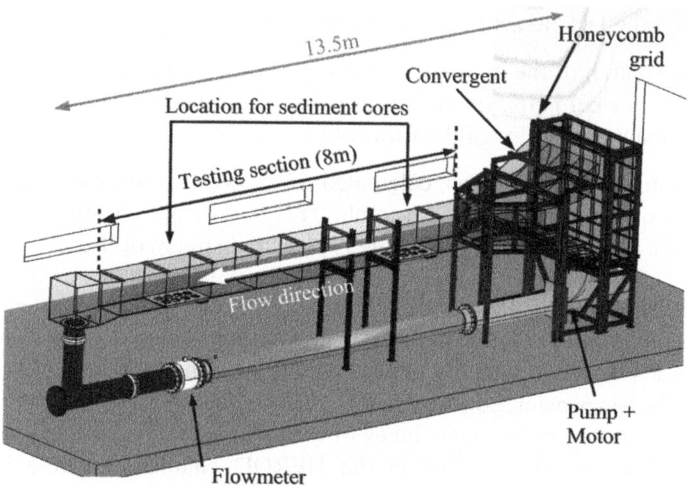

Figure 9.16. *HERODE channel*

The main characteristics of the channel are as follows:

– total length: 13.5 m;

– total width: 2 m;

8 Institut de Recherche sur les Phénomènes Hors-Équilibre.
9 Institute for Radiological Protection and Nuclear Safety.
10 Centre Océanologique de Marseille.
11 HEROD is the acronym for Hydraulic for EROsion and Deposition.

– maximum water volume: 9 m^3 in closed configuration, 6 m^3 in free surface (water depth of 0.6 m);

– theoretical maximum output: 1,200 $m^3.hr^{-1}$.

The detailed characteristics of the main elements are as follows:

– The test section, 8 m long, consists of eight boxes (1 m long each) of constant cross section, square (0.6 m square on the side).

– One of the boxes has a double base that allows us to install eight sediment cores.

– The return duct is composed of three elements. The first is polyvinyl chloride (PVC) (diameter 0.35 m, 2 m long). The second of 6 m in length (including 4 m with a heat exchanger), in stainless steel, is doubled with a cylindrical wall, within which it is possible to circulate a refrigerated liquid, so as to control the temperature of water. Finally, the last element is the one containing the propeller and the engine.

– The flowmeter, of the cylindrical shape, joins the return duct between the PVC and the stainless steel parts. Its operation is based on electromagnetic principles. It gives the instantaneous value of the flow rate;

– The converging part was calculated in a way to minimize the residual turbulence at the outlet of the honeycomb. This nest is placed after the settlement chamber of the pumped water. The whole is placed upstream of the test section.

The first step is devoted to the collection of sediments in the natural environment and their installation in the channel (Figure 9.17). Sampling is conducted carefully: we maintain a volume of water over the core to limit the reshuffling of sediments, not to alter the suspension/bottom interface and maintain conditions of saturation. Samples are collected in Plexiglas tubes of 0.5 m long with an internal diameter of 0.15 m. Eight cores are installed in the HEROD channel within 3 hours after collection. The HEROD channel is then filled up to 0.16 m depth of water above the cores, the latter flushed to the bottom of the channel. Additional cores are used for supplementary analyses. One core is subsampled to determine the vertical profiles of moisture content and two cores are used for rheological and size distribution analyses (by Malvern sizing). From these samples whose volume is determined and total mass is known, moisture content is deduced from weighing and dessiccation in a dry-off oven.

The HEROD channel is equipped with turbidimeters that track the changes of the suspended load over time. The characteristics of the boundary layer developed above the cores to be eroded were previously determined by velocity measurements using laser Doppler velocimetry (LDV) or particle image velocimetry (PIV). During this qualification stage, a study was conducted on the acceleration phenomenon, resulting from the thickening of the boundary layers in the wetted section, revealing a negligible effect of secondary currents in the test area. Moreover, the impact of the presence of cores on the estimation of the friction stress is not taken into account. In fact, no experimental parameter (critical shear stress profile, erosion flux) has shown such susceptibility, which seems consistent given the magnitude of the sediment concentration in the channel (a few milligrams per liter). The law connecting the flow rate in the channel to the friction stress over the cores is thus known. The HEROD channel can generate a maximum friction stress of 1 $N.m^{-2}$. By applying the strategy of increasing friction stresses, the erosion of the surface of the cores is initiated and maintained. We then obtain a growing trend in the concentration of the suspension. These signals are analyzed using the method presented by El Ganaoui *et al.* [ELG 04] in order to characterize erosive properties of the different layers *i* (Figure 9.18).

Figure 9.17. *(a) Coring in the Rhône river and (b) implementation of the cores in the HERODE laboratory channel*

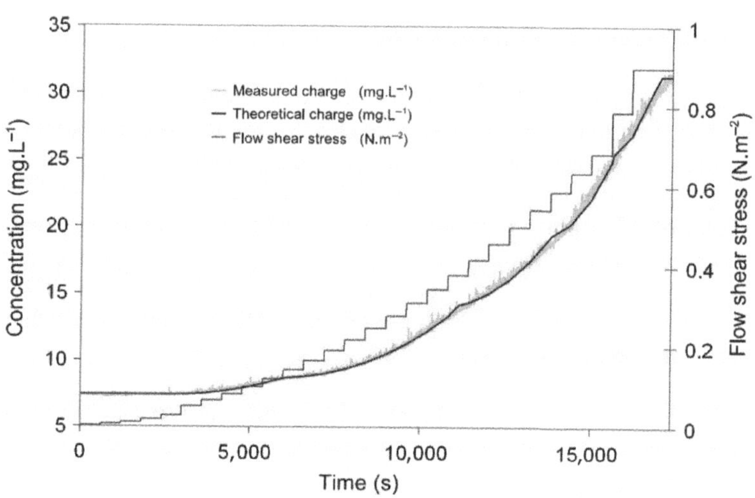

Figure 9.18. *Example of shear stress signals – suspended load and associated modeling*

The method is based on the use of the Partheniades erosion law (equation [9.13]), by considering that the bottom is eroded as the friction stress of the flow exceeds the critical erosion stress $\tau_{ce,i} < \tau_f$. Model calibration is performed gradually, erosion layer by erosion layer, by adjusting the values of the critical erosion stress, the erosion rate, and the initial concentration of the cores C_0^i in order to fit the time evolution of the suspended matter concentration provided by the model to the measurements (Figure 9.18). Table 9.3 shows the characteristics of the sediment column eroded in eight successive layers. The first four layers show erosion critical stresses that are increasing functions with depth, which would be characteristic of type I erosion [PAR 85]. Stabilizations of both $\tau_{ce,i}$ and erosion rate e_i between layers 3 and 4 highlight the presence of layers presenting type II erosion[12].

N	$\tau_{ce,i}$ ($N.m^{-2}$)	e_i ($kg.m^{-2}.s^{-1}$)	$[m]_i$ ($kg.m^{-2}$)
0	0.0275	0.0029	18
1	0.12	0.014	75
2	0.34	0.07	75
3	0.54	0.25	85
4	0.68	0.25	85

Table 9.3. *Identification of layers and their erosion parameters ($[m]_i$ is the surface density of the sediments)*

9.5. Example of operational application

9.5.1. *Simplified approach by a box model*

The general equations shown in the previous section are three-dimensional equations that describe in detail the transfers within the water column and the sediment column. A river may extend over several hundred kilometers and study periods may span over several days, months, or years. At these scales, a very accurate level of description is supposed to be costly in terms of parameterization and computational time. To meet these operational restrictions, it is customary to simplify the equations first by vertical integration (two-dimensional (2D) model), then by lateral integration (one-dimensional (1D) model), and, finally, by longitudinal integration (box model).

These models are well suited to large scales of time and space. The river is then described by a succession of reaches consisting of a water column and a sediment

12 This type of erosion, also known as surface erosion, is limited by the rehydration of the sediment surface before its erosion.

column that can, for example, be composed of three layers (an interface layer, an active layer, and a passive layer) to reflect the main properties related to vertical porosity profiles [MON 03, WAN 03, BOY 05].

Modeling is based on four levels, which are the representations of the hydrography of the river, hydraulics, sediment dynamics, and pollutant transfers [BOY 05].

The hydrographic model describes the river by a succession of reaches of trapezoidal shape and allows a spatial discretization based on any space step desired. For each reach, the variables are the length, width, angle of the banks, and the bed slope.

Hydraulic modeling is based on a semi-permanent approach based on the balance equation of water mass and the relationship of Manning–Strickler. It allows the estimation of the spatial and temporal variations of the average characteristics of the water column (flow rates, velocities, water volumes, water depths, and shear stresses).

The sediment model may take into account several classes of particles according to their hydro-sedimentary dynamics and their physicochemical affinities. It calculates the deposits and fluxes of matter in the water column and three sedimentary layers (Figure 9.19).

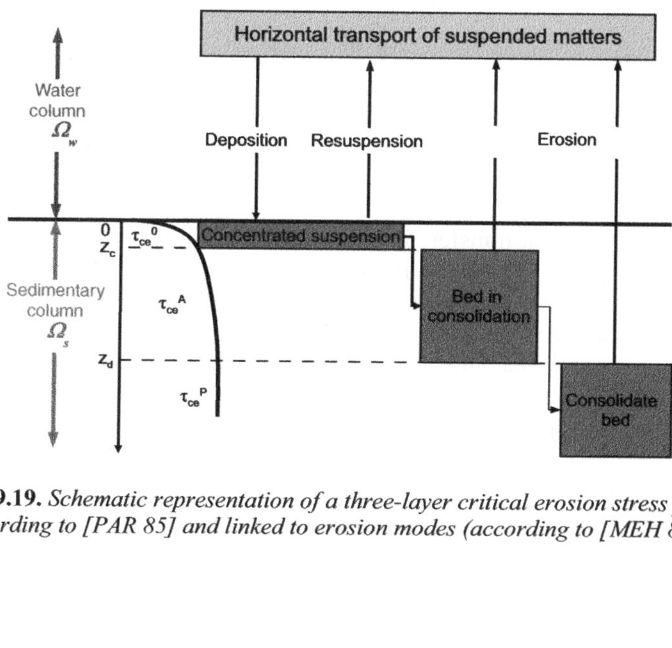

Figure 9.19. *Schematic representation of a three-layer critical erosion stress profile, according to [PAR 85] and linked to erosion modes (according to [MEH 82])*

The interface layer between the water column and bottom sediments is a thin layer formed by the recent deposits. Even if its particles are of very small diameter, they retain a non-cohesive behavior with regard to the erosion because this layer is neither compacted nor consolidated. Thus, the interface layer plays an important role on the transfers between the water column and the sediment column [MAA 98] because it determines the equilibrium concentrations of suspended matter classes [ELG 04]. Given its thinness (a few grain diameters) and high porosity ($\approx$1), it can be assumed that the concentrations of pollutants in the interface layer are in equilibrium with those of the water column.

The active layer is a layer undergoing compaction within which the cohesive effects become significant. The erosion does not depend on individual particles, but generally applies to the matrix consisting of whole solid grains. As noted before (section 2.3.2.1), taking into account the Van der Waals forces allows us to model the cohesive effects by involving only the variations of porosity and size distribution. In general, other factors, such as the saturation state, pH, or salinity of the water, as well as the impact of flora and benthic fauna, should also be considered. The concept of the active layer is related to the fact that the porosity of this layer remains sufficiently large to allow the pore water and dissolved phase of pollutants to transfer themselves easily toward the water column.

The passive layer results from the compaction of the active layer when the porosity becomes low enough to limit interstitial transfers. It corresponds to a form virtually stabilized across several years.

In the model, these three layers are characterized by their moisture contents, the maximum thickness of the interface and the active layers, and the coefficients of consolidation of the active and passive layers. The auto-consolidation model applies mainly to the interface layer and consolidation layer. The consolidated layer is assumed to have a constant porosity.

Modeling of pollutant transfers is partly based on hydraulic and sediment fluxes which drive the spatial and temporal variations of liquid and solid forms within the water column and the three sedimentary layers. For each pollutant, the parameters are those of the chemical reactions considered and inputs described by their kinetics and locations. In particular, the solid–liquid transfers are treated with an approach based on a reference K_d weighted in relation to the size of the particles using a geometric method [ABR 96].

9.5.2. *Application to the radioactive contamination of the Techa River (Russia)*

To illustrate the implementation of this model, this section describes its application, in the framework of IAEA's EMRAS Programme [KRY 09], to radioactive contamination of the Techa River, which is located in Russia in the southern part of the Ural Mountains (52° N, 61° E). This river is part of the watershed of the Kara Sea. It measures 207 km long and its watershed covers a surface of 7,600 km^2. The water depth varies from 0.5 to 2 m, its average width gradually changing from 15 to 30 m in a downstream direction, and its slope is, on average, 0.0006 $m.m^{-1}$. In its final part, there is an average flow rate of 7 $m^3.s^{-1}$, the maximum output of 29 $m^3.s^{-1}$ being reached in April.

During the 1949–1952 period, the military Mayak nuclear complex produced about 10^{17}Bq in the form of radioactive liquid waste of which approximately 95% of the waste was discharged into Lake Karachai upstream of the Techa River between March 1950 and November 1951. From this date, the radioactive waste has decreased drastically and the contamination of Lake Karachai has been controlled by a dam and a channel system. Since then, sources of contamination of the river are due to transfers from the floodplain area and due to seepage under the dam and the containment channels [KRY 09].

The application of the model consists of simulating the transfers of ^{90}Sr, ^{137}Cs, and $^{239,240}Pu$ within this hydrosystem for the period between 1950 and 1996. The input data are monthly averages of flow rates and suspended loads of nine grain size classes, the waste discharge of these three radionuclides from March 1950 to November 1951, and their average annual flow of infiltration and runoff from the end of this period [KRY 09].

The calculation results are assessed from measurements of activities carried out at different parts of the Techa River in 1996. Given the available data, the model was applied as follows:

– Annual chronicles of average monthly flow rates and suspended loads are repeated throughout the period between 1950 and 1996.

– In the absence of data on the texture of bottom sediments, it is not possible to determine the erosion parameters of active and passive layers. Fortunately, for the interface layer, the only data on suspended loads allow us to perform this characterization because it is assumed that the particles deposited in this layer have a non-cohesive behavior [ELG 07]. Thus, for each grain size class, the erosion parameters of the interface layer can be determined using the Shields diagram.

– In these circumstances, the erosion of the active and passive layers has been disabled. The active layer is subjected to the process of compaction and interstitial diffusion and the passive layer is subjected to only compaction. The parameters of the model are then the water contents (water mass/solid mass) of the three layers and the thickness of the interface and active layers.

The thickness of the interface layer is taken equal to the diameter of the larger suspended particle, that is 0.1 mm. The active layer corresponds to the upper portion of the sediments capable of maintaining transfers by diffusion with the water column. In practice, it corresponds to the dissolved oxygen zone that is a few centimeters thick. By default, this thickness is fixed at 20 mm. Finally, the values of moisture contents of passive, active, and interface layers are, in the absence of additional data, attributed to their default values, that is 66%, 150%, and 400%, respectively.

With these assumptions, the calibration of the model is carried out in two steps. The first step applies to the hydraulic model, which provides, among other things, the bed shear stresses. This calibrates the friction conditions through the adjustment of the Strickler coefficient. This is determined so that the water levels calculated by the model in relation to flow rate conditions correspond to the height vs. flow rate relationships provided by the scenario. As the flow rate increases stream-wise, an extra input of flow rate had to be included across the borders of the rivers to match the monthly measurements between PK44 and PK180. This adjustment converges to a uniform Strickler coefficient over the entire river that is equal to 15 $m^{1/3}.s^{-1}$. This value is consistent with the characteristics of the Techa River (Table 9.4), which shows an abundant aquatic vegetation.

Strickler coefficient ($m^{1/3}.s^{-1}$)	Characteristics
70–80	Coated bed
30–50	Natural bed, river not too winding and well calibrated
10–30	High vegetation, crowded bed, flood plain area

Table 9.4. *Variability of the Strickler coefficient according to [VIO 03]*

The second calibration step applies to the radio ecological model by determining the coefficient of liquid–solid partition for each radionuclide. These parameters are adjusted in order to match the computing activities of the water column and the sediments to the measurements performed in 1996. The values obtained after this step are shown in Table 9.5, where it is verified that they fit well within the range of variation available in the literature [IAE 10].

The calculations are performed on the whole of the river (207 km) using an interval of 500 m and a time step of 0.5 day for a total of 17,200 days (≈ 47 years). Figure 9.20 shows the comparison curves of the calculations of spatial changes of the levels of radioactive elements in the water column and sediments to the measurements performed in 1996.

Radionuclide	Coefficient K_d ($m^3.kg^{-1}$)	IAEA
^{90}Sr	0.3	0.23 < 1.2 < 6.3
^{137}Cs	30	1.6 < 29 < 520
$^{239,240}Pu$	50	10 < 240 < 5,200

Table 9.5. *Liquid–solid partition coefficients for the three radionuclides studied*

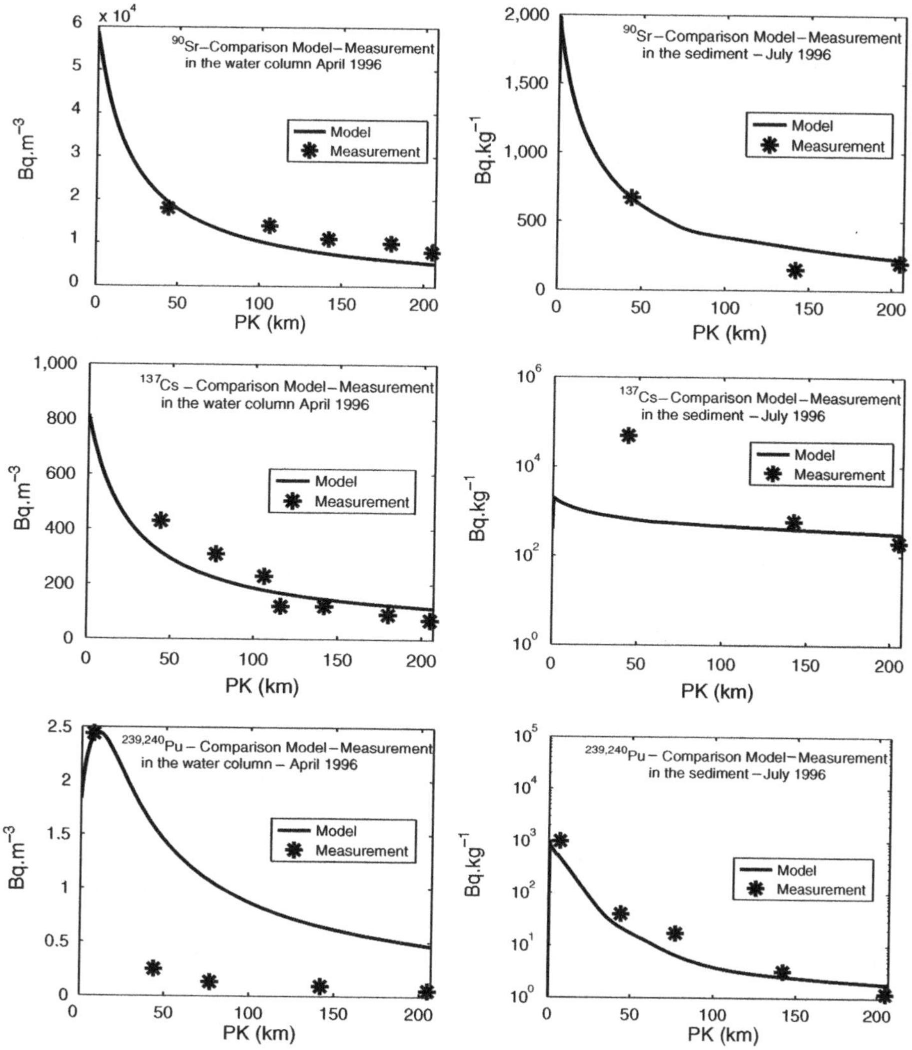

Figure 9.20. *Spatial distribution of concentrations of radioactive pollutants in the Techa River (Russia): comparison model – measurements*

These results show large variations in concentrations along the river, often in more than two orders of magnitude. These amplitudes are consecutive to the longitudinal variations of the flow rates and to the grain size discrimination of the resulting deposits. The multiclass approach is well suited to represent this type of situation. In practice, the more mobile are the elements, the more efficient is the adjustment of the model. Thus, the best results are obtained for ^{90}Sr.

For the $^{239,240}Pu$, the difficulty in obtaining consistency between calculations and measurements is mainly linked to the uncertainty in the quantification of the inputs between the years 1950 and 1951. While these inputs are precisely quantified for ^{90}Sr and the ^{137}Cs, this is not the case for $^{239,240}Pu$, for which it is noted that they are less than 1% of the total inputs. Here, the calculations were carried out by fixing these inputs at 1% of the total inputs.

From these results, it is possible to provide an estimate of the chronology of transfers throughout the simulated period, which extends from 1951 to 1996 (Figure 9.21). These results indicate that the purification of the medium is first carried out for strontium and cesium, and finally for plutonium. This illustrates the difference in the behavior of radionuclides in relation to their affinity to suspended particles. Thus, the balance between the concentrations in the sediments and those in the water column is reached faster when this affinity, characterized by the value of K_d, is lower.

For the $^{239,240}Pu$, which is the least mobile of the three, it appears that the state of contamination observed in 1996 remains highly correlated with the initial inputs. This observation, which confirms the role of uncertainty on the initial source term in the difficulty of modeling this radionuclide, suggests that the initial inputs were probably less than the 1% rate used for these calculations. In fact, a calculation with lower initial rejects will give lower concentrations in the water column and will require the solid–liquid partition coefficient of $^{239,240}Pu$ to be increased to maintain consistency with concentrations observed in the sediments. Both these changes go well in the direction of the difference between calculations and measurements and properties of this radionuclide (Table 9.5).

This analysis emphasizes the relevance of a multi-class sediment approach. The latter takes into account the different affinities of radionuclides in relation to the size of the particles; that is, the finer the particles are, the stronger the affinity. It also highlighted that in the long term, linkage between transfers and hydro-sedimentary processes plays an important part in that the pollutants show stronger affinities with the suspended particles.

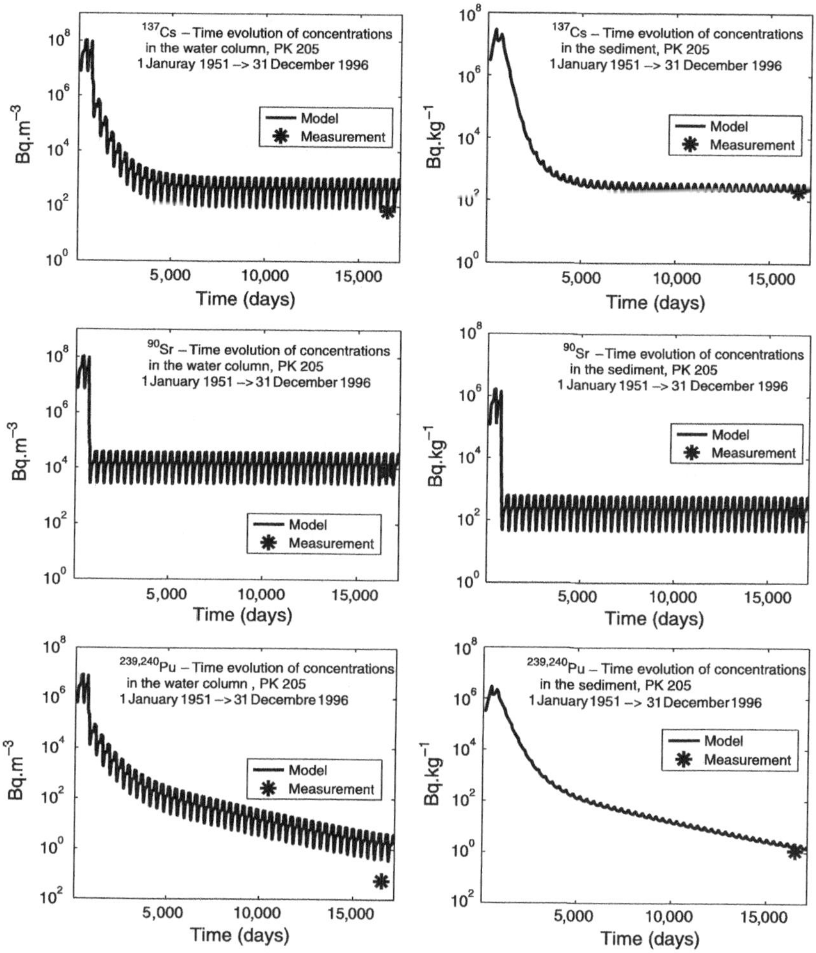

Figure 9.21. *Estimation of the time evolutions of radioactive pollutant concentrations at the exit of the Techa between 1951 and 1996*

9.6. Conclusion

9.6.1. *Review*

Pollutants may accumulate in sediment beds by the deposit of contaminated suspended matter. They then pass intermittently during periods of erosion during which large quantities can be remobilized. The erosion phenomenon appears to be a key mechanism of the management of water quality of rivers.

The modeling discussed in this chapter is not only based on the coupling between transport equations of contaminants and matter in the water column and sediment bed related by transfer fluxes whose main parameters are the settling velocity of particles, the erosion rate and the critical stress. New parameters are introduced in this original approach, which now accounts for the porosity and deformability of the sediments. The particular interest in fine particles led us to deal with the problem of erosion of cohesive sediments, subject to burial and aging, representative of auto-consolidation. The study of microscopic properties of the interactions between clay particles allowed the development of a cohesion model, based on the attractive component of the Van der Waals force, which was incorporated into a model of erosion threshold. This force depends on the state of compaction of the sediment bed and its physicochemical properties. The erosion threshold therefore depends on an additional variable: the porosity, itself described as a function of time and depth by the consolidation model. Based on experimental observations, a description in layers of the sediment bed simplified the modeling of transfer fluxes across the interface between the water and sediment columns. Finally, the affinities between fine particles and most pollutants have allowed us to simplify the transport model in a sedimentary way by considering classes of settling particles.

These modeling elements reduce the number of unknowns to be addressed to solve the problem, namely by reducing them to accessible parameters by an experiment. The key parameters for the closure of the global model are determined through experiments that were discussed: the artificial erosion in the HERODE channel, the decantation in the TASLISMEN tank.

With these simplifications, we see that it is possible to simulate the scenario of actual pollutant inputs as shown in the case of a contaminated river in the carly 1950s. In this example, the sediments act as a storage compartment and are clearly a key to water quality management of rivers.

9.6.2. *Outlook*

The theme of the water quality of rivers involves the interaction of several scientific fields such as physics, chemistry, and biology, for which several points revealed are interesting and could be developed.

The deposit fluxes of sediments are determined by the intensity of flow turbulence, also called transport capacity, which is often described by the Meyer–Peter formula. The equilibrium concentration proposed in the multiclass approach shows a connection with the transport capacity of the flow (see the thesis of C. Leupi [LEU 06], p. 104). This theme is similar to the phenomenon of segregation of particles according to the level of turbulence in the flow. The use of tools such as laser anemometers (Phase Doppler Anemometer – PDA) should allow us to study this phenomenon more closely and in particular to monitor the size distribution of the particles during erosion.

Further in-depth research is required for the study of erosion parameters, concerning, *inter alia*, the consolidation. The effect of parameters, such as pH, salinity of water, and the presence of a biofilm, on erosion should also be finely analyzed.

Diffusion phenomena at the interface between the water column and the sediment column remain unclear. On the one hand, the spread is still quite difficult to model but may contribute to the longer-term storage toward the sediment column of the dissolved or adsorbed pollutant on the very fine particles. On the other hand, when the non-consolidated or weakly consolidated sediments have been eroded, the erosion flux is no longer linear and appears saturated or limited by another slower process, linked to cohesion, despite significant friction.

Finally, other interactions may sometimes contribute to cohesion, such as the capillarity in unsaturated sediments, as we may find on the banks of rivers. Is this work transferable to such a context?

Is the state of suspended particles after erosion comparable to that at the time when they were deposited? Investigations using PDA technology would also allow us to learn about transformation of particles during their stay in the sediment bed.

Finally, concerning the erosion rate, there is currently no theory that can model the experimentally observed values.

9.7. Bibliography

[ABR 96] ABRIL J., FRAGA E., "Some physical and chemical features of the variability of Kd distribution coefficient for radionuclides", *Journal of Environmental Radioactivity*, vol. 30, pp. 253–270, 1996.

[AND 07] ANDERSEN T., FREDSOE J., PEJRUP M., "In situ estimation of erosion and deposition thresholds by Acoustic Doppler Velocimeter (ADV)", *Estuarine, Coastal and Shelf Science*, vol. 75, pp. 327–336, 2007.

[BLA 02] BLACK K., TOLHURST T., PATERSON D., HAGERTHEY S., "Working with natural cohesive sediments", *Journal of Hydraulic Engineering*, vol. 128, no. 1, pp. 2–8, 2002.

[BOR 00] BOROVEC Z., "Element in size-fractionated bottom sediment of the Elbe River in its Czech part", *Aquatic Sciences*, vol. 62, pp. 232–251, 2000.

[BOU 97] BOUDREAU B., *Diagenetic Models and Their Implementation: Modelling Transport and Reactions in Aquatic Sediments*, Springer, 1997.

[BOY 05] BOYER P., BEAUGELIN-SEILLER K., TERNAT F., ANSELMET F., AMIELH M., "A dynamic box model to predict the radionuclide behaviour in rivers for medium and long-term periods", *Radioprotection*, vol. 40, no. 1, pp. S307–S313, 2005.

[BRA 06] BRACH-PAPA C., BOYER P., TERNAT F., AMIELH M., ANSELMET F., "Settling classes of fine suspended particles", *Comptes Rendus de Mécanique*, vol. 334, no. 8–9, pp. 560–567, 2006.

[BÜR 98] BÜRGER R., CONCHA F., "Mathematical model and numerical simulation of the settling of flocculated suspensions", *International Journal of Multiphase Flow*, vol. 24, pp. 1005–1023, 1998.

[BUR 05] BURGOS R., CONCHA F., "Further development of software for the design and simulation of industrial thickeners", *Chemical Engineering Journal*, vol. 111, no. 2–3, pp. 135–144, 2005.

[BUS 99] BUSTOS M., CONCHA F., "Settling velocities of particulate systems: 10", *International Journal of Mineral Processing*, vol. 57, pp. 183–203, 1999.

[CAR 03] CARROL J., LERCHE I., "Sedimentary quantification using radionuclides", *Journal of Environmental Radioactivity*, Series radioactivity in the environment, vol. 5, pp. 97–113, 2003.

[CIT 04] CITEAU L., Etude des colloïdes naturels dans les eaux gravitaires de sols contaminés: relation entre nature des colloïdes et réactivité vis-à-vis des métaux (Zn, Cd, Pb, Cu), Université de Paris VI, INRA Versailles, 2004.

[CON 87] CONCHA F., BUSTOS M., "A modification of the Kynch theory of sedimentation", *American Institute of Chemical Engineers*, vol. 33, no. 2, pp. 312–315, 1987.

[DEY 99] DEY S., "Sediment threshold", *Applied Mathematical Modelling*, vol. 23, pp. 399– 417, 1999.

[ELG 04] EL GANAOUI O., SCHAAFF E., BOYER P., AMIELH M., ANSELMET F., GRENZ C., "The deposition and erosion of cohesive sediments determined by a multi-class model", *Estuarine, Coastal and Shelf Science*, vol. 60, no. 3, pp. 457–475, 2004.

[ELG 07] EL GANAOUI O., SCHAAFF E., BOYER P., AMIELH M., ANSELMET F., GRENZ C., "Erosion of the upper layer of cohesive sediments: characterization of some properties", *Journal of Hydraulic Engineering*, vol. 133, no. 9, pp. 1087–1091, 2007.

[GAR 00] GARRIDO P., BÜRGER R., CONCHA F., "Settling velocities of particulate systems: 11. Comparison of the phenomenological sedimentation – consolidation model with published experimental results", *International Journal of Mineral Processing*, vol. 60, pp. 213–227, 2000.

[GEL 05] GELARD D., Identification et carctérisation de la cohésion interne du matériau terre dans ses conditions naturelles de conservation, Doctoral Thesis, Institut National Polytechnique de Grenoble, 2005.

[GIB 67] GIBSON R.E., ENGLAND G.L., HUSSEY M.J.L., "The theory of one-dimensional consolidation of saturated clays 1: finite non-linear consolidation of thin homogeneous layers", *Geotechnique*, vol. 17, pp. 261–273, 1967.

[GRA 84] GRAF W.H., *Hydraulics of Sediment Transport*, Water Resources Publications, LLC, Highlands Ranch, Colorado, 1984.

[HOL 81] HOLTZ R.D., KOVACS W.D., *An Introduction to Geotechnical Engineering*, Ecole Polytechnique de Montréal, 1981.

[IAE 10] IAEA, Handbook of parameter values for the prediction of radionuclide transfer in terrestrial and freshwater environments, Technical reports series no. 472, Vienna, 2010.

[ISR 85] ISRAELACHVILI J.N., *Intermolecular and Surface Forces*, Academic Press, 1985.

[KAM 83] KAMPHUIS J.W., HALL K.R., "Cohesive material erosion by unidirectional current", *Journal of Hydraulic Engineering*, vol. 109, no. 1, pp. 49–61, 1983.

[KEL 81] KELLY W.E., GULARTE R.C., "Erosion resistance of cohesive soils", *Journal of the Hydraulics Division*, vol. 107, no. HY10, pp. 1211–1224, 1981.

[KRO 62] KRONE B., Flume studies of the transport of sediments in estuarial shoaling processes, Final report, Hydraulic Engineering Laboratory and Sanitary Engineering Research Laboratory, University of California Berkeley, 1962.

[KRY 09] KRYSHEV I., BOYER P., MONTE L., BRITTAIN J., DZYUBA N., KRYLOV A., KRYSHEV A., NOSOV A., SANINA K., ZHELEZNYAK M., "Model testing of radioactive contamination by ^{90}Sr, ^{137}Cs and ^{239,240}Pu of water and bottom sediments in the Techa River (Southern Urals, Russia)", *Science of the Total Environment*, vol. 7, no. 407, pp. 2349–2360, 2009.

[LAD 67] LADD C.C., KINNER K.B., The strength of clays at low effective stress, Report no. 3-101, U.S. Army Engineer Waterways Experiment Station, January 1967.

[LEE 68] LEE I.K., *Soil Mechanics: Selected Topics*, Butterworths, London, 1968.

[LEE 04] LEE C., "Automated sediment erosion testing system using digital imaging", *Journal of Hydraulic Engineering*, vol. 130, no. 8, pp. 771–782, 2004.

[LEH 06] LE HIR P., CANN P., JESTIN H., BASSOULLET P., "Instrumentation légère pour la mesure de l'érodabilité des sédiments vaseux ou sablo-vaseux", *IXèmes Journées Nationales Génie Civil – Génie Côtier*, 2006.

[LEU 06] LEUPI C., Numerical modeling of cohesive sediment transport and bed morphology in estuaries, Doctoral Thesis, École Polytechnique Fédérale de Lausanne, 2006.

[MAA 98] MAA J.P.-Y., SANFORD L., HALKA J.P., "Sediment resuspension characteristics in Baltimore Harbor, Maryland", *Marine Geology*, vol. 146, pp. 137–145, 1998.

[MEH 82] MEHTA A.J., "Resuspension potential of cohesive sediment bed: estuarine comparisons", *Sixth Biennial International Estuarine Research Conference*, Gleneden Beach, Academic Press, pp. 591–601, 1982.

[MIG 68] MIGNIOT C., "A study of the physical properties of different very fine sediment and their behaviour under hydrodynamic action", *La Houille Blanche*, vol. 7, pp. 591–620, 1968.

[MON 03] MONTE L., HAKANSON L., "Radioactivity in lakes and rivers", in SCOTT M.E. (ed.), *Modelling Radioactivity in the Environment*, Elsevier, 2003.

[PAR 65] PARTHENIADES E., "Erosion and deposition of cohesive soils", *Journal of Hydraulics Division, ASCE*, vol. HY1, no. 91, pp. 105–139, 1965.

[PAR 85] PARCHURE T.M., MEHTA A.J., "Erosion of soft cohesive sediment deposits", *Journal of Hydraulic Engineering*, vol. 111, no. 10, pp. 1308–1326, 1985.

[PIL 01] PILOTTI M., MENDUNI G., "Beginning of sediment transport of incoherent grains in shallow shear flows", *Journal of Hydraulic Research*, vol. 39, no. 2, pp. 115–124, 2001.

[POS 67] POSTMA H., "Sediment transport and sedimentation in the estuarine environment", in Estuaries, LAUFF G.H. (ed.), AAAS, Washington D.C., vol. 83, pp. 158–179, 1967.

[SAN 01] SANFORD P.L., MAA J.P.-Y., "A unified formulation for fine sediments", *Marine Geology*, vol. 179, pp. 9–23, 2001.

[SIG 01] SIGG L., BEHRA P., STUMM W., *Chimie des milieux aquatiques: chimie des eaux naturelles et des interfaces dans l'environnement*, Dunod, Paris, 2001.

[SUN 56] SUNDBORG A., *The River Klaralven: A Study of Fluvial Processes*, Esselte Aktiebolag, Stockholm, 1956.

[SZT 04] SZTABA K., "Influence of grain size upon its falling velocity", *Physicochemical Problems of Mineral Processing*, vol. 38, pp. 207–220, 2004.

[TER 07] TERNAT F., Erosion of self-weight consolidated cohesive sediments, PhD Thesis, Aix-Marseille II Mediterranean University, 2007.

[TER 08] TERNAT F., BOYER P., ANSELMET F., AMIELH M., "Erosion threshold of saturated natural cohesive sediments: modeling and experiments", *Water Resources Research*, vol. 44, no. W11434, pp. 1–18, 2008.

[TIL 81] TILLER F.M., "Revision of Kynch sedimentation theory", *American Institute of Chemical Engineers*, vol. 27, no. 5, pp. 823–829, 1981.

[VER 09] VERNEY R., LAFITE R., BRUN-COTTAN J.-C., "Flocculation potential of estuarine particles: the importance of environmental factors and of the spatial and seasonal variability of suspended particulate matter", *Estuaries and Coasts: Journal of the Estuarine Research Federation*, vol. 32, no. 4, pp. 678–693, 2009.

[VIO 03] VIOLLET P.-L., CHABARD J.-P., ESPOSITO P., LAURENCE D., *Mécanique des Fluides Appliquée*, Edition Presses de l'Ecole Nationale des Ponts et Chaussées, p. 367, 2003.

[WAN 03] WANG H., APPAN A., GULLIVER J., "Modeling of phosphorus dynamics in aquatic sediments: I – model development", *Water Research*, vol. 37, no. 16, pp. 3928–3938, 2003.

[WIB 87] WIBERG P.L., SMITH J.D., "Calculations of the critical shear stress for motion of uniform and heterogeneous sediments", *Water Resources Research*, vol. 23, no. 8, pp. 1471–1480, 1987.

[WIN 04] WINTERWERP J.C., VAN KESTEREN W.G.M., *Introduction to the Physics of Cohesive Sediment in the Marine Environment*, Elsevier, New York, 2004.

[ZIE 03] ZIERVOGEL K., BOHLING B., "Sedimentoloical parameters and erosion behaviour of submarine coastal sediments in the south-western Baltic Sea", *Geological Marine Letter*, vol. 23, pp. 43–52, 2003.

Chapter 10

Sediment Transport and Morphodynamics in Nearshore Areas

10.1. Introduction

This chapter is an introduction to the description and modeling of the physical phenomena that trigger the dynamics of sediments, sand, and pebbles in coastal and nearshore. The first part of this chapter deals with the characterization of sediments, the conditions that need to be met to set these sediments into motion under the effect of currents and waves, as well as with some general aspects regarding transport equations. Also, this chapter proposes a few models for sediment transport and for the morphodynamics that accompany it. These models are illustrated using examples ranging from natural sites, or sites that were less affected by human intervention (e.g. sandy beds and beaches), to morphological evolutions induced by artificial structures. In particular, we introduce examples of local growth or erosion, in structures such as pipelines, vertical cylinders, and coastal dikes. Some behavioral models are presented for the study of shoreline erosion and its intensification, which is linked to shoreline landscaping and climate change.

By examining "sediment transport", we pay, *a priori*, particular attention to the particles that are likely to sediment and to be set in motion by a moving fluid; these particles may range from fine particles to pebbles. There are two regimes that characterize the movement of particles: the "Brownian" regime for the small particles, whose sedimentation period tends toward infinity, and the hydrodynamic regime where the movement of particles is a composition between the settling

Chapter written by Vincent REY and Damien SOUS.

velocity and the advection velocity. The settling time is characterized separately from the fluid flow considered in its settling velocity; the latter is an essential parameter for characterizing particles, together with size and density.

Whereas the formulations for the suspended transport are derived from the classical transport equations of passive contaminants, the formulas for the transport of coarse particles are much more empirical. It is noteworthy that the dissolved specimens can also be adsorbed by particles (substratum), and thus sediment transport contributes to the transfer or the deposit of dissolved contaminants.

This chapter is an introduction to the dynamics of shoreline sediments and does not constitute an exhaustive description of the great variety, and oftentimes the great complexity of the physical processes that come into play. For every phenomenon that is examined, relevant bibliographical references are given to allow the reader to follow the matter in more detail. It is equally important to note that we limit ourselves to studying the transport of passive sediments, and that the biological and chemical problematics exceed the purely physical scope that we have limited our studies to. Unless otherwise specified, all the measurable quantities used henceforth are expressed following the International System (SI) of Units.

10.2. Marine sediments

10.2.1. *Characteristics of sediments*

We can distinguish between two large categories of sediments: cohesive and non-cohesive sediments. We generally consider sediments to be cohesive when their diameter D is less than 64 μm. Called "lutite" or "pelite", they are the equivalent of what we usually call "mud". These sediments have physicochemical properties that lead to aggregation or flocculation phenomena that can significantly modify their behavior in suspension. Their behavior in a fluid can be strongly dependent on salinity, due to the physicochemical effects that they have on the surface. For the non-cohesive sediments, we may distinguish between "arenite" ($0.064 < D < 2$ mm) and "rudite" ($D > 2$ mm). Rudite is essentially made of rock debris, and arenite of grains of quartz or (mineral or organic) limestone.

When suspended, the behavior of sediments becomes dependent on their size. For the small particles (called "Brownian"), the effects of the thermal agitation are significant and their diffusion coefficient D_t (agitation effect) is given by [GUY 92]:

$$D_t = \frac{k_B T}{6\pi\mu R} \qquad [10.1]$$

where R is the radius of the particle, μ the dynamic viscosity of the fluid, T the temperature, and k_B the Boltzmann constant.

For the coarser particles, the hydrodynamic effects are predominant. The relative significance of these two effects is measured by the Péclet number, Pe:

$$Pe = \frac{UL}{D_t} \qquad [10.2]$$

with U being the characteristic velocity and L the characteristic length, of the order D, diameter of the particle. If $Pe \ll 1$, the particle has a Brownian behavior, whereas if $Pe \gg 1$, it has a non-Brownian behavior. If we consider that the most important parameter is the velocity gradient of the particle G throughout its displacement over a distance of the order of its radius R, and if $G = U / R$, then $Pe \sim 6\pi\mu GR^3/(k_B T)$ is proportional to R^3 and thus strongly dependent on the size of the particle. Knowing that $D_t = 10^{-6}$ m^2/s, if we choose $G = 1$ s^{-1}, we can deduce from here that the non-Brownian particles have a diameter that is more than $R = 1$ µm.

Given that the size of the grains is not uniform, we can define the statistic parameters of distribution by considering D_k as the diameter, insomuch as k% of the grains have a diameter that is lower than D. It is noteworthy that we can also find D_k defined as the diameter insomuch as k% of the weight of the grains have a diameter that is lower than D, the coarse particles being thus heavier than the small particles. We can thus define the average characteristics of the sediment, such as the median diameter 1 D_{50}, the geometrical average $D = (D_{84.1}D_{15.9})^{0.5}$, etc., and characterize the distribution according to its variance, its coefficient of dissymmetry, its flattening etc. For the non-cohesive sediments, the distribution is often Gaussian.

We define a form factor SF to characterize the geometry of the grains by $SF = c(ab)^{-0.5}$ with a, b, and c being the lengths of the main axes. Generally, the shape indeed influences the settling velocity. The density of particles plays an important role, especially in the settling velocity and the inertial effects in the direction of the flow.

The settling velocity of a particle is an essential parameter in the formulation of sediment transport equations in a fluid. It depends on the characteristics of the fluid (volume mass ρ, kinematic viscosity ν), and of the particle (volume mass ρ_s, diameter D, shape and the condition of the surface).

10.2.2. *Settling velocity of sediments*

For the Reynolds numbers of the order 1 (particles of a size less than approximately 100 µm), the calculated drag for the spherical particles of a diameter $D = 2R$ is given by $F = 6\pi\mu R W_s$, where W_s is the settling velocity. It is also called the Stokes velocity. It is opposed to the relative weight of the particle in a permanent regime. The intensity of settling velocity can thus be expressed as follows (Stokes law):

$$W_s = \frac{rgD^2}{18\,\nu}, \quad r = \frac{\rho_s}{\rho} - 1 \qquad [10.3]$$

For the small particles, of a cohesive type (theoretically $D < 64$ µm), we can observe the effects of aggregation or separation. For non-spherical particles and/or for a set of polydispersed particles, it is sufficient to replace D with a characteristic diameter d, oftentimes the median diameter d_{50}. The expression of W_s evolves with the square of the diameter of the particle, hence the transition between cohesive diameters and non-cohesive sediments is quite fast.

For flows with larger Reynolds flownumbers (larger particle, more significant velocity), the dependence of the settling velocity on the diameter of the particles varies [MIG 77]: $W_s \sim d$ for $1 < Re < 1{,}000$, and $W_s \sim d^{0.5}$ for $1{,}000 > Re$ (turbulent flows). More generally, we can write the drag force F depending on the friction coefficient C_d and of the master couple of the particle S (surface):

$$F = \frac{1}{2}\rho C_d W_s^2 S \qquad [10.4]$$

The settling velocity can be obtained by equaling the entrainment force and the reduced gravitational force:

$$W_s = \sqrt{\frac{4\,rgd}{3\,C_d}} \qquad [10.5]$$

However, the friction coefficient depends on the Reynolds number, and the expression is not the expression that is used in the models of shoreline transport. The usual formulations are given depending on the size of particles:

$$W_s = \begin{cases} \dfrac{rgd^2}{18\,\nu} & \text{if } d \le 100\ \mu\text{m} \quad \text{(Stoke's Law)} \\ \dfrac{10\,\nu}{d}\left[\left(1+\dfrac{0.01rgd^3}{\nu^2}\right)^{0.5}-1\right] & \text{if } 100\ \mu\text{m} \le d \le 1{,}000\ \mu\text{m (Zanke's Law)} \\ 1.1(rgd)^{0.5} & \text{if } d \ge 1{,}000\ \mu\text{m} \quad \text{(Van Rijn's Law)} \end{cases}$$

For the small, cohesive particles (i.e. theoretically, $D < 64$ µm), aggregation effects, or separation effects, may appear when the characteristics of the medium vary. This happens particularly in the vicinity of estuaries or outlets where salinity varies significantly. Even though, for reasons of convenience, transport models are frequently used as the hypothesis for non-cohesive particles (i.e. for a behavior that is independent of other particles when transported), we observe significant variations in their settling velocity, depending on salinity. When the particles are not spherical, the settling velocity decreases. This decrease is all the more important when the Reynolds number is large. If the particles have elongated shapes (e.g. in the case of pellets), their characteristic size is their length. Their settling trajectory in the absence of the flow, which is strictly vertical for spherical particles, is usually neither vertical nor rectilinear [HER 96, CHH 99].

In fact, the settling velocity of the particles depends on the concentration of suspended particles. The variation in the settling velocity can have a significant effect on the suspension distribution of the water column. If we consider, for example, the case of the two particles of a radius R, distant from a, and settling at a velocity of W_s, the ratio W_{s0}/W_s, where W_{s0} is the settling velocity for an isolated particle, has a value of $1 - 0.75R/a$ if the particles fall side-by-side, and $1 - 1.5R/a$ if they fall one after the other [GUY 92]. Therefore, two particles fall quicker than a single particle, as a consequence of the entrainment phenomenon. However, the effect of the concentration rather has the tendency to decrease the settling velocity because of the increase in the effects of the viscosity; the velocity is thus given by $W_s = W_s(1 - 6.55c)$, where c is the suspension concentration [GUY 92].

A frequently used alternative expression was proposed by [RIC 54]: $W_s = W_t(1 - c)^n$, where W_t is the terminal settling velocity of a particle in an infinite fluid and n an empirical exponent that varies from 4.65 to 2.4 for the increasing Reynolds numbers. For the natural sands that have a high concentration ($c > 0.3$), the settling velocity, which differs significantly from the theoretical prediction for the spherical particles, can be divided by more than 5 in relation to the free settling velocity [BAL 04]. For a very concentrated suspension, the behavior of the water/sediment mixture tends toward the behavior of a pileup, where the spheres are fixed in relation to one another and because of that, it is as if the particles fell as a block; this leads us to considering the flow of a fluid throughout a porous medium.

In the case of cohesive sediments, the settling velocity increases up to a concentration of the order 10–20 g/l because of the effects of flocculation; then it decreases as a consequence of the obstructions in the high concentrations. Salinity, which enhances the flocculation phenomenon, increases the settling velocity. It is thus a phenomenon that may prove significant when near the outlets or the estuaries, as it has been observed on the Rhône and is shown in Figure 10.1.

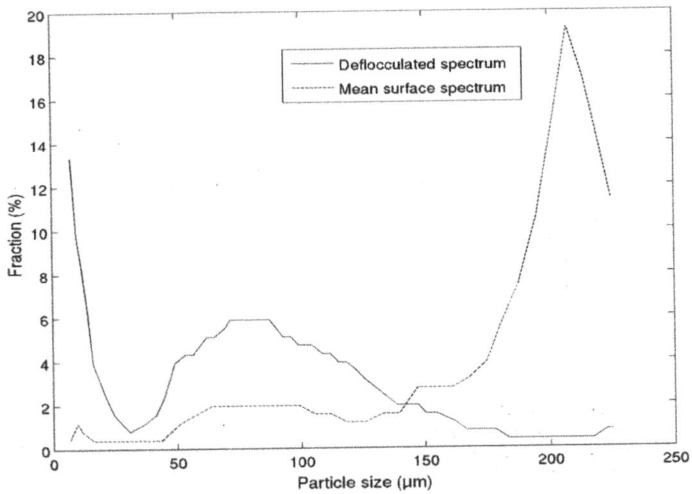

Figure 10.1. *Grain size distribution of suspended particles at the mouth of the river Rhône [DUF 08]; the deflocculated particles were filtered and exposed to sonic waves to break down the aggregates*

10.2.3. *Rheology of suspensions and deposits*

The diluted suspensions behave like homogeneous Newtonian fluids, of a viscosity of $\mu > \mu_0$, where μ_0 is the viscosity of the fluid in the absence of suspension [BAT 67]. Einstein [EIN 05] showed that $\mu = \mu_0(1 + 2.5c)$ for the low concentrations in Brownian or non-Brownian spherical particles. The concentrated suspensions behave like non-Newtonian fluids. This happens particularly near the bottom where the viscosity is thus more significant, which tends to diminish the turbulence.

At the bottom, the deposits have the tendency to press down and compact (Figure 10.2), which influences the conditions of resetting into suspension. The compaction of the deposit of non-cohesive sediments is very quick, whereas the compaction of cohesive sediments is slow. The rheological behavior of the mud corresponds to a viscoplastic Bingham model, the stress tensor τ follows the law $\tau = \tau_0 + \mu_v \dot{\gamma}$, where μ_v is the viscosity of the mud, τ_0 the rigidity threshold that depends on the size of grains and on their concentration.

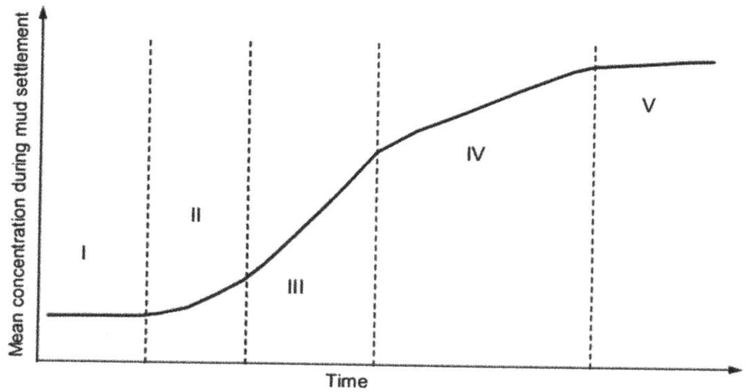

Figure 10.2. *Different phases of mud compaction (cohesive sediment). **I**: particles deflocculated by agitation; **II**: flocculation, settling velocity obstructed by the flocs; **III**: phase 1 of the compaction, crushing the flocs, eliminating interstitial water; **IV**: phase 2 of the compaction, evacuating the water by means of porosity and draining wells; **V**: phase 3 of the compaction, evacuating the water by means of compression [MIG 68]*

10.3. Sediment transport

10.3.1. *Suspending the particles, the notion of threshold*

Because of the scattering of the sizes, shapes, and positions of sediment particles as well as of the velocity fluctuations in space and time, setting the particles into motion must be considered as a statistical process.

The suspension of solid particles depends, on the one hand, on hydrodynamic conditions and, on the other hand, on the nature and size of particles. A particle that was deposited at the bottom is subject to a relative weight $\vec{P}$, a hydrodynamic force $\vec{F}$, and a contact force (or a resistance force) $\vec{R}$. The particle will be set into motion as soon as the resulting force is directed upward (suspension) or as soon as these forces allow the particle to rotate around one of its fulcrums (rolling motion).

The hydrodynamic force results from the stress exerted by the fluid upon the bottom, which is called bottom or bed shear stress τ_0. It depends on the type of boundary layer (laminary or turbulent). According to the different "schools" or according to the different types of flow, there are three general expressions:

– $\tau_0 = \rho g R_H J$ in fluvial hydraulics, where R_H is the hydraulic radius, and J the slope of the line of total energy;

– $\tau_0 = \rho u_*^2$ for a turbulent flow in permanent regime, with u_* being the friction velocity;

– $\tau_0 = 0.5 \rho f_w U_m^2$ for an oscillating flow, where U_m is the average velocity (or the maximum velocity) of the oscillation of the fluid at the bottom under the effect of waves, and f_w the friction coefficient at the bottom, depending on the Reynolds number and the rugosity.

If the bearing force is higher than its apparent weight, the particle will be uplifted under the effect of this shear force. We thus define non-dimensional numbers, which compare the hydrodynamic force with the reduced gravitational force. For the stationary flows, we define the *Shields number* θ as:

$$\theta = \frac{\tau_0}{(\rho_s - \rho) g D} = \frac{u_*^2}{rgD} \qquad [10.6]$$

For alternate flows (swells), we define the *mobility number* Ψ as:

$$\Psi = \frac{U_m^2}{rgD} = \frac{(A\omega)^2}{rgD} \qquad [10.7]$$

with A being the amplitude of the wave at its surface, and ω the pulsation.

The Shields number θ and the mobility number Ψ are related as follows:

$$\theta = \frac{1}{2} f_w \Psi \qquad [10.8]$$

General remarks

– The analyses carried out on the Shields number or the mobility number are generally restricted to non-cohesive sediments. For the fine particles for which the aggregation effects become significant, the cohesion forces must be taken into account to describe the dynamics of particles.

– The shear stress τ at the bottom is the sum of two terms, the drag due to the pressure gradients τ'' and the friction on the surface of the grain τ'. The most destabilizing influence comes from τ', and so we define a Shields parameter for the surface friction $\theta' = \tau' / [(\rho_s - \rho)gD]$. This parameter is frequently used for determining the threshold of setting sediments into motion; τ' is called *effective shear*.

– The field measurements as well as those carried out in the laboratory show the difficulty of obtaining a unified view of the threshold for setting into motion cohesive sediments. The proportion of these cohesive sediments in the considered sediment stock has a very strong influence. For the fine sediments of the sand or mud type ($D < 300\,\mu\text{m}$), van Rijn [RIJ 07a] suggests that we define the bed shear stress $\tau_{\text{cr,bed}}$ as follows:

$$\tau_{\text{cr,bed}} = \begin{cases} \dfrac{c_{\text{gel}}}{c_s}\left(\dfrac{d_{\text{sand}}}{d_{50}}\right)^{\gamma_a} \tau_{\text{cr, 0}} & \text{if } D<64\ \mu\text{m} \\ (1+P_{c,s})^3 \tau_{\text{cr, 0}} & \text{if } D>64\ \mu\text{m} \end{cases} \qquad [10.9]$$

where $d_{\text{sand}} = 64\ \mu\text{m}$, γ_a is an empirical exponent between 1.5 and 2, c_{gel} the volume concentration of the consolidation of the bed (where "consolidated" corresponds to a rigid bed, or a viscoplastic bed), c_s the maximum concentration of a pure sediment (which can be assimilated to the compactness = 1 – porosity), $\tau_{\text{cr,0}}$ the critical stress of non-cohesive particles, and $P_{c,s}$ the mass proportion of clay in a sample of the bed.

– The granulometry scattering of the grains of natural sand changes the conditions of the setting into motion threshold considerably, in relation to a monodispersed sand. Thus, we consider $k_s = 2.5\ D_{50}$ to calculate the relative rugosity that the Shields number depends on, therefore marked as $\theta_{2.5}$.

– For practicality, we represent the Shields parameter or the mobility parameter according to the non-dimensional diameter of particles D_*, which is defined as follows:

$$D_* = \left(\frac{rg}{\nu^2}\right)^{\frac{1}{3}} D \qquad [10.10]$$

which may itself be expressed according to the particle Reynolds number $Re_* = u_* D/\nu$. The experimental curves [MAN 55, BAG 56] show a strong dispersion in the experimental points around the extrapolation curve of the setting into motion threshold. We can also observe in Figure 10.3 the differences in evaluating these suspending thresholds, according to different authors.

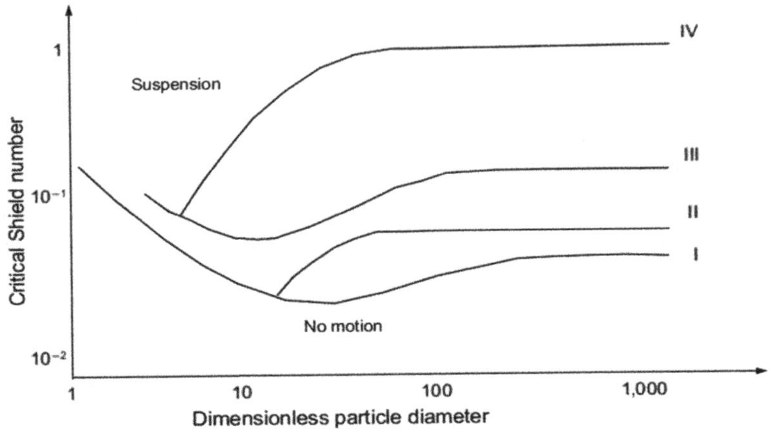

Figure 10.3. *The thresholds for setting into motion and suspending the particles, under the effect of a current, depending on the reduced diameter.* ***I****: Yalin–Shields diagram,* ***II****: Engelund diagram,* ***III****: van Rijn diagram, and* ***IV****: Bagnold diagram [RIJ 89]*

10.3.2. *Modes of transport*

We distinguish between two main types of transport: the bedload transport, and the suspended transport. The bedload transport is the result of intragranular forces; it regards the particles that can be found near the bottom, and it is the dominant mode of transport for the coarse particles. We can also distinguish the sliding transport (on a smooth bottom, therefore a rather rare situation) by rolling and by saltation. Note that the latter is an intermittent suspended transport.

The suspended transport is due to the hydrodynamic entraining forces. The suspended transport concerns the small particles, whose settling velocity is sufficiently low (the characteristic sedimentation time is much higher than the transport characteristic time). When the suspension is balanced (or slightly unbalanced), a sediment concentration profile is established between the bottom and the free surface. In the case of a sediment intake around the surface (river mouth, outlets), the particles have the tendency to sediment, but they are not likely to be suspended again. We then talk about "wash load".

We have just observed that there may be a balance, or an imbalance, between the suspending movement and the deposit. Thus, there are more favorable areas for the deposit, i.e. calm areas (weaker, less turbulent flow), or the mixture between waters of different kinds (physicochemical phenomena).

10.3.3. *Equations and transport models*

Several approaches have been proposed for estimating the sediment transport; they were divided into two major schools, which we briefly present below: the first approach aims to calculate the solid sediment transport in a balanced state (*a priori* rather more adapted to bedload transport) and the other approach is founded on determining the advection/diffusion of the matter. These two approaches use empirical or quasi-empirical parameterizations. For a more detailed look at the main formulas of total transport that were used in a shoreline medium, the reader may refer to [DYE 86], [ENG 67], [RIJ 85], [BIJ 71], [BAG 66], [BAI 81], and [SOU 97].

The transport of suspended particles is due to hydrodynamic forces. It can be expressed starting from the classic advection–diffusion equation, which includes the settling velocity $\vec{W}_s = -W_s\vec{e}_3$, supposed to be independent from the flow velocity:

$$\frac{\partial c}{\partial t} + div\left[\left(\vec{u} + \vec{W}_s\right)c\right] = D_t\nabla^2 c \qquad [10.11]$$

where c is the concentration of particles and D_t the molecular diffusion coefficient. Generally speaking, the flow is turbulent, and the turbulent diffusion is predominant in relation to molecular diffusion. By writing $c = C + c'$ and $u_j = U_j + u'_j$ (breaking it down into average and fluctuation), equation [10.11] becomes:

$$\frac{\partial C}{\partial t} + U_j\frac{\partial C}{\partial x_j} = W_s\frac{\partial C}{\partial x_3} - \frac{\partial \overline{u'_j c'}}{\partial x_j} \qquad [10.12]$$

where $\overline{u'_j c'}$ represents the turbulent mass flow of particles with x_j, $j = 1.2$ horizontal coordinates and x_3 as the ascending vertical. This flow is generally expressed in relation to average fields.

The observations regarding the vertical distribution of the concentration generally indicate the presence of strong concentrations around the bottom, gradually decreasing toward the free surface, which depend on the effective settling velocity. We obtain vertical distributions of the average concentrations over a wave period for an advection–diffusion equation.

$$cW_s + \varepsilon_{s,cw}\frac{dc}{dz} = 0 \qquad [10.13]$$

where $\varepsilon_{s,cw}$ is the mixture coefficient of the sediment for a combined flow of wave and current [RIJ 07b].

The modeling of the transport requires us to master the set of on-site hydro-sedimentary parameters (sediment size, settling velocity, flocculation, stratification, salinity, etc.) and to recalibrate the model using on-site measurements.

In the case of bedload transport, the approach is generally different. It is the consequence of granular interactions that take place around the bed, forced by hydrodynamics. It is expressed starting from empirical formulas that represent the transport capacity, from which we can deduce the local flow mobility number Ψ and the critical value of this number Ψ_c:

$$Q_b(\vec{x},t) = \text{sediment transport rate} = a(\Psi - \Psi_c)^n \quad [10.14]$$

where a and n are constants. This is a threshold law. Several formulations were proposed for permanent flows (analogous to the flows of rivers) [MIG 77], or for flows in the presence of waves [HOR 88]. It is noteworthy that a deviation of a factor of 10 may appear between the different formulations.

10.4. Sediment structures and morphodynamic structures

The observations carried out in nearshore areas reveal the presence of sediment structures over really varied scales, ranging from a few cm to several km, constantly evolving due to different driving forces (tide, waves, currents, alluvia, winds, etc.). The formation and migration of these sediment structures is a very complex phenomenon. Sediment structure plays a defining role in the evolution of shorelines, estuaries, and fluvial beds. The hydraulic facilities, such as pipelines, structures, and piers, that are installed on sediment bottoms can be damaged or weakened as a consequence of sand silting or erosion, which takes place around it. Furthermore, because of this, the breakwaters may become inefficient, and the navigation canals can become sand silted. The quantitative studies carried out on these undermining phenomena are mainly experimental; the results are often expressed according to the characteristic measurable quantities of the obstacles, and the non-dimensional numbers that characterize these flows.

10.4.1. *Sediment structures outside the breaking area*

10.4.1.1. *The formation of ripples and sand dunes*

The observations carried out in nature or in the laboratory have emphasized a large variety of sediment structures, whose classification is difficult to establish.

These structures interact strongly with the hydrodynamics by modifying the characteristics of the flow. In what follows, we examine the formation of ripples and sand dunes under the effect of a current or a swell, as well as the formation of sand bars due to the spatial modulations in the amplitude of the swell. We must bear in mind that several other physical processes can be at the origin of the development and migration of sediment structures. In particular, on an average scale, the unbalance in sediment transport, connected to flow sediment gradients (e.g. estuaries) or hydrodynamic perturbations (e.g. wakes), could often cause a variety of morphological structures.

10.4.1.2. *Under the effect of current*

For simplicity and following an analogy with natural situations, most of the studies on the sediment structures formed under the effect of a current have concerned unidirectional flows on a free surface. We generally distinguish between ripples and sand dunes, according to the height and the wavelength of the structures compared with the thickness of the boundary layer. On the one hand, the ripples never exceed the boundary layer and their effects are limited to the near-bed region. On the other hand, the dunes are characterized by wavelengths that are more significant and that cause accelerations and deccelerations throughout the entire height of the flow, up to the free surface. We can also observe the megaripples and the antidunes [RIJ 07b].

Three regimes can be usually distinguished:

– the slow regimes with plain beds, ripples, or dunes (the "lower flow regime" in Figure 10.4);

– the transitional regimes with "washed" dunes; and

– the fast regimes with plain bed and antidunes ("upper flow regime" in Figure 10.5).

The distinction between these different regimes is made by the Froude number [ENG 67] or by the suspension parameter (the inverse of the Rouse number) coupled with the Reynolds number of the particle [LIU 57].

Figure 10.5 shows the presence of current ripples formed in a feeder of a rip channel on the beach of Truc Vert (Aquitaine) when the tide is low.

Estimating friction on the bottom often requires defining an equivalent rugosity. The equivalent rugosity k (non-dimensional), related with the presence of these structures, is estimated by empirical formulas.

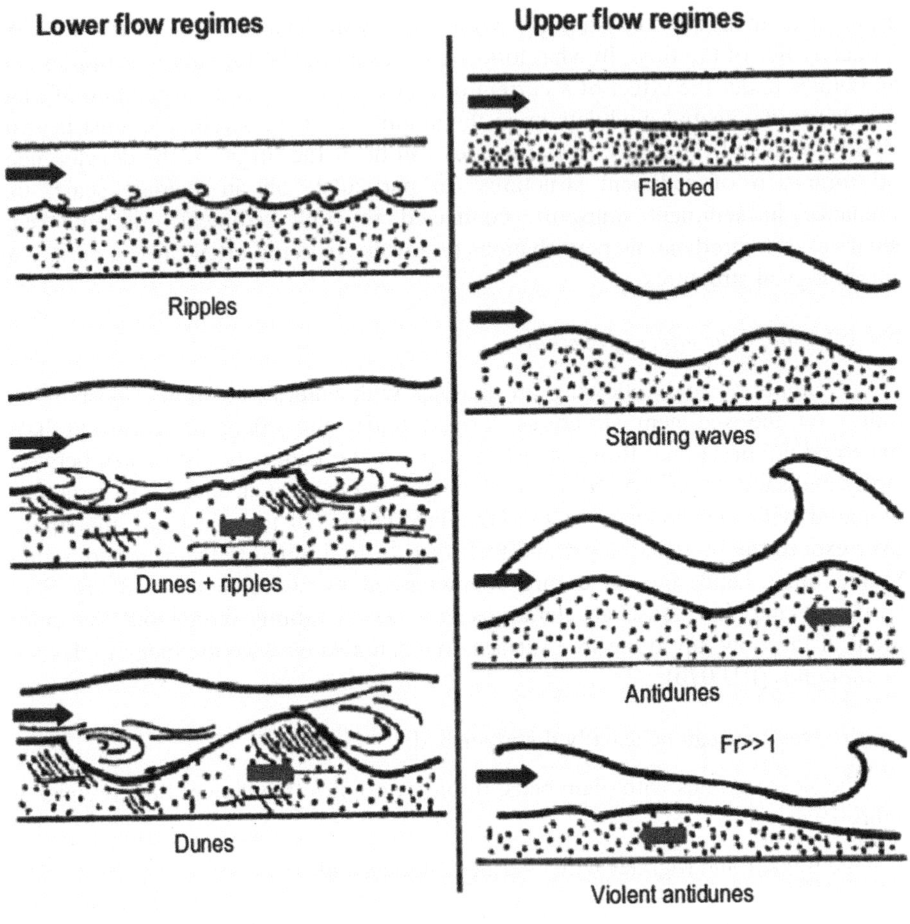

Figure 10.4. *Sequence of sediment forms developed by a unidirectional current, for a flow with increasing intensity [SIM 65]*

For ripples, we generally assume that:

$$k = \begin{cases} 150 f_{cs} D_{50} \quad \text{if } \Psi<50 \text{ (low wave-current regime, shallow)} \\ (182.5 - 0.652\psi) f_{cs} D_{50} \quad \text{if } 50<\Psi<250 \text{ (transitional regime)} \\ 20 f_{cs} D_{50} \quad \text{if } \Psi>250 \text{ (strong wave-current regime, sheetflow)} \end{cases}$$

where $f_{cs} = \text{Max}[1, (0.25 D_{\text{gravel}} / D_{50})^{1.5}]$, and $D_{\text{gravel}} = 2\,\text{mm}$.

Figure 10.5. *Current ridges formed by the descending tide, Truc Vert beach*

For the dunes, the formulas are the following:

$$k = \begin{cases} 0.0004 f_{fs} \Psi h & \text{if } \Psi < 100 \\ (0.048 - 0.00008\Psi) f_{fs} \Psi h & \text{if } 100 < \Psi < 600 \\ 0 & \text{if } \Psi > 600 \end{cases}$$

where $f_{cs} = \mathrm{Max}[1, D_{50} / (1{,}5 D_{sand})]$, and $D_{sand} = 64$ µm.

10.4.1.3. *Under the action of waves*

When the swell energy is sufficient, the sand particles are set in motion and form small sediment structures called ripples. Their more symmetrical aspect generally distinguishes them from current ripples. Their mode of formation depends on the value of the mobility number Ψ associated with the local flow. If Ψ is hardly higher than the critical number Ψ_c, the sand grains roll, forming small ripples (*rolling grain ripples*) that do not perturb the flow above. For the values that are higher than Ψ, the ripples are of a larger size, and the flow above the ripples is perturbed, forming a turbulence eddy (hence the name of *vortex ripples*). In fact, we can observe, in a first stage, the formation of small ripples, whose wavelength L_r and amplitude η_r subsequently increase because of the eddies generated nearby. For a partially stationary swell, we can observe an initial formation and increase around the nodes.

Numerous laboratory and on-site studies have led us to propose laws that regard the bending of the ripples according to the shear stress. Nielsen [NIE 89] proposed an empirical law for the bending of the ripples:

$$\frac{\eta_r}{L_r} = 0.182 - 0.24\left(\tfrac{1}{2} f_w \Psi\right)^{\frac{3}{2}} \qquad [10.15]$$

with

$$f_w = \exp\left[5.213\left(\frac{2.5D}{A}\right)^{0.194} - 5.977\right] \qquad [10.16]$$

The height of the mature ripples may be expressed according to the semi-amplitude A of the oscillation of the fluid at the bottom [NIR89]:

$$\frac{\eta_r}{A} = 0.275 - 0.022\ \Psi^{\frac{1}{2}} \qquad [10.17]$$

A linear expression is proposed in [FRE 92] for the length of the ripples depending on A

$$L_r = 1.2\,A \qquad [10.18]$$

which corresponds to a size of the order 0.6 times the horizontal excursion of the fluid.

10.4.1.4. *Formation of sand bars*

We have seen that locally the solid particles are suspended or transported by bedload. If this local deficit is compensated by bedload transport and by alluvial deposit, the bathymetry no longer evolves. If there is an imbalance, then accretion and erosion start affecting certain areas. From a qualitative point of view, we can say that the calm areas (weaker flow, less turbulent) will correspond to an accumulation of sediments, whereas the more agitated areas are the most prone to erosion. In fact, we cannot limit ourselves to such a simplistic reasoning, since, for example, sediment bars may form rippling areas during tempests, whereas these areas are the areas where the agitation is at its maximum. This demonstrates the current difficulties for correctly modeling the bathymetric evolutions, which are not linked to the quantity of solid matter transported, but to the gradients of its transport. Besides, the bathymetric modifications quickly induce evolutions in the breaking zone, and thus modifications in hydrodynamic driving forces.

This next part is limited to coupled evolutions without wave breaking, and it is based on studies carried out in wave flume [REY 91, REY 95, DUL 00]. Its goal is to show the impact of spatial modulations in the amplitude of gravity waves on their evolution, combined with the bottom modulations, whose topography presents modulations called bars. These considerations were recently applied to theoretical as well as numerical studies in bathymetric evolutions [OST 93, YU 00], where the transport in the bottom boundary layer is taken into account for the bedload transport.

In reality, such bars are however rarely observed because they suppose a certain regularity in the swell fields over quite a significant period of time. However, they were observed in Escambia Bay in Florida [LAU 73].

For monochromatic linear waves, the only possible spatial modulations are due to the presence of a partially stationary wave, generated either by a non-negligible reflection of the shoreline wave (significant slope, or shoreline structure), or by a reflection over a bar. The experiments shown in Figure 10.6 were carried out in a small swell canal (length 470 cm, width 40 cm, depth 6 cm). The bathymetric developments as well as the developments of the surface wave are measured between the initial time ($t = 0$) and the final time ($t = 3$ hours) over a sedimentary bed (local depth 4.75 cm) initially flat, and an incidental frequency wave $f = 1.5$ Hz with an amplitude $H = 8$ mm.

At first, we can observe quite a regular modulation of the surface wave along the canal, which corresponds to the signal of a partially stationary wave, whose deflected free surface is expressed as:

$$\eta = \eta_- + \eta_+ = a_- \sin(\omega t - kx) + a_+ \sin(\omega t + kx + \varphi) \qquad [10.19]$$

where the indexes "–" and "+" correspond to the incident wave and the reflected wave. The reflection coefficient is defined by:

$$R = \left| \frac{a_+}{a_-} \right| \qquad [10.20]$$

It may be calculated by:

$$R = \frac{a_{\max} - a_{\min}}{a_{\max} + a_{\min}} \qquad [10.21]$$

where $a_{\max}$ is the maximum amplitude of the wave and $a_{\min}$ its minimal amplitude. Two successive maxima (or minima) are separated by approximately half a wavelength. The reflection coefficient R has an initial value of 10%, and ends up being 23% in the final stage, as shown in Figure 10.6.

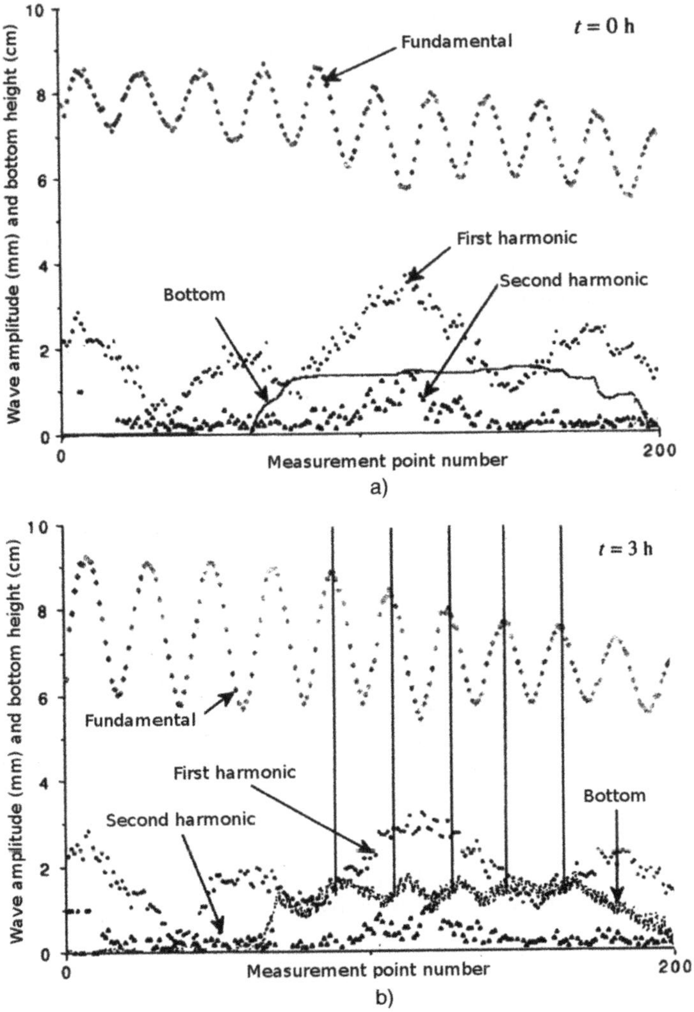

Figure 10.6. *Coupled evolution for a partially stationary wave: initial time (t=0 s) and final time (t=3 h) [REY 95]*

Furthermore, the bottom that was initially plain presents, after the erosion, a spatial periodicity of around half of a wavelength. We can observe that the accretion areas correspond to the positions of the antinodes of the surface wave. This means that the predominant mode of transport is the suspension mode. We detail the effects of the two different modes in the next section. This bathymetric evolution is coupled with the evolution of the wave, which means an increased reflection of the incident swell. This phenomenon bears the name "Bragg resonance", by an analogy with the phenomena of interference that are well known in the case of electromagnetic waves, as well as in the physics of solids. This phenomenon was largely examined over the past 20 years in the field of shoreline oceanography, since it is related to a self-protection phenomenon of the beaches that are subject to swells, by reflecting a part of their energy toward the high sea, by a bathymetric adaptation. In fact, this study represents an ideal situation. In reality, the nonlinear effects are significant and the beaches do not have a null slope. Considering the nonlinearity of the slope and of the beach requires further developments, which are beyond the scope of this chapter.

10.4.2. *Sediment structures in breaking zone*

10.4.2.1. *Conceptual classification of sediment beaches*

Several conceptual models were created to understand the morphological variations of beaches according to environmental conditions; these conditions are mainly: the height of the breaking wave, the wave period, the tidal regime, the settling velocity of the sediment, and the local slope of the beach. Starting from these parameters, these beaches were classified as dispersive, intermediary, or reflective [WRI 84]. Whereas the extreme regimes (dispersive and reflective) can be clearly discerned with the aid of environmental parameters, the intermediary regime, where we can find complex morphologies (beaches with sand bars), is more difficult to characterize. In particular, the dynamic of beaches with multiple bars (generally two or three) remains poorly understood. The variability of the environmental conditions as well as the difficulty in obtaining satisfying morphological, on-site measurements – these are factors that limit the comparison with the laboratory models (canal or swell basin) as well as with the numerical models.

10.4.2.2. *Breaking bars*

There are two main hypotheses, which can be combined, that are usually advanced for explaining the formation of breaking bars [MAS 03]. The first hypothesis suggests that the bars form close to the breaking point by the convergence of the sediment fluxes. The second hypothesis links the formation of a bar to the spatial modulations of surface waves (long stationary wave).

There are several experimental and numerical studies that were carried out for identifying the influence that environmental conditions have on the formation and migration of these bars. We present here the results of a study carried out in a swell canal under the effect of the size of sediments for identical hydraulic conditions [DUL 00]. The final bathymetric profile for each of these grain size distributions is presented in Figure 10.7.

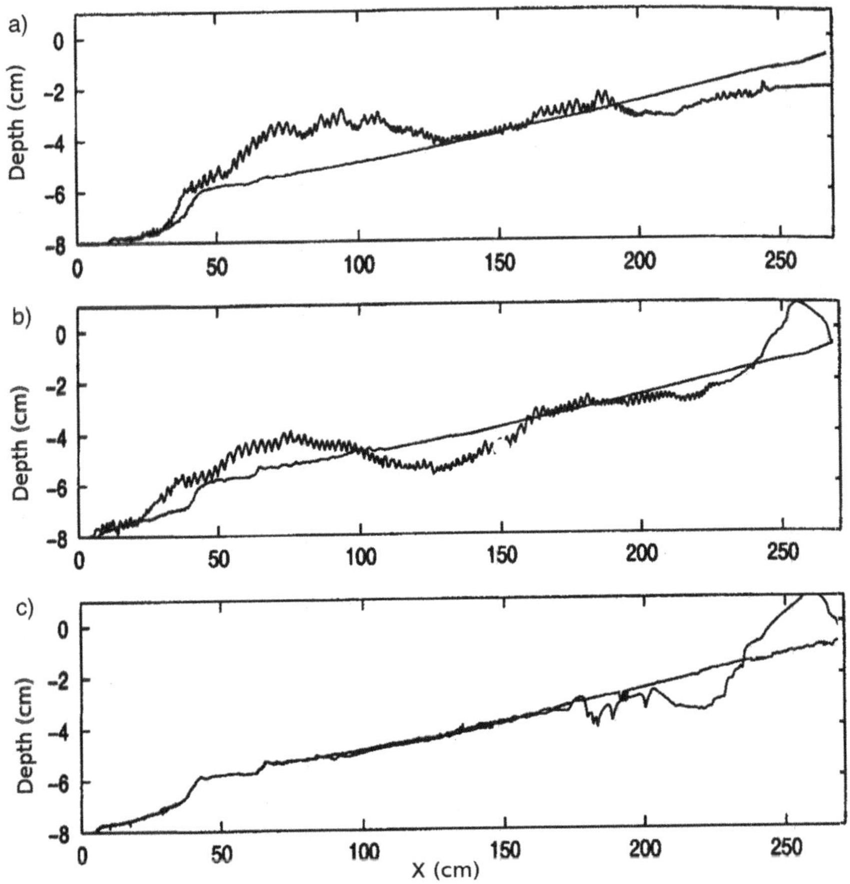

Figure 10.7. *Final bathymetric profiles for $D_l = 80\mu m$ and $\Psi_l = 34.3$: (a) $D_l = 120\mu m$ and $\Psi_l = 22.9$; (b) $D_l = 200\mu m$ and $\Psi_l = 13.7$; (c) $H_0 = 1.9 cm$, $H_b = 2.5 cm$; and $h_b = 2.9 cm$ [DUL 00]*

For a low-mobility number Ψ, the transport is made by bedload and in the direction of the propagation of the wave. The shoreline advances, and the beach is in accretion (Figure 10.7(c)). For a high-mobility number, the transport is mainly carried out by suspension, and a large bar forms offshore. The direction is thus opposite from the propagation of the swell, the shoreline recedes, and the beach is eroded (Figure 10.7(a)). In the case of an intermediary mobility number, the transport is mixed, the offshore bar is smaller, and the top of the beach is affected by accretion. These different modes of transport emphasize the influence of the average flows, which are induced by swell and the wave breaking. The particles can thus be transported in opposite directions according to the altitude at which they are dragged in the fluid. The coast is likely to become eroded in conditions of strong agitation, and to recompose itself in conditions of moderate agitation.

Therefore, the evolution of beaches, which regards an almost constant grain size distribution of sediments, is linked to the energy of the swells that determine the predominant types of transport. In the aim of studying the evolutions of the shoreline by behavioral models (see, for example, [VRI 93]), we can consider temporal scales that are higher than the temporal scales of storms.

10.4.2.3. *Bars induced by wave modulations*

The main idea here is to demonstrate, starting from modeling experiments, that the breaking of the waves may cause upstream wave modulations, whether it is through its "linear" effects (partially stationary wave), or through its nonlinear, wave–wave interactions, thus leading to bounded infra-gravitational waves. In what follows, we take the example of the interaction between a monochromatic wave and a bottom sediment.

The experimental studies presented here were carried out on the device presented in Dulou *et al.* [DUL 00]. The simultaneous evolutions of the surface wave and the bottom of the water are presented in Figures 10.8 and 10.9. The height of the generated wave is $H_0 = 1.35$ cm, and the sediment diameter $D = 80$ µm. The results are presented in more detail in Dulou *et al.* [DUL 00]. Initially, the wave is progressive; no energy can be observed in the second harmonic, and the breaking is regular, with an almost linear decrease in the height of the wave after the breaking point, situated at $x = 200$ cm. After one hour of the action of the swell, the bottom is rippled, the decrease in the wave height becomes less regular; furthermore, it is partially reflected toward deep water, probably by the breaking bar that it has formed. This reflection increases with time, and after three hours it reaches 15%; spaced sediment bars, which are half the length of the wave, are also observed (Figure 10.8). Furthermore, the energy in the second harmonic of the wave becomes significant. The evolution of the beach in the breaking area induces a decrease in the height of the irregular swell, showing the presence of two successive breakings.

After 12 hours of action, the "Bragg" bars have decreased in number and in amplitude, and a modulation on a larger scale becomes apparent. The breaking point cannot be localized as long as the wave is perturbed in that area. After 76 hours of action, the bars that are half the size of the wavelength gave way to bars of a much larger spatial expansion. Thus, in this study, we can observe the successive apparitions of the sediment bars that were caused by spatial modulations of the surface wave. The first modulation is due to the partially stationary character of the wave, reflected in the flooding bar (Figure 10.9). We can also note that this reflection is itself accentuated by Bragg resonance with the same bars. The second modulation, which appears over longer periods of time, corresponds to the oscillations generated by nonlinear wave–wave interactions.

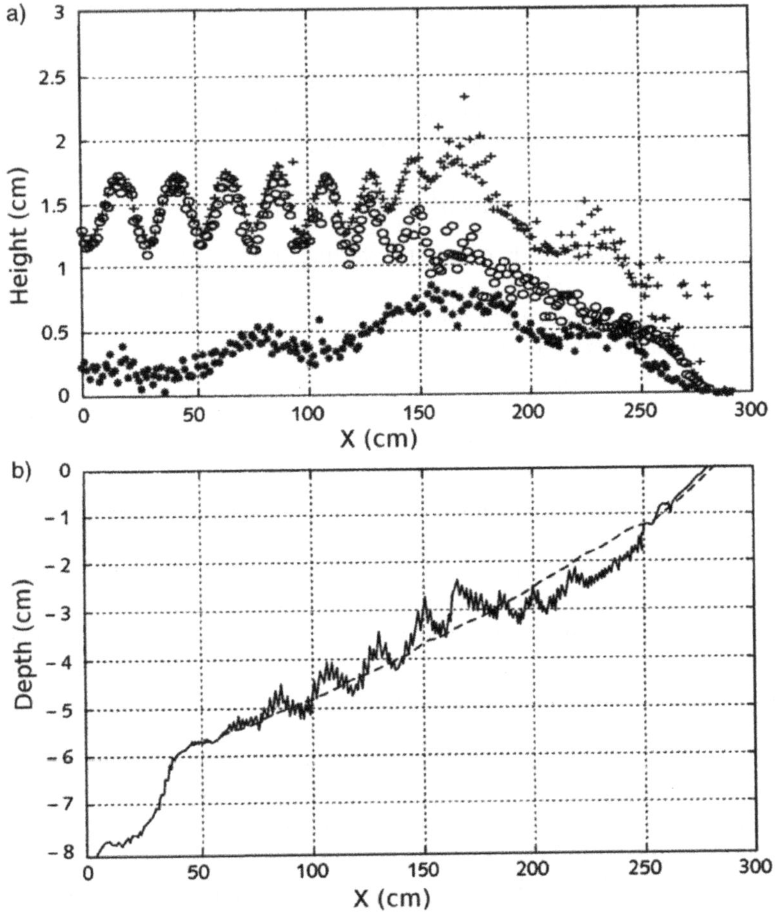

Figure 10.8. *(a) Envelopes of the total wave (+) and of the first (o) and second (*) harmonics at 12 hours above the bathymetric profiles (–) initial and (---) at 12 hours*

Therefore, we can find the two types of modulations observed outside the breaking area: the first modulation because of the partially stationary character of the wave and the second modulation caused by nonlinear interaction between a regular wave and a free wave. We can therefore conclude that, because of instability effects, in combination with the fact that the breaking point is not completely fixed, the breaking may cause nonlinear driving forces; in the beginning, these forces were concerned only with a monochromatic regular wave. These observations could lead us to conclude that we can generate infra-gravitational waves by a temporal evolution of the breaking point. Let us recall that the oscillation observed here is identified with a bounded subharmonic wave (or an infra-gravitational wave, since it is linked to low-frequency modulations).

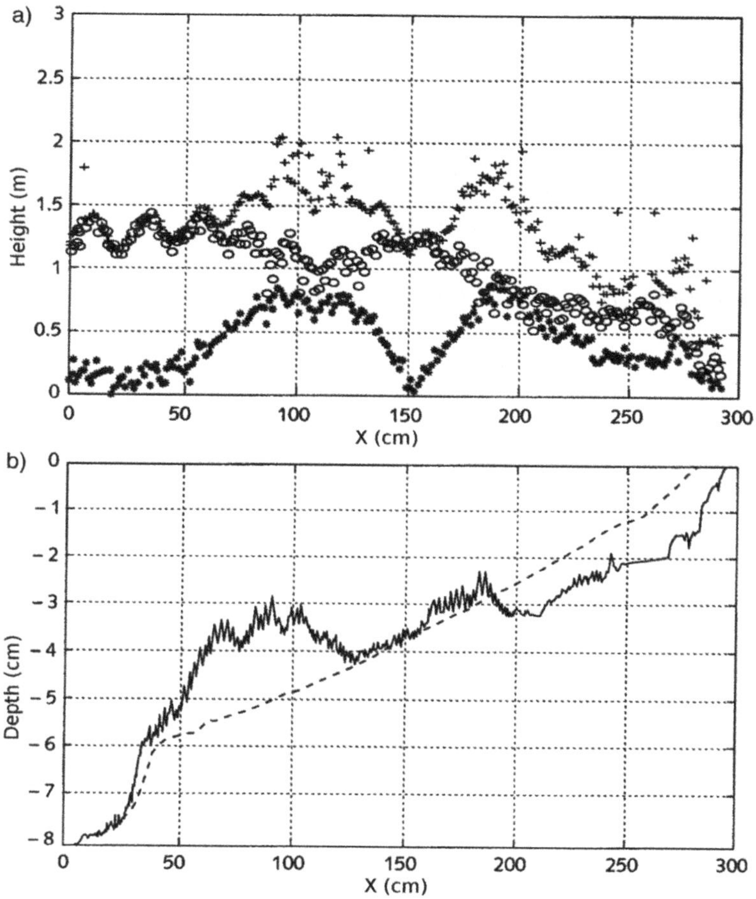

Figure 10.9. *(a) Envelopes of the total wave (+) and of the first (o) and the second (*) harmonics at 76 hours above the bathymetric profiles (–) initial and (---) at 76 hours*

10.4.2.4. *Sediment structures along the coast*

On-site studies show that the bathymetry can present modulations in the direction perpendicular to the coast. This is, for example, the case of Sète beach (see Figure 10.10), on the Mediterranean Sea, or the case of Omaha beach (see Figure 10.11) on the Normandy coast.

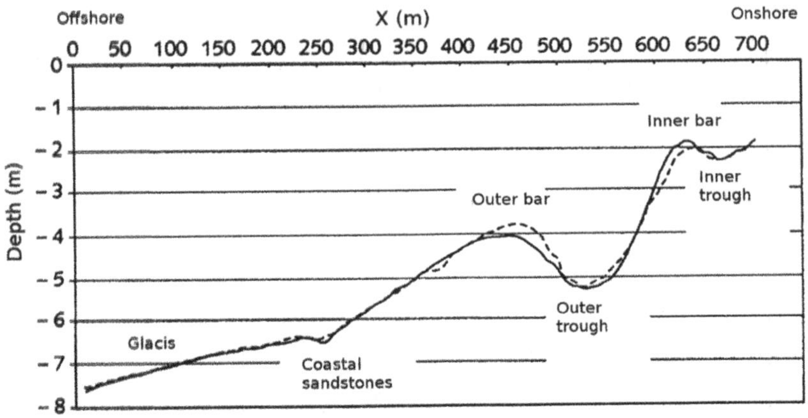

Figure 10.10. *Sète beach, beach profiles on 13 and 23 January 1999 [PNE 00]*

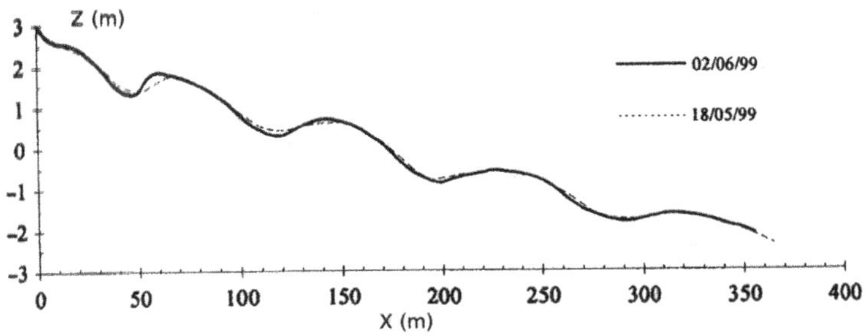

Figure 10.11. *Omaha beach, beach profiles on 18 May and 2 June 1999 [PNE 00]*

These structures can therefore be observed independently of the tide regime, i.e. microtidal in the Mediterranean and macrotidal along the Normandy coast. As mentioned above, their generation and migration are still very poorly understood, the attempts at explaining it can either advance the role of the waves from the point of view of their spatial modulations (subharmonic spatial modulations of incident waves by wave–wave interaction, partial reflection of infra-gravitational waves toward the high sea), or from an energetic point of view in the area of the breakers.

On Sète beach, for example, the internal bar – being more sensitive to the stormy episodes than the external bar, since they are situated at a lower depth – is thus erodible even by the moderate swells, whereas only the strong storms affect the external bar. At Omaha beach, the bars on the top of the beach are very mobile and may disappear, whereas the bars on the lower area of the beach are much more stable.

The beaches are not generally rectilinear and the bars rarely have, in reality, the regular profiles suggested by the ideal theories. More or less regularly spaced structures may be observed underwater and at the level of the shoreline. We still lack quantitative knowledge of the formation and the mobility of these morphological structures, despite the fact that numerous qualitative studies were dedicated to these structures.

The edge waves, the rip currents (narrow return currents which are intense and that traverse the surface area), or the processes of self-organization are frequently considered to be responsible for their formation, but no obvious conclusion can be reached explaining the formation of several types of observed rhythmical topographies, whose characteristic size may vary from several meters to several kilometers.

10.4.2.5. *Subtidal and intertidal bars*

The case of the bars formed parallel to the coast has been presented above. In fact, in reality, we can observe that the one-dimensional character of these structures is always only local, and starting from a certain spatial, observation scale along the coast, the sediment structure is two-dimensional. From a practical point of view, it is however useful to be able to isolate the perpendicular evolutions of the shore (cross-shore) and those along the shore (alongshore). This is often justified morphodynamically by much shorter evolution periods in the "cross-shore" direction than in the "alongshore" direction. However, because the rhythmical structures (or pseudoperiodical structures) observed along the shore are of a characteristic size comparable to the distance between the area of the breakers and the shore, a completely two-dimensional study should be carried out.

The two-dimensional submerged bars have shapes that depend essentially on incident swells, and on their direction. They are given in Figure 10.12 for these swells of normal and oblique incidence. The presence of rip currents attenuates the wave energy in these channels, whereas it breaks over the submarine structures, feeding the rip currents by the current generated in the breaking zone. This two-dimensional shape of the beaches appears to be quite stable until significant storms, through their energy and their chaotic aspect, make the profile of the beach uniform. Such structures are observed in France along the Aquitaine shoreline (see Figure 10.13); the rip currents are called "baïne" currents, the "baïnes" being the trench separating two sedimentary bars. These bars are strongly remodeled during winter storms, and they remain stable during the summer period.

10.4.2.6. *Shoreline: the formation of cusps*

At the beachface, we can frequently observe quasi-periodic structures on the shoreline in the shape of cusps. Their regularity allows us to define an average wavelength as well as amplitude, defined in the horizontal by the distance between its peak and its base (which can be "materialized" by the line traced by the swash). Their regular geometry is the reason why they are called "sand waves" by several authors. They are categorized according to their sizes, and called beach cusps, storm cusps, or giant cusps according to their characteristic size: 10, 100, or 1,000 m. They can move along the shore, regardless of their size, which may pose significant problems for development works or shore protection.

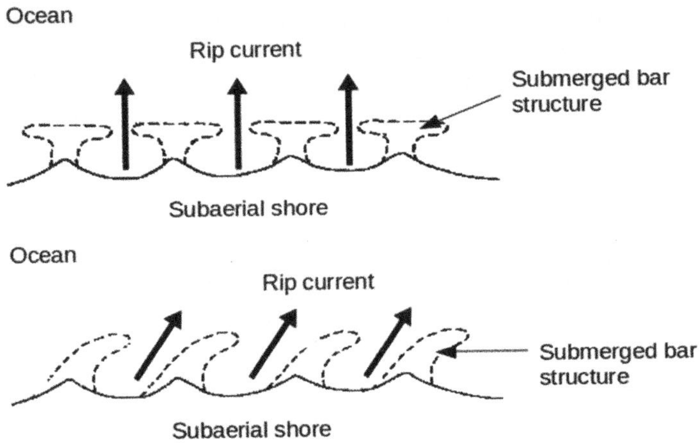

Figure 10.12. *Typical forms of undersea sediment bars after swells [COA 00]*

Figure 10.13. *Sediment structures revealed by the wave breaking on the shallow parts of the bathymetry, Truc-Vert, Aquitaine coast (courtesy of Natalie Bonneton)*

An example of the evolution of a shoreline on a Japanese beach is given in Figure 10.14 [SON 69]. The author emphasizes the possibility of superposing rhythmic systems of very different characteristic sizes (as we have observed for the bathymetry with the superposition of rips and bars) and, on the other hand, their evolution over time.

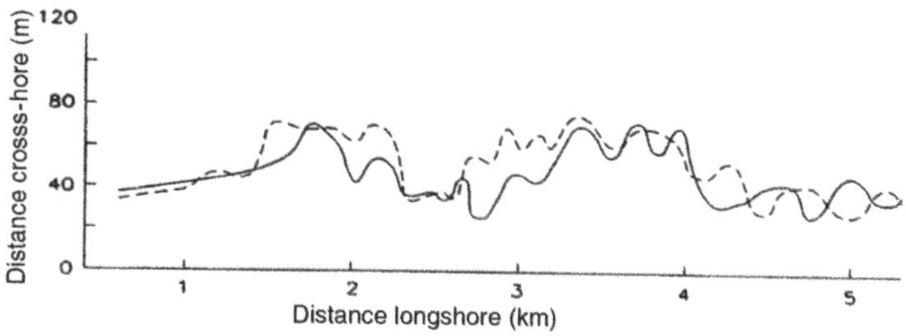

Figure 10.14. *Fluctuations of the coastline, Tokai beach, Japan [SON 69]. August (dotted line) and October (full line)*

These rhythmical structures of the shore are sometimes associated with the undersea bars in the shape of cusps (*crescentic bars*). One explanation for their existence, as was given by Bowen and Inman [BOW 71], is the presence of stationary edge waves, of a pulsation ω and with a wave number k_n:

$$\eta(x,y,t) = \sum_n a_n L_n(2k_n x)e^{-k_n x}\cos(k_n y)\cos(\omega t) \qquad [10.22]$$

Equation [10.22] reveals for every mode n several nodes (or antinodes) spaced by $\lambda_n / 2 = \pi / k_n$. Their presence can also account for the presence of rhythmical structures along the beach (sand waves). Let us bear in mind that the on-site observations equally show the presence of swash currents, which are stronger at the trough of the sand waves; this, in turn, leads to more caution in interpreting the phenomenon only by the action of the wave, knowing that the instantaneous currents induced by this wave remain very inferior to those of the swash. However, the spatial modulations of the currents caused by the edge waves follow the direction of the formation of these sedimentary structures.

10.5. Local effects: scouring around the structures

Even the presence of artificial structures such as pipelines, piers, and offshore or shoreline structures may lead to erosion and accretion phenomena around their immediate vicinity. The quantitative studies on these scouring phenomena are mainly experimental, the results are often expressed according to the characteristic sizes of the obstacles, and to the non-dimensional numbers that characterize the flows.

10.5.1. *Scouring around the pipelines*

The scouring problems around the pipelines play a significant role in their design because of the resonant oscillations that may appear under the action of the movement of the surrounding fluid. Studies were recently carried out on the effects of the scouring of pipelines, as well as on their protection against swells and currents – regardless whether this phenomenon is natural or imposed around their installations.

In shallow water, the drag and lift forces are quite significant. The flow that oscillates around the pipeline can be characterized by the next two dimensionless numbers:

$$Re = \frac{U_m D_p}{\nu} \text{ (Reynolds number concerning the pipeline)} \qquad [10.23]$$

$$KC = \frac{U_m T}{D_p} \text{ (Keulegan–Carpenter number)} \qquad [10.24]$$

where U_m is the orbital velocity of the fluid above the bottom, $D_p = 2R$ the diameter of the pipeline, ν the kinematic viscosity of the fluid, and T the wave period. The parameter that must be determined is the maximum scouring depth $S = (H - R)$ underneath the pipelines (H is the maximum scouring depth found at the center of the pipeline). Several studies tend to show that the relation between this depth and the diameter of the pipeline depends only on the Keulegan–Carpenter number KC [SUM 90] according to which $S/D_p = 0.1\ KC^{0.5}$. More recent studies have shown that they also depend on the frequency parameter $\beta = Re/KC = D_p^2 T/\nu$ [CEV 97].

As shown in Figure 10.15, the scouring depth can be the size of the diameter of the pipeline. Its dependence of KC and β is given by the empirical formulas $S/D_p = 0.045\ KC^{0.76}$ and $S/D_p = 15.79\ \beta^{-0.55}$. Under certain conditions of waves and currents, the pipelines can sand silt by themselves. Certain geometries (the presence of fins) were tested for increasing this sand silting [HUL 84]. The goal is a better stability of the pipelines at the bottom. The critical sand silt depth e_{crit} was studied by Sumer and Fredsoe [SUM 91]. It depends only on the Keulegan–Carpenter number $e_{crit} = 0.1 \ln(KC)$ for regular waves.

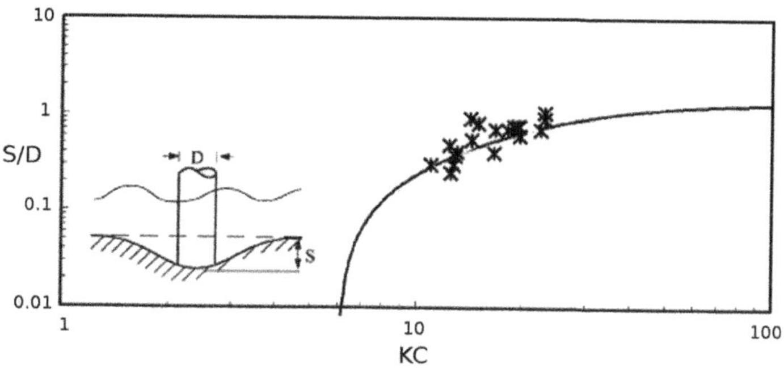

Figure 10.15. *Maximum scouring depth around a pipeline [SUM 91]*

10.5.2. *Scouring around vertical cylinders*

The presence of vertical cylinders generates eddies, which set the sediments into motion. The scouring around the structures under the action of the swell depends on their size, compared with the wavelength of the swell. The obstacles are of a "large size" if their diameter is higher than a tenth of the wavelength or else they are qualified as "small size". Therefore, scouring depends on the Keulegan–Carpenter number. The shape of the cylinders has an insignificant influence on scouring: a column that has an angular section has the same effect as a circular cylinder of a section of the same surface. These geometry effects are detailed in Herbich [HER 91]. Experiments carried out on scouring around structures for a flow that oscillates have shown that after a transition period during which the scouring is initiated on the lateral sides of the cylinder and a sand silt forms both upstream and downstream, a state of dynamic equilibrium is reached. It has the shape of an inverse cone for a low *KC*, of a slope that is equal to the deposit angle of the sediments.

10.5.3. *Scouring upstream from nearshore structures*

Shoreline structures can cause the waves to break, and/or reflect a significant part of their energy. In the vicinity of these structures, the scouring is sometimes significant, particularly because of the turbulent character of the flow. The maximum depth of the scouring can be estimated experimentally according to the number of wave periods (i.e. the time) for different slopes of the structure.

We show that the influence is more important for the low slopes (<30°) than for the steep slopes, for which the reflection of the structure remains almost constant. The maximum depth of scouring is generally around $\lambda/4$. Far upstream from the structure, and even if the waves do not break, several modulations will emerge. The accretion and erosion regions are linked to the position of nodes and antinodes of the partially stationary wave. The sediment bars form at the level of the nodes when the bedload transport is predominant, and at the level of the antinodes when the suspension is predominant. These modulations of spatial periodicity, which is around half of the wavelength of the surface wave, increase the reflection of the wave toward the high sea via Bragg resonance, as shown in Figure 10.16. These modulations were studied extensively because they contribute to the protection of the shoreline, through their capacity to partially reflect incident swells (see [REY 95] for detailed information).

Figure 10.16. *Scouring around a vertical cylinder (V.Rey, BGO)*

10.6. Long-term morphodynamics of beaches

For calculating the evolution of bathymetries over weeks, months, or years, the hydrodynamic models do not allow us to correctly predict these evolutions; on the one hand, that is because of the numerical limitations, and on the other hand, it is due to physical parameters (winds, waves, currents, etc.) that are not sufficiently known and/or physically modeled extensively. We can then turn to behavioral models, which temporally "integrate" the short-term processes in the balance equations for the conservation of the mass of sediments.

10.6.1. *Evolution models of the profile of the beaches*

The medium-term evolutions (a few days) mainly regard the direction perpendicular to the shore (beach profile) and are largely dominated by the action of the different swell regimes. The sediment transport is not homogeneous, and the result is a slow evolution of the profile. This evolution can be modeled by a diffusion equation:

$$\frac{\partial y}{\partial t}-\frac{\partial}{\partial z}\left[D(z)\frac{\partial y}{\partial z}\right]=S(x,z,t) \qquad [10.25]$$

where y is the translation of the direction perpendicular to the profile shore in the shore z in relation to a balance profile. $S(x,z,t)$ is a source term, which depends and represents the intakes or the losses due to longitudinal transport. The expression of the diffusion coefficient $D(z)$ often remains difficult to establish.

This type of model does not seek to detail the physical processes that come up in the shoaling area (amplification of the wave through bathymetric effect before breaking), breaking area and on the beach setup (sheared currents, sediment transport). It simply "models" the evolution of a beach profile subjected to these driving forces, particularly the return to a state of balance after a stormy episode. Figure 10.17 illustrates a typical profile of normal balance, and a typical profile of a storm.

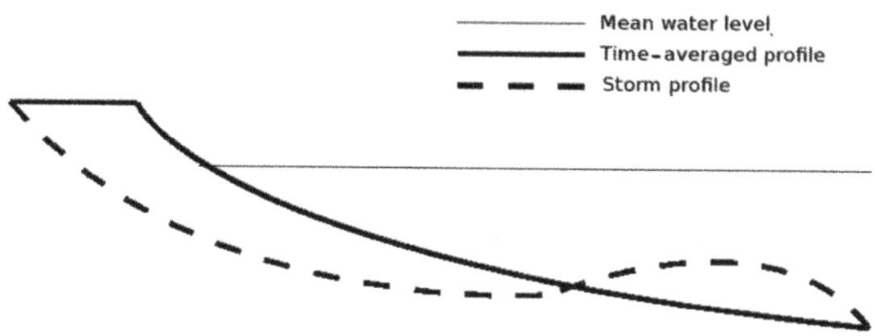

Figure 10.17. *Typical profiles of a natural beach*

Bruun [BRU 54] has actually shown that the beaches presented a balance profile of the form $h(x)=Ax^{2/3}$ where h represents the depth of the water at a distance x from the shore. The constant A depends on D, sediment diameter, as well as on the wave characteristics that we can define starting from the dimensionless parameter $H_b/(W_sT)$ where H_b is the wave height at the breaking point, T the wave period, and W_s the settling velocity of the sediments.

In practice, the low slope profiles correspond to large waves with short periods, as well as to small particles; the profiles of an abrupt slope correspond to small waves with long periods, as well as to coarse sediments.

The major interest of the hypothesis of this balance profile is that it allows us to treat the longitudinal transport (alongshore) in a manner that is separated from the transport that is perpendicular to the shore (cross-shore); this is justified by the fact that the characteristic evolution times along the shore are much higher than those associated with the cross-shores processes.

However, this still remains a hypothesis, supposing, for example, that there is enough sand on the dune, that the shoreline is homogeneous enough, etc.; and also, as any other hypothesis, it can be proven wrong by development works such as dikes or slopes built on the seafront.

10.6.2. *Shoreline evolution models*

The behavior of the coast is the result of the large number of processes and mechanisms that act and interact on different time and spatial scales [VRI 93]. Calculation of the evolution of the position of the coast line can be carried out following two approaches [VRI 91]:

– a medium-term approach that uses the wave data, which is either measured or calculated by numerical models. One equation for the conservation of the sediment mass is completely solved, since the sediment transport is known. We particularly show that the conservation equation can be reduced to a diffusion equation for the oblique swells with a low incidence [PEL 56];

– a more "long-term" approach, i.e. for the periods of time where the wave data and particularly the data on bathymetric evolutions are absent, but where the evolution of the shoreline over several decades can be known (satellite images or aerial photography). By analogy with the classical conservation equations, we may suppose that the conservation equation of the sediment mass can be described using a diffusion equation or, more generally, a diffusion–advection equation [BER 00]. The coefficients can then be calculated using an interpolation method or an assimilation method [DEV 98], and the future evolution of the shoreline can be projected from historical data.

The evolution model of the shoreline uses an equation based on the conservation of the volume of the sediment. We use an orthogonal direct Cartesian mark where the axis x is directed parallel to the shoreline (we assimilate the curvilinear abscissa s of the shoreline to the axis x), the y axis is directed toward the high sea, representing the position of the shoreline (see Figure 10.18).

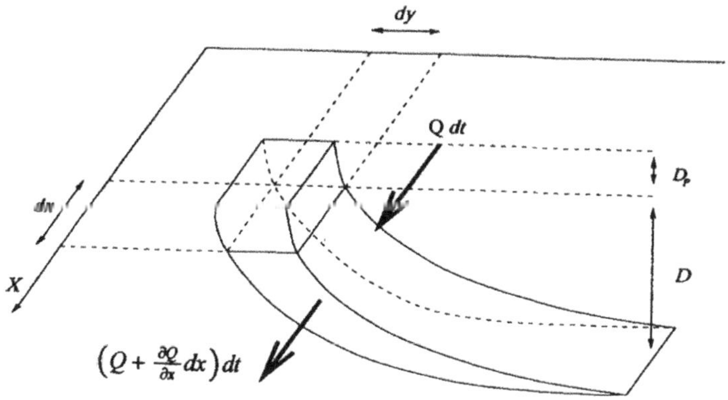

Figure 10.18. *A perspective view at beach profiles*

We may suppose that the beach profile is in balance (Bruun profile), and that it can support only translation movements following the axis *Oy*. The variation in the position of the shoreline *dy*, of a *dx* section of the beach, during an interval *dt* is equal to the difference between the input quantity of the sediment, hence the following conservation equation:

$$\frac{\partial y}{\partial t} = -\frac{1}{h_s}\left(\frac{\partial Q}{\partial x}\right) + S(x,t) \qquad [10.26]$$

where $h_s = h + h_p$, with h and h_p being the closure depth beyond which the effect of the sediment transport on the morphodynamics is considered to be insignificant, and the height of the beach where the sediment transport actually takes place, and Q is the sediment flow along the coast. $S(x,t)$ represents the source terms (bringing or extracting sediments), or still the fact that the beach profile can evolve (see previous section, storm profile and balance profile).

Having studied that data on beach profiles subjected to waves, Hallermeier [HAL 81] has suggested for h the following expression:

$$h = H_0\left(2.28 - 10.9\frac{H_0}{L_0}\right) \qquad [10.27]$$

where H_0 and L_0 are the height and the wavelength of the swell at sea.

The transport Q depends on the characteristics of the breakers. In this study, we limit ourselves to the transformation of the swells in a shoaling area, for regular and slowly variable sea bottoms. The transformation model of the swell in finite depth is a refraction mode; for linear waves (Airy waves), it is based on the theory of rays. By taking into account the shoaling, the height of the waves H_B at the breaking point is given by:

$$H_B = K_R K_S H_0 \text{ with } K_R = \sqrt{\frac{\cos(\alpha_0)}{\cos(\alpha_b)}} \text{ and } K_S = \sqrt{\frac{C_{g0}}{C_{gB}}} \qquad [10.28]$$

where K_R and K_S are the refraction and shoaling coefficients. α_0, α_b and C_{g0}, C_{gB} are the incidence angles and the group velocities of the offshore and breaking waves, respectively. The angle α_b is obtained starting from the Descartes–Snell law. The breaking point corresponds to the water depth h_B, calculated from the breaking criterion. For the more complex bathymetries, refraction–diffraction models are the most widely used.

Quite a simple expression of the sediment transport Q was proposed by the Shoreline Engineering Research Center [COA 84]:

$$Q = (H_B^2 C_{gB}) a_1 \sin(2\alpha_{bs}) = Q_0 \sin(2\alpha_{bs}) \qquad [10.29]$$

where H_B is the significant height of the waves at the breaking point, C_{gB} the group velocity of the waves at the breaking point, and α_{bs} the local angle between the wave crests and the shoreline at the breaking point. The coefficient a_1 depends on the physical characteristics of the water and the sediments:

$$a_1 = \frac{K_1}{16r(1-p)(1.416)^{\frac{5}{2}}}, \quad r = \frac{\rho_s}{\rho} - 1 \qquad [10.30]$$

where ρ_s is the volume mass of the sediment ($\rho_s = 2{,}650$ kg/m^3), ρ the volume mass of the water ($\rho = 1{,}003$ kg/m^3), and p the porosity of the sediment at bottom level (settled at $p = 0.4$). The coefficient K_1 is empirical and it is used for adjusting the model.

For the evolutions of beaches in the presence of shoreline structures (dikes, piers), a more complex formulation in the aim of accounting for the diffraction of waves behind the structures was proposed by [HAN 89]:

$$Q = H_B^2 C_{gB} \left[a_1 \sin(2\alpha_{bs}) - a_2 \cos(\alpha_{bs}) \frac{\partial H_b}{\partial x} \right] \qquad [10.31]$$

where tan β is the average slope of the beach profile and

$$a_2 = \frac{K_2}{8r(1-p)\tan\beta(1{,}416)^{\frac{7}{2}}} \quad [10.32]$$

If we limit ourselves to the equation [10.29], the sediment flux is of the form

$$Q = Q_0 \sin(2\alpha_{bs}) \quad [10.33]$$

The incidence angle at the breaking point is given by

$$\alpha_{bs} = \alpha_b - \alpha_s = \alpha_b - \tan^{-1}\left(\frac{\partial y}{\partial x}\right) \quad [10.34]$$

where α_b is the incidence angle of the swell in relation to the axis of x and α_s is the angle between the coastline and the axis x. For the beaches with a low slope, the angle α_{bs} is weak, and we may deduce from it the equations [10.33] and [10.34]:

$$Q = 2Q_0\left(\alpha_b - \frac{\partial y}{\partial x}\right) \quad [10.35]$$

If we ignore the source terms, and we assume the long shore transport Q_0 and the angle α_{bs} as being independent of x and t, having used the equation [10.35], the equation [10.26] becomes a diffusion equation:

$$\frac{\partial y}{\partial t} - a\frac{\partial^2 y}{\partial x^2} = 0 \quad [10.36]$$

More generally, this evolution can be considered to pursue a law of the advective–diffusive type. This is particularly what we find in the academic case of waves having larger height in the offshore advances of the shoreline [BER 00]. Under these conditions, the conservation conditions become:

$$\frac{\partial y}{\partial t} + U\frac{\partial y}{\partial x} = a\frac{\partial^2 y}{\partial x^2} \quad [10.37]$$

10.7. Conclusion

The aim of this chapter was to present the sediment transport and the associated morphodynamics in nearshore areas. It cannot be exhaustive since the problematic is so large and complex. Rather, this chapter aimed at giving a description of the approaches and formulations that are generally used, by underlining the more significant aspects that bear not only on the modeling of sediment transport but also on the morphological evolution of the seabed and the beaches.

More specifically, the difficulty with modeling the transport, be it physical or numerical, comes from the two categories of sediments that have very different properties (cohesive and non-cohesive), and the two modes of transport, suspension and bedload – the former taking place all over the entire water column, and the latter in the boundary layer of the bottom. The sediment transport formulas proposed in the relevant literature are numerous and they lead to very different results. This proves the complexity of the problem and its sensitivity to hydrodynamic parameters and sediment parameters, which are represented and considered in a very diverse manner.

The use of physical models (tests carried out in canals and basins), widely used for establishing not only erosion laws and transport laws (setting into motion thresholds, maximum scouring depth, bedload transport, etc.) but also for studies of morphological evolution must also be treated cautiously. These model studies are often characterized as not respecting the conditions of hydrodynamic similitude (the Froude number is most often preserved and is preferred over the Reynolds number) and by a distortion of the scale for the grain size distributions (for example, a weaker grain size distribution scale for maintaining the conditions of non-cohesive particles). Despite their potential inconveniences, the physical models are still an indispensable tool for studying the impact that hydraulic structures might have on sediment dynamics, whether in the vicinity of the structures or at a considerable distance.

Finally, whereas sediment transport can be observed on a hydrodynamic scale, the morphological evolutions become slower as their spatial scale grows larger, and as their transport gradients get lower. For this reason, the studies carried out on erosion/accretion phenomena most often turn to integrated models of sediment mass conservation rather than to the balances of sediment transport, forced by hydrodynamic models.

10.8. Bibliography

[BAG 46] BAGNOLD R.A., "Motion of waves in shallow water. Interaction between waves and sand bottoms", *Proceedings of the Royal Society of London, Series A*, vol. 187, no. 1008, pp. 1–18, 1946.

[BAG 66] BAGNOLD R.A., *An Approach to the Sediment Transport Problem from General Physics*, United States Department of the Interior, U.S. Geological Survey, 1966.

[BAI 81] BAILARD J.A., "An energetics of total load sediment transport model for plane sloping beach", *Journal of Geophysical Research*, vol. 86, pp. 10938–10954, 1981.

[BAL 04] BALDOCK T.E., TOMKINS M.R., NIELSEN P., HUGHES M.G., "Settling velocity of sediments at high concentrations", *Coastal Engineering*, vol. 51, pp. 91–100, 2004.

[BAT 67] BATCHELOR G.K., *An Introduction to Fluid Dynamics*, Cambridge University Press, 1967.

[BER 00] BERTHOT A., REY V., FRAUNIÉ P., "Modélisation de l'évolution long terme du traît de côte", *VIèmes Journées Nationales Génie Civil-Génie Côtier*, 17–19 May 2000.

[BIJ 71] BIJKER E.W., "Longshore transport computations", *Journal of Waterways, Harbour and Coastal Engineering*, vol. 99, 1971.

[BOW 71] BOWEN A.J., INMAN D.L., "Edge waves and crescentic bars", *Journal of Geophysical Research*, vol. 76, no. 36, pp. 2569–2577, 1971.

[BRU 54] BRUUN P., Coast erosion and the development of beach profiles, PhD Thesis, Beach Erosion Board US Army Corps of Engineers, Waterways Experiment Station, Vicksburg, MS, 1954.

[CEV 97] CEVIK E.O., YUKSEL Y., "Local scour around pipelines under wave action", *Proceedings of the Seventh International Journal of Offshore and Polar Engineering*, 25–30 May 1997.

[CHH 99] CHHABRA R.P., AGARWAL L., SINHA N.K., "Drag on non-spherical particles: an evaluation of available methods", *Powder Technology*, vol. 101, 1999.

[COA 84] COASTAL ENGINEERING RESEARCH CENTER, *Shore Manuel Protection,* 4th ed., U.S. Army Corps of Engineers, Washington, 1984.

[COA 98] COASTAL ENGINEERING RESEARCH CENTER, *Coastal Engineering Manual*, U.S. Army Corps of Engineers, Washington, 1998.

[COA 00] COASTAL ENGINEERING RESEARCH CENTER, *Coastal Engineering Manual*, U.S. Army Corps of Engineers, Washington, 2000.

[DEV 98] DEVENON J.L., FRAUNIÉ P., REY V., "Une méthode d'assimilation de données pour identifier le coefficient de diffusion d'un modèle d'évolution long terme de traît de côte", *Vèmes Journées Nationales Génie Civil-Génie Côtier*, 13–15 May 1998.

[DUF 08] DUFOIS F., Modélisation du transport particulaire dans le Golfe du Lion en vue d'une application au devenir des traceurs radioactifs issus du Rhône, PhD Thesis, Sud Toulon-Var University, 2008.

[DUL 00] DULOU C., REY V., BELZONS M., "Etude expérimentale de l'évolution spatiotemporelle du couplage onde-fond sédimentaire en zone de déferlement", *Vèmes Journées Nationales Génie Civil-Génie Côtier*, 17–19 May 2000.

[DYE 86] DYER K.R., *Coastal and Estuarine Sediment Dynamics*, John-Wiley, 1986.

[EIN 05] EINSTEIN A., "On the motion of small particles suspended in liquids at rest, required by the molecular-kinetic theory of heat", *Annalen der Physik*, vol. 17, pp. 49–560, 1905.

[ENG 67] ENGELUND F., HANSEN E., *A Monograph on Sediment Transport*, Technisk Forlag, Copenhague, Denmark, 1967.

[FRE 92] FREDSOE J., DEIGAARD R., *Mechanics of Coastal Sediment Transport*, World Scientific, Singapore, 1992.

[GUY 92] GUYON E., HULIN J.P., PETIT L., *Hydrodynamique Physique*, Editions du CNRS, 1992.

[HAL 81] HALLERMEIER H., "A profile zonation for seasonal sand beaches from wave climate", *Coastal Engineering*, vol. 4, pp. 253–277, 1981.

[HAN 89] HANSON H., KRAUS N.C., GENESIS: Generalized model for simulating shoreline change, Technical Report CERC, 1989.

[HER 91] HERBICH J.B., *Handbook of Coastal and Ocean Engineering*, vol. 2, Gulf Publishing, 1991.

[HER 96] HERZHAFT B., GUAZZELLI E., MACKAPLOW M.B., SHAQFEH E.S.G., "Experimental investigation of the sedimentation of a dilute fiber suspension", *Physical Review Letters*, vol. 77, pp. 290–293, 1996.

[HOR 88] HORIKAWA, *Nearshore Dynamics and Coastal Processes*, Tokyo Press, 1988.

[HUL] HULSBERGEN C.H., "Stimulated self-burial of submarine pipelines", *16th Annual OTC in Houston*, 7–9 May 1984.

[LAU 73] LAU J., TRAVIS B., "Slowly varying stokes waves and submarine longshore bars", *Journal of Geophysical Research*, vol. 78, no. 21, pp. 4489–4497, 1973.

[LIU 57] LIU H.K., "Mechanics of sediment ripple formation", *Journal of Hydraulic Engineering, ASCE*, vol. 83, no. HY2, 1957.

[MAN 55] MANOHAR M., "Mechanics of bottom sediment movement dur to wave action", *U.S. Army Corps of Engineers, B.E.B Tech. Memo.*, vol. 75, p. 121, 1955.

[MAS 03] MASSELINK G., HUGHES M.G., *An Introduction of Coastal Processes and Geomorphology*, Oxford University Press, 2003.

[MCN 50] MCNOWN J.S., MALAIKA J., "Effects of particle shape on settling velocity at low Reynolds numbers", *Transactions of American Geophysical Union*, vol. 31, no. 1, pp. 74–83, 1950.

[MIG 68] MIGNIOT F., "Etude des propriétés physiques de différents sédiments trés fins et de leur comportement sous des actions hydrodynamiques", *La Houille Blanche*, vol. 7, pp. 591–620, 1968.

[MIG 77] MIGNIOT F., "Action du vent, du courant et de la houle sur les sédiments", *La Houille Blanche*, vol. 1, pp. 9–47, 1977.

[NIE 89] NIELSEN P., *Coastal Sediment Transport*, World Scientific Publishing, Singapore, 1989.

[NIE 92] NIELSEN P., *Coastal Bottom Boundary Layers and Sediment Transport*, World Scientific Publishing, Singapore, 1992.

[OST 93] OSTENDORF D.W., MADSEN O.S., "Sand bar evolution beneath partially standing waves: laboratory experiments and model simulations", *Continental Shelf Research*, vol. 13, no. 11, pp. 1149–1181, 1993.

[PEL 56] PELNARD-CONSIDERE R., "Essai de théorie de l'évolution des formes de rivage des plages de sable et de galets", *4èmes Journées de l'hydraulique, les énergies de la mer, question III*, 1956.

[REY 91] REY V., Propagation d'ondes de gravité au dessus de fonds solides ou constitués de sédiments: Application à l'interactions dynamiques ondes-sédiments, PhD Thesis, Thèse de l'Université de Provence, 1991.

[REY 95] REY V., DAVIES A.G., BELZONS M., "On the formation of bars by the action of waves on an erodible bed: a laboratory study", *Journal of Coastal Research*, vol. 11, no. 4, pp. 1180–1194, 1995.

[RIC 54] RICHARDSON J.F., ZAKI W.N., "Sedimentation and fluidisation: part 1", *Transactions of the Institution of Chemical Engineers*, vol. 32, pp. 35–53, 1954.

[RIJ 85] VAN RIJN L.C., "Sediment transport in combined currents and waves", *Euromech 262 – Sand Transport in Rivers, Estuaries and the Sea*, 1985.

[RIJ 89] VAN RIJN L.C., *Handbook of Sediment Transport by Current and Waves*, Delft Hydraulics, Delft, the Netherlands, 1989.

[RIJ 07a] VAN RIJN L.C., "Unified view of sediment transport by current and waves. I: initiation of motion, bed roughness and bed-load transport", *Jorunal of Hydraulic Engineering*, vol. 133, no. 6, 2007.

[RIJ 07b] VAN RIJN L.C., "Unified view of sediment transport by current and waves. II: suspended transport", *Journal of Hydraulic Engineering*, vol. 133, no. 6, 2007.

[SIM 65] SIMONS D.B., "Open channel flow", in CHORLEY R.G. (ed.) *Introduction to Fluvial Processes*, Methuen, London, 1965.

[SON 69] SONU C.J., "Collective movement of sediment in littoral environment", *Proceedings of Eleventh Conference of Coastal Engineering*, 1969.

[SOU 97] SOULSBY R., *Dynamics of Marine Sands*, Thomas Telford, London, 1997.

[SUM 90] SUMER B.M., FREDSOE J., "Scour below pipelines in waves", *Journal of Waterway, Port, Coastal and Ocean Division*, vol. 116, no. 3, 1990.

[SUM 91] SUMER B.M., FREDSOE J., "Onset of scour below a pipeline exposed to waves", *Proceedings of the International Journal of Offshore and Polar Engineering*, vol. 1, no. 3, 1991.

[VRI 91] VRIEND H.J.D., "Mathematical modelling and large-scale coastal behaviour", *Journal of Hydraulic Research*, vol. 39, no. 6, 1991.

[VRI 93] VRIEND H.J.D., CAPOBIANCO M., CHESTER T., SWART H.E.D., STIVE M.J.F., "Approches to long-term modelling of coastal morphology: a review", *Coastal Engineeging*, vol. 21, 1993.

[WRI 84] WRIGHT L.D., SHORT A.D., "Morphodynamic variability of surf zones and beaches: a synthesis", *Marine Geology*, pp. 93–118, 1984.

[YU 00] YU J., MEI C.C., "Formation of sand bars under surface waves", *Journal of Fluid Mechanics*, pp. 315–348, 2000.

List of Authors

Muriel AMIELH
CNRS
Marseille
France

Fabien ANSELMET
Ecole Centrale de Marseille
France

Pascale AUSSILLOUS
Aix-Marseille University
France

Bruno ANDREOTTI
Paris Diderot University and ESPCI
France

Rémi BÉGUIN
GeophyConsult
Chambéry
France

Nadia BENAHMED
Irstea
Aix en Provence
France

Ahmed BENAMAR
Le Havre University
France

Stéphane BONELLI
Irstea
Aix en Provence
France

Patrick BOYER
IRSN
Cadarache
France

Bernard CAMBOU
Ecole Centrale de Lyon
France

Christophe CHEVALIER
IFSTTAR
Paris
France

Philippe CLAUDIN
CNRS and ESPCI
Paris
France

Yves-Henri FAURE
University of Grenoble
France

Jean-Jacques FRY
EDF-CIH
Chambéry
France

Frédéric GOLAY
Sud Toulon Var University
France

Elisabeth GUAZZELLI
CNRS
Marseille
France

Cyril GUIDOUX
GeophyConsult
Chambéry
France

Didier MAROT
University of Nantes
Saint-Nazaire
France

Fabienne MERCIER
Irstea
Aix en Provence
France

Yannick PEYSSON
IFP Energies Nouvelles
Rueil Malmaison
France

Pierre PHILIPPE
Irstea
Aix en Provence
France

Nadège REBOUL
University of Lyon I
France

Vincent REY
Sud Toulon Var University
France

Paul ROYET
Irstea
Aix en Provence
France

Damien SOUS
Sud Toulon Var University
France

Fabien TERNAT
Aix-Marseille University
France

Eric VINCENS
Ecole Centrale de Lyon
France

Index

A, B

advection, 67, 194, 202, 290, 326, 335
bar, 248, 337, 340, 341, 343-345, 348, 349, 351, 354
bedload transport, 128, 144, 187, 254, 265, 269, 276, 334-336, 340, 341, 354, 360

C

clogging, 40, 44, 58-61, 63, 71, 73-75, 110, 129, 134, 145, 149, 150, 255
coefficient
  diffusion, 64, 67, 234, 288, 326, 335, 355
  friction, 31, 137, 224, 227, 243, 257, 258, 266, 270, 271, 276, 328, 332
  of erosion, 17, 34, 35
cohesive sediments, 254, 281, 291, 292, 306, 318, 326-328, 330-333
compaction, 6, 7, 14, 17, 180, 260, 290, 293, 295-297, 299, 301, 312, 313, 318, 331
confinement, 43, 51, 271
constriction, 59, 63, 82-85, 87, 89, 91-98, 101, 103, 105-108, 111, 125, 142
cracks, 7-10, 13-15, 25, 33, 81
current, 165, 169, 267, 307, 335, 337, 339, 340, 349

D

dam, 4-11, 16, 20, 32, 113, 143, 182, 313
diluted suspension, 193
discrete element, 84, 85
dune, 237, 240, 241, 246, 356

E

embankment, 1, 2, 5, 8, 35, 115, 143
energy, 2, 13, 25, 26, 31, 33-35, 55-58, 75, 85, 181, 204, 205, 229, 254, 281, 332, 339, 343, 345, 349, 353
erosion
  contact, 12, 13, 21, 22, 25, 32, 115, 120, 188
  surface, 19, 44, 45, 51-55, 58, 120-123, 128, 130, 141, 144, 145, 148, 152, 203, 310
erosion kinetics, 11, 17, 170, 172, 201-203, 207-209

F

fictitious domain, 206, 207, 211, 213, 214, 219
finite element, 107
finite volume method, 207
flood, 81, 117, 118, 183, 314, 345
flow
  laminar, 187, 188, 194, 201, 202, 207, 210-212, 241, 261, 265-269, 275
  local, 256, 336, 339
  turbulent, 188, 194, 202-204, 207-210, 219, 224, 241, 265, 266, 328, 332, 334
friction, 14, 17, 25-28, 30, 31, 35, 91, 123, 135-137, 140, 147, 177, 224, 227, 243, 257, 258, 263, 270, 271, 275, 276, 285, 292, 309, 310, 214, 319, 328, 332, 333, 337
  angle, 140, 147, 258

G, H

granulometry, 286, 296, 297, 333
hole, 10, 11, 17, 19, 20, 35, 51, 52, 145, 156-158, 162-167, 169, 179, 183, 184, 216, 218, 219
Hole Erosion Test (HET), 17, 19, 34, 35, 51, 52, 145, 156, 157, 160, 165, 170, 178, 179, 182-184, 216, 219
hydraulic fracture, 10, 155

I, J

instability, 3, 41, 50, 59, 99, 240, 245-248, 346
interstitial pressure, 32-34, 59, 75
Jet Erosion Test (JET), 19, 34, 215, 216, 219

L, M

law of erosion, 16, 17, 35, 107, 158, 189, 207, 291
migration, 33, 44, 60, 67, 99, 336, 337, 343, 348
morphodynamic, 336

N

number
  Froude, 136, 337, 360
  Mobility, 23, 199, 201, 281, 332, 333, 336, 339, 344, 349
  Reynolds, 25, 26, 28-31, 59, 123, 124, 126, 137, 143, 170, 172, 199, 201, 204-206, 217, 225, 259-264, 269, 293, 328, 329, 332, 333, 337, 352, 360
  Shields, 6, 20, 23, 27, 30, 31, 34, 51, 126, 127, 136-140, 143, 145, 152, 224, 225, 230, 248, 255-265, 274-276, 292, 297, 306, 313, 332, 333

P, R

pore, 41, 43, 44, 50, 52, 61, 63, 67, 69, 71, 73, 75, 84, 88, 89, 99, 101-103, 121, 132, 136, 145, 147, 150, 284, 290, 294, 301, 312
porous medium, 1, 3, 28, 30, 39-41, 55, 59-73, 120, 121, 124-128, 131, 136-138, 144, 147, 152, 192, 273, 295, 300, 301, 330
pressure drop, 51, 56, 71, 163, 164, 170, 199, 209, 210, 214, 217, 218, 253
radionuclide, 314, 316
ripple, 129, 144, 228, 229, 240, 244–249, 335–340, 345
roughness, 26, 35, 226, 232, 238, 243, 248, 266

S

salinity, 73, 290, 312, 319, 326, 329, 330, 336
scouring, 211, 351-353, 360
sedimentation, 60, 62, 63, 65, 116, 127-129, 131, 134, 142, 149, 248, 260, 299, 325, 334
stress
  critical, 26, 30, 35, 41, 51, 131, 197, 215, 265, 291-294, 302, 310, 318, 333
  hydraulic, 125, 128, 131, 137, 146
  yield, 148, 209
suffusion, 11-13, 16, 22, 23, 32, 39-56, 58, 59, 74, 75, 116, 118, 119, 149, 202
swell, 337, 339, 341-345, 353, 355, 357, 359

T

transport capacity, 142, 187, 318, 336
turbidimeter, 160, 163, 165, 166, 305, 308
turbidity, 134, 141, 142, 145, 149, 163-166, 168, 183, 305
turbulence, 16, 25, 26, 34, 35, 56, 123, 124, 137, 160, 203-205, 209, 214, 288, 308, 318, 319, 330, 339
two-phase, 143, 169, 189, 190, 192, 195, 198, 199, 203, 219, 253, 254, 267, 268, 272, 275

V, W

velocity,
  critical, 21, 23, 52, 135-138, 140, 147
  of erosion, 141
  flow, 59, 67, 70, 81, 106, 149, 201, 219, 226, 247, 260, 263, 335
wave, 240, 245, 246, 275, 332, 335, 339-347, 349, 351-356, 358
wind, 226-229, 236-239, 244, 248
work, 2, 4, 17, 21, 22, 31, 40, 82, 105, 119, 131, 137, 139, 147, 150, 249, 255, 259, 275, 276, 319